Remote Sensing
of Turbulence

Remote Sensing of Turbulence

Victor Raizer

CRC Press
Taylor & Francis Group
Boca Raton London New York

CRC Press is an imprint of the
Taylor & Francis Group, an **informa** business

First edition published 2022
by CRC Press
6000 Broken Sound Parkway NW, Suite 300, Boca Raton, FL 33487-2742

and by CRC Press
2 Park Square, Milton Park, Abingdon, Oxon, OX14 4RN

Library of Congress Cataloging-in-Publication Data

Names: Raizer, Victor Yu, author.
Title: Remote sensing of turbulence / Victor Raizer.
Description: First edition. | Boca Raton : CRC Press, 2022. | Includes
  bibliographical references and index.
Identifiers: LCCN 2021020836 (print) | LCCN 2021020837 (ebook) | ISBN
  9780367469788 (hardback) | ISBN 9781032108902 (paperback) | ISBN
  9781003217565 (ebook)
Subjects: LCSH: Turbulence--Remote sensing. | Turbulence--Data processing.
  | Geophysics.
Classification: LCC QA913 .R35 2022 (print) | LCC QA913 (ebook) | DDC
  551.46/2--dc23
LC record available at https://lccn.loc.gov/2021020836
LC ebook record available at https://lccn.loc.gov/2021020837

ISBN: 978-0-367-46978-8 (hbk)
ISBN: 978-1-032-10890-2 (pbk)
ISBN: 978-1-003-21756-5 (ebk)

DOI: 10.1201/9781003217565

Typeset in Times
by SPi Technologies India Pvt Ltd (Straive)

Contents

Preface

"It would be possible to describe everything scientifically, but it would make no sense; it would be without meaning, as if you described a Beethoven symphony as a variation of wave pressure."

Albert Einstein

"There is no subject so old that something new cannot be said about it."

Fyodor Dostoevsky

Turbulence is all around us – in air, in water, in technology, in life. It is hard not to be intrigued by a subject which pervades so many aspects of our world. Scientists tend to view turbulence as complex chaotic process (flow) with unpredictable behavior in space and time. By this means, they create significant challenges and uncertainties in mathematical description and prediction of turbulence. These quotations above just reminder us an analogy – turbulence has no ultimate explanation but it inspires endless explorations and insights, continuing the last 100 years. The major contribution however remains the Navier–Stokes equations (1822), which are a key mathematical tool in fluid mechanics. Today these equations provide flexible computational framework in a vast array of industries. Yet turbulence is regarded as the last major unsolved problem of classical physics.

Many researchers in different fields, their activities and efforts have inspired me to consider and maybe elaborate a novel approach, based on space observations. These absolutely undeniable methods give us more immediate and more beneficial results in our daily routine than abstract theories or models. In fact, satellite data are an exceptional information resource needed for monitoring, control, and surveillance of our environment.

Specifically, this book is about remote sensing of turbulence. The objective is to establish concepts, tools, and methods of turbulence research from space. The book provides a broad view on turbulence science and summarizes current problems and future challenges. The selected material intended to emphasize a novel geophysical application – detection and recognition of complex turbulent flows in the ocean and atmosphere.

The book is organized in classical manner – physics, technology, application, and observational data. The book falls into five parts. The book begins with an introduction to the physics of turbulence that ensures the use of remote sensing technology and then offers a thorough review of electromagnetic wave propagation in turbulent media. A brief review of remote sensing principles and miscellaneous aspects is given as well. Finally, the book describes a methodology of turbulence observations and discusses purposefully available satellite data. The main attention, however, is paid here not to the concrete calculations, but rather to the expostulation of the principal ideas of turbulence observations for a variety of purposes.

Ultimately, the book explores consecutively remote sensing capabilities for detection and recognition of turbulent events, their signatures, and variability. Taking a multi-disciplinary approach for using remote sensing technology, the book offers a single comprehensive resource for many researches and specialists working in various fields of applied science – geophysics, hydrodynamics, oceanography, hydrology, meteorology, atmospheric science, environmental engineering, and related disciplines. Because the book covers many research areas, the reader may realize as many interesting details as possible. This is a material for future research, but hopefully the physics, tools, and data presented in this book will also be useful for ongoing projects.

The present volume, I believe, can be used as an introductory textbook on the foundation of remote sensing of turbulence. It will be comprehensible and helpful not only to remote sensing community but also for physicists and computer science specialists who are interested to turbulence problems including numerical studies as well. Such, at least, was the aspiration of the author.

I wish to thank CRC Press team and editors, especially Irma Britton, for her kind support and continued encouragement over a number of years. I am also grateful to my wife, Elena, for her outstanding patience and prudence during this difficult time of pandemic.

Fairfax, Virginia
March 2021

Author

Victor Raizer, PhD, DSc, physicist and researcher, has 40 years of experience in the field of electromagnetic wave propagation, radiophysics, hydrophysics, microwave radiometer/radar and optical techniques, geosciences, and remote sensing. He was graduated from the Moscow Institute of Physics and Technology in 1974, Department of Aerophysics and Space Research (FAKI). He earned a PhD in experimental physics in 1979 and a DSc in major physics and mathematics in 1996. He worked with the Space Research Institute (IKI), the Russian Academy of Sciences, Moscow (1974–1996), and then he joined NOAA and Zel Technologies, Virginia, USA (1997–2016) as a senior scientist.

Dr. Raizer provided a broad spectrum of scientific research, including developments of multisensor observation technology, nonacoustic detection capabilities, modeling, simulation, and prediction of complex remotely sensed data. He is Senior Member of IEEE and the author of two recent books: *Advances in Passive Microwave Remote Sensing of Oceans* (2017, CRC Press) and *Optical Remote Sensing of Ocean Hydrodynamics* (2019, CRC Press). He is the coauthor with I. V. Cherny of the notable book *Passive Microwave Remote Sensing of Oceans* (1998, Wiley, Chichester, UK).

List of Acronyms

ABL	Atmospheric boundary layer
ABW	Antarctic bottom water
AMSR-2	Advanced microwave scanning radiometer
AMSU	Advanced microwave sounding unit
ANN	Artificial neural network
ASAR	Advanced synthetic aperture radar
ASTER	Advanced spaceborne thermal emission and reflection radiometer
ATOVS	Advanced TIROS operational vertical sounder
AVHRR	Advanced very high resolution radiometer
CAT	Clear air turbulence
CFD	Computer fluid dynamics
CIMSS	NOAA's Cooperative Institute for Meteorological Satellite Studies
DLR	Deutsches Zentrum für Luft- und Raumfahrt (German Aerospace Center)
DMSP	Defense meteorological satelliteprogram
DNS	Direct numerical simulation
DST	Dynamical systems theory
EnMap	Environmental mapping and analysis program satellite, Germany
ENVISAT	Environmental satellite, ESA
EO-1	Earth observing-1 satellite
EOS	Earth observing system
ERS	European remote sensing satellite
ESA	The European Space Agency
GEO	Geostationary orbit
GIFOV	Ground-projected instantaneous field of view
GIS	Geographic information system
GFD	Geophysical fluid dynamics
GLONASS	Global orbiting navigation satellite system
GMI	Global precipitation measurement microwave imager
GMS	Geostationary meteorological satellite, Japan
GNSS	Global navigation satellite system
GOES	Geostationary operational environmental satellite
GPS	Global position system
GSD	Ground sample distance
GTO	Geostationary transfer orbit
HEO	Highly elliptical orbit
HS	Hyperspectral
FOV	Field of view
ICO	Intermediate circular orbit
IFOV	Instantaneous field of view
IR	Infrared
IRS	Indian remote sensing satellite

ISS	International space station
JPSS	Joint polar satellite system
K41	Kolmogorov theory of 1941
KARI	Korea Aerospace Research Institute
KIOST	Korea Institute of Ocean Science and Technology
KO62	Kolmogorov-Obukhov theory of 1962
LEO	Low earth orbit
LES	Large eddy simulation
LWIR	Longwave infrared radiation
MEO	Medium low earth orbit
METEOSAT	Geostationary meteorological satellites
MODIS	Moderate resolution imaging spectroradiometer
MS	Multispectral
MSG	Meteosatsecond generation satellite
MTSAT	Multifunctional transport satellites, Japan
MTVZA-GY	Imaging/sounding microwave radiometer (module for temperature and humidity sounding of the atmosphere and "GY"' are the initials of the prominent Russian space system designer G. Ya. Guskov); Meteor satellite instrument
MWIR	Middle wave infrared
NADW	North Atlantic deep water
NASA	National Aeronautics and Space Administration
NIR	Near infrared
NIW	Near-inertial wave
NOAA	National Oceanic and Atmospheric Administration
NPOESS	National polar-orbiting operational environmental satellite system
NPP	National polar-orbiting partnership
OBL	Upper-ocean boundary layer
OLI	Operational land imager
OTF	Optical transfer function
PAN	Panchromatic
PBL	Planetary boundary layer
pdf	Probability density function
PER	Polarization extinction ratio
PO	Polar orbit
POES	Polar operational environmental satellites
PSF	Point spread function
RANS	Reynolds-averaged Navier–Stokes equations
RIT	Radiative transfer theory
RTE	Radiative transfer equation
SAR	Synthetic aperture radar
SGS	Subgrid scale stress tensor
SLAR	Side-looking aperture radar
SMMR	Scanning multi-channel microwave radiometer
SMOS	Soil Moisture and Ocean Salinity satellite
SPOT	Satellite Pour l'Observation de la Terre (French)

SSM/I	Special sensor microwave imager
SSMIS	Special sensor microwave imager/sounder
SSO	Sun synchronous orbit
SSS	Sea surface salinity
SST	Sea surface temperature
SVN	Support vector machine
SWIR	Short-wave infrared
TIR	Thermal infrared
TIROS	Television infrared observation satellite
TKE	Turbulence kinetic energy
TRMM	Tropical rainfall measuring mission satellite
UNDEX	Underwater explosion
USGS	U.S. geological survey
UV	Ultraviolet
VHT	Vortical hot tower
VIIRS	Visible infrared imaging radiometer suite
VIS	Visible
VNIR	Visible and near-infrared
WKE	Wave kinetic equation

1 Turbulence
Introductory Overview

We realize thus that: big whirls have little whirls that feed on their velocity, and little whirls have lesser whirls and so on to viscosity – in the molecular sense.

Lewis F. Richardson, 1922

The turbulence world is around us – in nature, in technology, in life. Turbulent motions are very common in nature: in the oceans, in the atmosphere, in rivers, even in stars and galaxies. *"Turbulence is the last great unsolved problem of classical physics."* This quotation, attributed to Richard Feynman, is being in almost all scientific books on turbulence. One reason to consider this statement truthful is that turbulence actually exhibits unpredictable chaotic behavior in space and time. Most turbulence phenomena, observed in nature, still require theoretical explanation and/ or proper interpretation. Turbulence is a subject of continuous and extensive study in physics, mathematics, technology, and engineering.

There are two acceptable ways for exploring turbulence – fundamental physics and applied engineering, and their methods and goals differ. The first provides scientific (mostly theoretical but sometimes without experimental verifications) description of turbulent phenomena, the second focuses on technological aspects and practical solutions of the problem. The "bridge" between these two ways is very woozy and based only on the equations of classical fluid dynamics, which give us just mathematical foundations. Physicists attempt to find some rigorous relationships and/or relevant approaches, whereas engineers and applied mathematics use direct *numerical* methods (*solvers) to obtain practical* solutions of various turbulence problems. These efforts are not necessarily complementing each other.

A rapidly growing interest to study of turbulence has produced a huge amount of scientific and technical literature. Various theories, approximations, models, and semi-empirical relationships, describing the phenomenon of turbulence, have been suggested over the years. We even do not try survey these works in any way; however, we have to mention several textbooks, which provide fundamental knowledge to the subject matter. First of all, we refer to the two-volume encyclopedic book by Monin and Yaglom (1971, 1975), which is detailed historical account of the theory of turbulence. Excellent monographs and books (Batchelor 1953; Hinze 1959; Kadomtsev 1965; Loitsyanskii 1966; Tennekes and Lumley 1972; Frost and Moulden 1977; Landau and Lifshitz 1987; McComb 1990, 2014; Chorin 1994; Frisch 1995; Lesieur et al. 2001; Lesieur 2008; Davidson et al. 2013; Davidson 2015; Bailly and Comte-Bellot 2015; Ting 2016) cover various aspects of turbulence science and applications. We encourage the reader to save time for searching relevant references on this subject. Someway, we did this work instead he/she.

DOI: 10.1201/9781003217565-1

This chapter is a brief overview of turbulence with a short history and introduction to the problem. It highlights the main milestones, physical essence, and bibliography including classical point of view on the theory. The subject of turbulence is extremely complicated and vast and we will consider only those aspects that, in our opinion, would be relevant for our interest – geosciences and remote sensing.

1.1 HISTORICAL REMARK

The first usage of word "turbulence" dates back to early 15th century and has its roots in Latin word *turbulentus* originated from the word *turba*, which means disorder, tumult, crowd. As a scientific term, "turbulence" is used to describe certain complex and irregular motions of a fluid, gas, and plasma, both in nature and in technical devices. This definition, however, is far from being perfect; it expresses just our perception of this mysterious phenomenon in natural science, from antiquity to present days.

The subject of turbulence has a very long history (Table 1.1). More than 2000 years ago, Lucretius described eddy motion in his *De rerum natura* (translated as *On the Nature of the Universe*): "...an external current of air from some other quarter may whirl them along in their course." For more meaningful, we mention only two fascinating examples of art masterpieces, created years before scientists could even begin to think about the theory.

Over five centuries ago, Leonardo da Vinci (1452–1519) was probably the first who used the word "turbulence" (in Italian *turbolenza*) in a scientific sense, observing, describing, and sketching the natural flow of the water (Figure 1.1). Leonardo illustrated turbulent flow of water as a field of various swirl structures of different form and sizes. He wrote (referring to Richter 1970):

Observe the motion of the surface of the water, which resembles that of hair, which has two motions, of which one is caused by the weight of the hair, the other by the direction of the curls; thus the water has eddying motions, one part of which is due to the principal current, the other to the random and reverse motion.

His sketches of turbulence can still be seen today.

Other example is the famous Vincent van Gogh's *The Starry Night*, painted in 1889 (Figure 1.2). The prominent effect of apparent flow is caused by *luminance*, the intensity of the light in the colors on the canvas. Recently, researchers analyzed digitally this painting and found that it has structural composition, closely matched turbulent properties of molecular gas clouds and fluid flow (Aragón et al. 2008). "*In summary, our results show that Starry Night, and other impassioned van Gogh paintings, painted during periods of prolonged psychotic agitation transmitted the essence of turbulence with high realism*" wrote Aragón and his associates. Van Gogh genially renders turbulent structure with such a mathematical precision. Visually, both Leonardo and Vincent art works remind us some kind of vortex/eddy formations (or "coherent structures" in the modern terminology) that can be characterized as *turbulent motions*.

Since the first Leonardo sketch, many famous scientists have worked on the problem of turbulence: Leonhard Euler (1707–1783), Hermann Ludwig Ferdinand von Helmholtz (1821–1894), William Thompson (Lord Kelvin) (1824–1907), Lord

TABLE 1.1
Important Contributions in Turbulence Research

1500	Leonardo da Vinci uses the term "la turbolenza" observing two states of flow
1822	C. Navier derives the momentum equation describing the behavior of viscous flow
1839	Rediscovery of two states of fluid motion by G. Hagen
1877	J. Boussinesq introduces the concept of eddy viscosity
1883	O. Reynolds investigates transition of flow and identifies of the Reynolds number
1887	Lord Kelvin introduces the notion of "turbulence"
1895	Reynolds decomposition
1909	D. Riabuchinsky invents the constant-current hot-wire anemometer
1912	J. Morris invents the constant-temperature hot-wire anemometer
1914	L. Prandtl demonstrates the transition to turbulence in tank experiments
1921, 1935	Statistical approach by G. I. Taylor
1922	L. Richardson's hierarchy of eddies *"big whirls have little whirls…"*
1924	L. Keller and A Friedman formulate the hierarchy of moments
1936	Theory of submerged turbulent jets of G. N. Abramovich
1937	T. von Kármán's concept of homogeneous shear flow
1938	G. I. Taylor discovers the prevalence of vortex stretching
1941	A. N. Kolmogorov theory, 2/3 and 5/3 power laws
1943	S. Corrsin discovers the sharp interface between turbulent and non-turbulent regions
1949	Discovery of intrinsic intermittence by G. Batchelor and A. Townsend
1951	Turbulent transition and spots of H. Emmons
1952	E. Hopf functional equation for turbulence
1954	Monin–Obukhov similarity theory
1962	Kolmogorov–Obukhov scaling hypothesis with intermittency correction
1963	Lorenz attractor and butterfly effect as a model of "deterministic chaos"
1967	Bursting phenomenon by S. Kline et al.
1967	S. Smale introduces the "horseshoe map" as the hallmark of chaos
1960s–1970s	Theory of wall-bounded turbulence
1971	Poincaré–Andronov–Hopf bifurcations
1971	D. Ruelle and F. Takens suggest "chaos to turbulence" scenario
1972	H. Tennekes and J. Lumley "A First Course in Turbulence," MIT. Must read book
1976	Recapitulation of large scale coherent structures by A. Roshko
1975, 1982	B. Mandelbrot introduces fractal geometry and explores self-similarity
1977	I. Prigogine develops concept of self-organization in large complex dynamic systems
1980s	Chaos/Dynamical systems theory and application to turbulence
1983	Theory of helicity turbulence of S. Moiseev et al.
1995	U. Frisch introduces the notion of "multifractal" for structure of turbulence region
1997	Barenblatt–Chorin–Prostokishin power scaling law for turbulent boundary layers
1980–1990	CFD codes for modeling and simulation of turbulent flow (DNS, LES, RANS)
2000s–	Commercial CFD packages for bluff body aerodynamics and hydrodynamics

(*Sources:* Thoroughly updated data from Tsinober 2009; Eckert 2019).

Rayleigh (1842–1919), and Andrei Nikolayevich Kolmogorov (1903–1987). We named only few of them, most known.

Many historical details and references are given in review (Lumley and Yaglom 2001).

1.1.1 THE REYNOLDS ERA

The modern era of turbulence research has begun with pioneering work of Osborn Reynolds (1842–1912). Reynolds explored the flow in pipes in his laboratory and found

FIGURE 1.1 Leonardo da Vinci's illustration of the swirling water flow. (The Royal Collection, Her Majesty Queen Elizabeth II.)

FIGURE 1.2 The Vincent van Gogh's *The Starry Night*. (Exhibited in the Museum of Modern Art in New York City.)

the existence of two general types of moving fluid structure; he named them as *laminar flow* and *turbulent flow*. Reynolds was the first who studied turbulent flow in systematic experiments and, as a result, introduced the non-dimension parameter known today as the *Reynolds number*, Re = UL/v, where U and L are the velocity and length scales and v is the kinematic viscosity of the fluid. The Reynolds number has different interpretation, but the most fundamental one is that this is the ratio of inertial forces to viscous forces. The value of the Reynolds number defines main flow regimes which are:

- **Laminar flow** at low Reynolds numbers (Re < 2,300), where viscous forces are dominant, and this flow is characterized by smooth, constant fluid motion.
- **Transition flow** at Reynolds numbers (2,300 < Re < 4,000); it is dominated by inertial forces, which tend to produce eddies, vortices, and other flow instabilities.
- **Turbulent flow**, which is qualified at high Reynolds numbers (Re > 4,000); the flow becomes chaotic and unpredictable (with *irregular variations of velocity*).
- **Subcritical** (Fr < 1), **critical** (Fr = 1), or **supercritical** (Fr > 1) flow is defined by the relationship between flow velocity U and flow depth d, i.e., by the Froude number Fr = U / $\sqrt{gd}$ (g is the acceleration of gravity).

More detailed specification has been developed for different vortex shedding regimes of fluid flow across smooth circular cylinders (Lienhard 1966) (Figure 1.3).

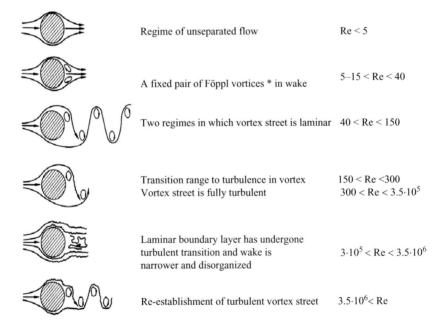

	Regime of unseparated flow	Re < 5
	A fixed pair of Föppl vortices * in wake	5–15 < Re < 40
	Two regimes in which vortex street is laminar	40 < Re < 150
	Transition range to turbulence in vortex Vortex street is fully turbulent	150 < Re < 300 300 < Re < 3.5·10⁵
	Laminar boundary layer has undergone turbulent transition and wake is narrower and disorganized	3·10⁵ < Re < 3.5·10⁶
	Re-establishment of turbulent vortex street	3.5·10⁶ < Re

---

* The region of nonlinear stability is called *Föppl equilibrium*. It is pointed out that a vortex pair properly placed downstream can overcome the cylinder and move off to infinity upstream

FIGURE 1.3 Schematic illustration of flow regimes at different Reynolds numbers. Overview of different vortex shedding regimes. (Based on Lienhard 1966).

These criteria clarify the fact that the flow will be laminar so long as the Reynolds number exceed some critical value Re_{cr}, while for $Re > Re_{cr}$ it will be turbulent. For example, the transition to turbulence in pipe flow can occur anywhere in the range $1,000 < Re < 50,000$ depending on pipe characteristics.

According to Lesieur (2008), the case of fully developed turbulence is specified as

> *a turbulence which is free to develop without imposed constrains. The possible constrains are boundaries, external forces, or viscosity.... So, no real turbulent flow, even at a high Reynolds numbers, can be fully developed in the large energetic scales. At smaller scales, however, turbulence will be fully developed if the viscosity does not play in the dynamics of these scales.*

From this viewpoint, it follows that the definition of the value Re_{cr} is tough question especially for transition regimes of the flow. It is also not possible to describe fully developed turbulent flow using methods of deterministic fluid dynamics (although some randomization procedures can be applied); therefore, it is imperative to apply statistical approach.

1.1.2 THE KOLMOGOROV ERA

In the 20th century, the Russian mathematician Andrey Nikolaevich Kolmogorov proposed a real theory. In 1941 and 1942, he published several papers (two of them reprinted in English: Kolmogorov 1941a, 1941b) that provided some of the most important and most-often quoted results in the history of turbulence science. The Kolmogorov statistical theory states that a turbulent flow poses dynamical process, called *energy cascade*, also known as *Richardson–Kolmogorov* energy cascade, introduced first by Richardson (1922). Cascade process is the division of large eddies into smaller eddies and a transfer of large scales to smaller scales (exactly as Richardson's "*Big whirls, little whirls...*"). This may be interpreted as a large eddy passing all of its kinetic energy unto smaller eddies within its life span. According to Richardson's energy cascade proposition, these smaller eddies subsequently pass all their energy unto even smaller eddies when making their revolution.

To control cascade process, Kolmogorov introduced the following three dimensionless scales (known as *Kolmogorov microscales*):

$$\eta = (v^3/\varepsilon)^{1/4} \qquad \text{length microscale}$$

$$\tau_\eta = (v/\varepsilon)^{1/2} \qquad \text{time scale}$$

$$u_\eta = (v\varepsilon)^{1/4} \qquad \text{velocity scale}$$

Here ε (m^2/s^3) is the average rate of dissipation of turbulence kinetic energy (abbreviated TKE) per unit mass. Physically, the TKE rate of dissipation is characterized by measured root-mean-square (rms) velocity fluctuations and can be defined experimentally or calculated based on a turbulence model. In some cases, the TKE dissipation rate is estimated by the Taylor dissipation law (1921) as $\varepsilon \approx C_\varepsilon U_L^3/L$, where L is commonly referred to as the integral length scale (which is a typical length scale of the flow), U_L is the characteristic velocity of the largest eddy, and C_ε is constant. These scales η, τ_η, and u_η are indicative of the smallest eddies present in the flow, i.e., the scales at which the energy is dissipated.

The Kolmogorov idea of microscales is based on a similarity hypothesis and the assumption that the smallest scales of turbulence eddy are universal (i.e., similar to each other in any turbulent flow) and depend only on ε and ν. Note that the Taylor's microscale λ_T is proportional to $\lambda_T/L \sim Re^{-1/2}$, while the Kolmogorov's length microscale η is proportional to $\eta/L \sim Re^{-3/4}$ (Tennekes and Lumley 1972).

Using dimensional analysis and scaling laws, Kolmogorov introduced famous formula for the energy spectrum of isotropic and homogeneous turbulence

$$E(k) = C_k \varepsilon^{2/3} k^{-5/3}$$

Here C_k is a universal constant known as the *Kolmogorov constant* ($C_k = 1.4 - 2.2$) and k is the wavenumber (or spatial frequency). The spectrum E(k) exists in the inertial range $1/\ell_0 \le k \le 1/L_0$, where L_0, ℓ, and ℓ_0 are the initial (outer), current, and inner turbulent eddy sizes, respectively. Kolmogorov assumed that in the *inertial subrange*, where $L_0 > \ell > \ell_0$ turbulent motions are both homogeneous and isotropic and energy may be transferred from eddy to eddy without loss, i.e., the amount of energy that is being injected into the largest structure must be equal to the energy that is dissipated as heat. Figure 1.4 illustrates schematically the Richardson–Kolmogorov energy cascade and the Kolmogorov turbulence spectrum within inertial subrange. This prediction, known as "K41 theory" or the "*Kolmogorov's five-thirds law*," has been tested in a variety of turbulent phenomena and in many cases it is extremely well satisfied (Frisch 1995).

Later in 1962 after "footnote remark" by Landau on the *non-universality* of the K41 theory, Kolmogorov and Obukhov revised the K41 theory and formulated new one including the effect of *intermittency*; this new theory is known as the "KO62 theory" (Kolmogorov 1962; Frisch 1995; Birnir 2013). Since that time, serious attempts have been made to develop full-scale stochastic theory of turbulence; citing Yaglom, "*the 20th Century becomes the century of the turbulence theory*" (Lesieur et al. 2001). Kolmogorov quantities are of fundamental significance in turbulence science for all time.

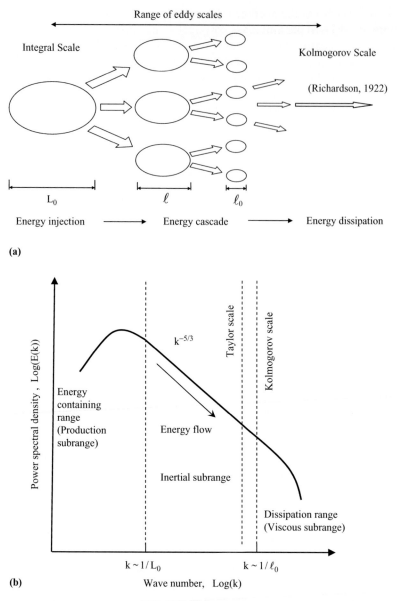

FIGURE 1.4 Schematic illustration of Kolmogorov K41 theory of turbulence. (a) Richardson–Kolmogorov energy cascade and (b) turbulence energy spectrum. Here input range $L_0 \leq \ell$, inertial range $L_0 < \ell < \ell_0$, and dissipation range $\ell \leq \ell_0$ (L_0, ℓ and ℓ_0 are the initial (outer), current, and inner turbulent eddy sizes).

1.1.3 The Computer Era

Eventually, practical analyses of turbulence inevitably invoke a large number of analytical and numerical models and techniques (Heinz 2003; Cardy et al. 2008; Durbin and Pettersson Reif 2011). Computer modeling of turbulence lifts off in 1980s on the basis of the Navier–Stokes equations. The developments of computer information technology (CIT) and "fast" algorithms led to the creation of the universal software product known as *Computational Fluid Dynamics* (CFD). CFD allow scientists and engineers to investigate various fluid dynamics and turbulence problems with lower costs and higher performance. In particular, CFD provide modeling and simulations of incompressible unsteady 2D or 3D Navier–Stokes equations with mixed boundary conditions. For example, deterministic Navier–Stokes equations and solutions describe deterministic laminar flows and the stochastic Navier–Stokes equations and solutions describe stochastic turbulent flows.

CFD include four main methods for numerical modeling and simulation, known as 1) direct numerical simulation (DNS), 2) large eddy simulation (LES), 3) Reynolds-averaged Navier–Stokes equations (RANS) method, and 4) hybrid RANS/LES method. These methods will be considered in more detail in Section 1.4.

Turbulence is a complex phenomenon that affects highly diverse environment. Many believe that numerical solvers are more efficient and informative than rigorous mathematical models and/or approximations (although in global statistical sense this viewpoint is questionable). Numerical methods have long been a subject, receiving greater attention than analytic methods or empirical evaluations. This tendency occasionally initiates *overprediction* of the phenomena and/or effects being studied with a huge amount of generated computer data. Unfortunately, computer data often remain without needed experimental or alternative verification. Sometimes users encounter issues like roundoff, overflow, or too high computation complexity that may evoke misunderstanding of the result.

Meanwhile, here is a short list of areas of interest in geosciences and remote sensing, where understanding of turbulence nature potentially leads to a significant scientific breakthrough:

- Geophysical fluid dynamics and climate theory.
- Ocean circulations.
- Ocean mixing.
- Boundary layer dynamics.
- Air-sea interactions.
- Variations of surface currents.
- Scaling of surface waves.
- Internal waves turbulence.
- Wave-wave interactions and energy transfer.
- Generation of wakes, jets, plumes, and multiphase flows.
- Variations of ocean temperature and salinity.
- Double-diffusive convection and turbulent spots.
- Fluctuations of wind.
- Cloud formation.
- Clear air turbulence.
- Air pollution.
- Thunderstorms and precipitation.
- Volcano eruption.
- Sedimentology.
- Magnetohydrodynamics.
- Superfluidity.

Turbulence also plays an important role in aeronautics and combustion, astrophysics, acoustics, medicine, the chemical industry, hydraulics, and many other fields of natural science and engineering. Nowadays, the theory of fluid dynamics is embodied into CIT, where classical equations of fluid mechanics are being "digitalized" and facilitated to provide numerical studies. However, the knowledge and expertise are necessary not only for digitalization of mathematical equations, but also for the understanding of their applicability for practical realization.

The next sections are devoted to fundamental aspects of turbulence science. We consider turbulence from a somewhat heuristic viewpoint, in particular, focusing on the importance of turbulence as a complex physical phenomenon and at the same time, as the relevant object for remote sensing. The main features of turbulence, models, and important results will be discussed, in the first place.

1.2 TURBULENT FLOW

According to common definition, turbulent flow refers to motions of fluid or gas, in which pressure and velocity at any point varies instantaneously in an irregular manner. One of the most popular textbooks on turbulence (Hinze 1959) gives the following, more extended definition: *"Turbulent fluid motion is an irregular condition of the flow in which the various quantities show a random variation with time and space coordinates, so that statistically distinct average values can be discerned."*

Geophysical and industrial turbulent flows are characterized by eddies, vortexes, and instabilities. At mathematical level, modeling and simulation of turbulent flows are usually relay to computations of the velocity and vorticity fields, geometry and structure of the flow at different regimes. The Reynolds number remains a common criterion, used to determine laminar, transition, or turbulent flow regime.

1.2.1 Introduction

As mentioned above, turbulence is a complex, nonlinear multiscale phenomenon, which poses the most challenging problems in physics and mathematics. The comment *"No complete quantitative theory of turbulence has yet been evolved"* (Landau and Lifshitz 1987) still stands. Nevertheless, over the last several decades, our understanding of physics of turbulence has significantly advanced; this progress has come a long way through combinations of theoretical models, numerical studies, experiments, and data analysis. The ability to model and predict turbulence has improved dramatically. Several excellent books (Townsend 1976; Zimin and Frik 1988; Voropayev and Afanasyev 1994; Deissler 1998; Pope 2000; Mathieu and Scott 2000; Piquet 2001; Bernard and Wallace 2002; Biswas and Eswaran 2002; Heinz 2003; Schiestel 2008; Yaglom and Frisch 2012; Nieuwstadt et al. 2016; Bernard 2019) provide comprehensive coverage of various problems of turbulence and demonstrate the progress. Leaving out details, we will consider only key features of the turbulence science.

1.2.2 Definition and Properties

In general, three (instead two famous) following categories of turbulent flow can be distinguished: 1) turbulent free shear flows, 2) wall-bounded turbulent flows, known also as boundary layer flows, and 3) convective turbulence flows (thermohaline convection, buoyancy-driven convection, etc.). Such a classification is based on the differences in the velocity-intermittency structure of natural turbulent flows that are inherent to atmospheric, oceanic, and/or fluvial (and industrial) flows.

All turbulent flows are also characterized by high level of vorticity fluctuations that distinguish turbulent flow from other fluid motions (e.g., ocean waves or atmospheric gravity waves). This important statement has been clearly formulated by Monin (1978):

> ...the chief criterion of turbulence is the chaotic, random nature of the spatial and temporal variations of the thermohydrodynamic properties of the flow. However, it is not useful to refer to every flow of this kind as turbulent; for a number of purposes it may be necessary to distinguish turbulent flows from other types of random liquid and gas motions that exhibit some degree of regularity.

Regardless of great variety of turbulent flows, they exhibit all of the following properties (Tennekes and Lumley 1972):

- Complexity, irregularity, vorticity.
- Disorganized, chaotic, seemingly random behavior.
- Non-repeatability and sensitivity to initial conditions.
- Extremely large range of spatiotemporal scales.
- Enhanced diffusion (mixing) and dissipation.
- 3D, time dependence and rotationality.
- Intermittency in both space and time.

Visualization of the flow is an important source of information and one of the main principles of turbulence detection. Many original images of various turbulent structures are collected in the classic *Album of Fluid Motion* (van Dyke 1982). Several pictures from this album are shown in Figure 1.5. These pictures have been used extensively for research purposes including verifications of physical models and numerical computations, describing turbulent flows of various structures. Formally, as assumed, the equations of fluid mechanics should be able to reproduce visualization samples.

1.2.3 EQUATIONS OF FLUID DYNAMICS

The fundamental equations governing fluid motion are equations of conservation of mass, momentum, and energy (Table 1.2). For an isothermal, incompressible flow of a Newtonian fluid, the Navier–Stokes equations and the continuity equation along with appropriate boundary conditions completely describe the flow. For compressible flows, the energy conservation equation and an equation of state for the fluid will be required additionally. The equations have been derived for a vanishingly small control volume and represent fundamental conservation laws and statements of laws of thermodynamics; thereby they should be valid for all flows irrespective of whether or not the flow is steady or turbulent or compressible.

The Navier–Stokes equations are the most important equations in fluid dynamics. The Navier–Stokes equations were formulated by a French engineer and scientist Claude-Louis Navier in 1822 and by Sir George Gabriel Stokes in 1845 on the basis of the Euler's differential equations. These equations describe motions of viscous

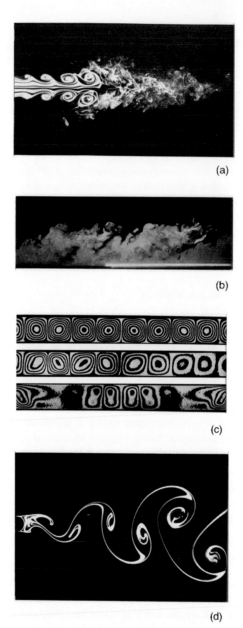

FIGURE 1.5 Illustrations of different turbulent flows. (a) Free shear, photo # 102 "Instability of an axisymmetric jet," (b) wall-bounded, photo # 158 "Turbulent boundary layer on a wall," (c) photo # 139 "Buoyancy-driven convection rolls," and (d) photo # 94 "Kármán vortex street behind a circular cylinder at R=140" (i.e., Re=140)

(*Source:* van Dyke 1982).

TABLE 1.2
Hydrodynamic Equations of Motions

Bernoulli's equation	$\dfrac{u^2}{2} + \Psi + \dfrac{p}{\rho} = 0$
Conservation of energy	$\dfrac{\partial}{\partial t}\left(\dfrac{1}{2}\rho u^2 + \rho e\right) = -\nabla \cdot \left[\rho v\left(\dfrac{1}{2}u^2 + w\right)\right], \ w = e + \dfrac{p}{\rho}$
Conservation of mass	$\nabla \cdot (\rho \mathbf{u}) = 0$
Euler equations	$\dfrac{\partial \rho}{\partial t} + \nabla \cdot \rho \mathbf{u} = 0$
Laplace equation	$\nabla^2 \phi = 0$ or $\nabla^2 u = 0$
Navier–Stokes equations	$\rho\left(\dfrac{\partial \mathbf{u}}{\partial t} + \mathbf{u} \cdot \nabla \mathbf{u}\right) = -\nabla p + \rho \mathbf{g} + \mu \nabla^2 \mathbf{u} + F$
	$\nabla \mathbf{u} = 0$

$\rho = \rho(p, T)$, ρ density, T temperature, p pressure, $\mathbf{u}$ velocity, μ dynamic viscosity, $\mathbf{g}$ body acceleration (e.g., gravity), ϕ velocity potential, e internal energy, w enthalpy, Ψ force potential ($\Psi = \rho gh$, h indicates elevation height), $\mathbf{F}$ body force, and t time.

incompressible fluid (or gas or plasma) practically at any scales from millimeters to astronomical units. However, the use of the Navier–Stokes equations in turbulence modeling is difficult task due to non-linearity of the equations and high sensitivity to initial and boundary conditions. Let's consider these equations in more detail.

The Navier–Stokes equations are a system of nonlinear partial differential equations given by

$$\frac{\partial \mathbf{u}}{\partial t} + \left(\mathbf{u} \cdot \nabla\right)\mathbf{u} = -\frac{1}{\rho}\nabla p + v\nabla^2 \mathbf{u} + \mathbf{F}\left(\mathbf{x}, t\right), \tag{1.1}$$

$$\nabla \cdot \mathbf{u} = 0, \tag{1.2}$$

$$\mathbf{u}\left(0, \mathbf{x}\right) = \mathbf{u_0}\left(\mathbf{x}\right), \ \mathbf{u}\left(\mathbf{x}, t\right)\big|_{\partial G} = 0, \ t > 0, \ \mathbf{x} \in G, \tag{1.3}$$

where $\mathbf{u}(\mathbf{x}, t)$ is the velocity field, $p(\mathbf{x}, t)$ is the pressure field, v is the kinematic viscosity, ρ is the density of fluid, $\mathbf{F}(\mathbf{x}, t)$ is the body force term, and t is time. **G** denotes the relevant domain of flow; the differential operators ∇ and ∇^2 are the gradient and Laplace operators, respectively. Equations (1.1)–(1.2) satisfy conservation of mass, momentum, and energy. In fact, these equations are continuum formulation of Newton's laws of motion. In classical form, Equations (1.1)–(1.2) are deterministic and may be extremely sensitive to initial and boundary conditions (1.3) that are related to so-called initial and boundary value problem. It is important to note that the Navier–Stokes equations are *invariant* under a particular change of time and space; the existence of so-called "scaling invariant spaces" provide a family of self-similar solutions of the Navier–Stokes equations that asymptotically approximate the typical behavior of a wide class of turbulent phenomena.

The principal difficulty in solving the Navier–Stokes equations arises from the presence of the nonlinear term $(\mathbf{u} \cdot \nabla)\mathbf{u}$. Because of the nonlinear structure of the equation, an analytical treatment is a formidable task. For instance, in the 3D case, a theorem for the existence of global solution for arbitrary value of viscosity ν is missing (the problem still belongs to list Millennium Prize problems, established by the Clay Mathematics Institute of Cambridge, MA in 2000 with $1 million prize for the solution (Carlson et al. 2006)).

Indeed, the stable solution can be found for slow flow at small Re $<< 1$ (i.e., so-called "microhydrodynamics"). The difficulties occur at very high values of Re $> 10^5 - 10^6$, for geophysical flow or flight aerodynamics, where solutions are very ambitious and often become unstable. A small perturbation in the single parameter, boundary or initial conditions may lead to completely different results.

While the Navier–Stokes equations are the mathematical core of turbulence, there are many cases when they are insufficient to represent the phenomena of interest. Additional terms and equations should be added to the set of fundamental equations of fluid dynamics. Moreover, numerical models require more information to close the system of equations: the number of unknowns is greater than the number of equations to be solved (system of equations is not closed). This common issue is called "closure problem of turbulence." Turbulence closure models are limited and use empirical formulas or parameterizations.

Meanwhile, few famous approximate calculations and proofs of the Navier–Stokes problem have been found; the most known are the following (see, e.g., Lemarie-Rieusset 2016; Tsai 2018):

1. Weak solution (Leray 1934) called also Leray–Hopf solution,
2. Smooth solution,
3. Proof of global unique solvability of the 2D problem (Ladyzhenskaia 1969),
4. Creeping flow approximation, known also as "Stokes Flow" or "Low Reynolds number flow,"
5. Sobolevski solution,
6. Galerkin truncation method,

and several others. Note that most analytical solutions are derived under simplified conditions; specifically they ignore nonlinear effects which have significant contributions in flow dynamics and turbulence in particular. These and other mathematical aspects of the Navier–Stokes equations are discussed in many books (Batchelor 1967; Doering and Gibbon 1995; Foias et al. 2001; White 2006; Tartar 2006; Kollmann 2019).

1.2.4 Instabilities

It is a common view that turbulence originates from the instability of flow(s). In fact, turbulence is a *state of continuous instability* (Tritton 1988). The relationship between instability and turbulence is complex; the problem is connected to nonlinear dynamics and chaos. In particular, *geophysical instabilities* or *instabilities in geophysical flows*, which are of our primarily interest, can be induced by many factors, e.g., stratified shear flow, the fluid–body interaction including flows around obstacles and/or actual physical boundary (known as *wall turbulence*), chaotic mixing, advection, diffusive convection, global atmospheric circulation (barotropic and baroclinic), thermohaline interleaving, etc. Here we pay attention to a number of most important types of instabilities that potentially can cause turbulence. These instabilities are known as the following:

- **Kelvin–Helmholtz Instability** occurs at the interface between two parallel flows with different velocities and densities. At certain values of the velocities and densities, the interface between the two flows begins to oscillate, indicating the onset of instability.
- **Taylor–Couette Instability** occurs in the annulus between differentially rotating concentric cylinders, e.g., when the outer cylinder is held fixed and the inner cylinder is rotating at some specified frequency. As the velocity of the inner cylinder increases, the flow becomes unstable and a new, qualitatively different steady flow arises.
- **Rayleigh–Bénard Convection** (also known as Rayleigh–Bénard instability) is a type of natural convection; it refers to the flow that develops when a stable fluid is heated from below. Such flow results from the development of the *convective instability*. The instability can also exhibit a sequence of transitions to other type of flow until fully developed turbulence occurs. One of these states is known as *convection cells* (or *Bénard cells*). Buoyancy and hence gravity are responsible for the appearance of convection cells.
- **Rayleigh–Taylor instability** occurs at an interface between two fluids of different densities when the lighter fluid is pushing the heavier fluid and/or when density and pressure gradients point in opposite directions. Any perturbation along the interface between the two fluids will grow. It is a fundamental fluid-mixing mechanism. The buoyancy-driven flow acts on fluids of different density that results in the variable-density Rayleigh–Taylor turbulent instability causing turbulence mixing. Moreover, the problem of Rayleigh–Taylor instability is a generic paradigm for convection.
- **The von Kármán Vortex Street** is a flow around a body (cylinder); it represents a repeating pattern in the form of a series of offset, counter-rotating vortex structures. It is a typical pattern of turbulent wake behind body moving through a fluid.
- **Holmboe instability** (1962) is exponentially growing instability occurring in unbounded stratified parallel shear flow due to resonant interaction between two or more progressive linear interfacial waves; this instability can lead to the formation of cusp-like waves. Beach cusps are one of the most commonly observed shoreline features of sandy and gravel beaches.
- **Richtmyer–Meshkov instability** is the fluid flow phenomenon that occurs when a shock wave impinges (in the normal direction) upon an interface separating two fluids (gases) having different densities. This instability arises due to the impulsive acceleration of an interface and the initial perturbations on the interface grow linearly in time in the beginning (that makes it different from the Rayleigh—Taylor instability).

Listed instabilities are usually referred to the first or primarily instabilities. More detailed investigations have demonstrated that laminar-turbulent transition and evolution of turbulent flow (e.g., spatiotemporal dynamics of 3D wake) consists of a cascade of successive instabilities; an important role in this process plays so-called *secondary and tertiary instabilities* that may involve (or lead to) *bifurcation* phenomena (the word "bifurcation" means *forked*). Cascades of higher-order instabilities and bifurcations (bifurcation sequences) ultimately lead to chaotic behavior of entire physical system, e.g., fully developed turbulence with absolute complexity and unpredictability. Relationship between instabilities and turbulence in fluid flow is described in several books (Chandrasekhar 1981; Moiseev et al. 1999; Drazin 2002; Tsinober 2009; Manneville 2010; Sengupta 2012, Gaissinski and Rovenski 2018; Smyth and Carpenter 2019).

1.3   DYNAMICAL SYSTEMS AND TURBULENCE

Dynamical system refers to a system exhibiting time-dependent behavior. A common mathematical formulation of such a system can be written as $dX/dt = f(X, t; \mu)$, where X may stand for any physical field, $f(\ldots)$ is a smooth function of X and of the vectors of parameters μ but does not depend explicitly on time.

Fundamental quantity that characterizes (in)stability of a dynamic system is known as *Lyapunov exponent*, introduced by the Russian mathematician Aleksandr Mikhailovich Lyapunov in 1892. Lyapunov's dissertation (PhD thesis) was translated into French in 1908, and much later in English (see book Lyapunov 1992). An essential aspect of the Lyapunov exponent is the fact that small differences in the initial values of a dynamic system over time can grow into very large differences. The exponent indicates the speed with which two initially close dynamic states diverge, if the exponent is positive, or converge, if the exponent is negative. A definition of the Lyapunov exponent is given in Appendix.

Chaotic behavior of dynamical systems is described by the bifurcation diagram of the logistic map with the representation of Lyapunov exponent values. Figure 1.6 shows typical example. A dynamical system with a positive Lyapunov exponent is called chaotic. This case is of particular interest. The paths of such a system are extremely sensitive to changes of the initial conditions. The Lyapunov exponents provide the most efficient mathematical framework for stability analysis of dynamical systems including the transition from deterministic chaos to a stochastic process (see, e.g., book Pikovsky and Politi 2016).

Over the last 40 years, with the discovery of chaos and strange attractors, dynamical systems gained considerable interest in many areas of natural science – mathematics (such as number theory and topology) and physics (such as electromagnetic and gravitational fields). Indeed, dynamical systems encompass a staggering range of geophysical and technological systems evolving motions in time.

The application of dynamical systems to analyses of turbulence is ubiquitous project. As known, key features here are multiscale dynamics, high dimensional phase space, nonlinear energy transfers, and intermittent instability. This bunch of properties can be accounted simultaneously using dynamical system approach only. Dynamic models can also facilitate an efficient computation of the complex turbulent

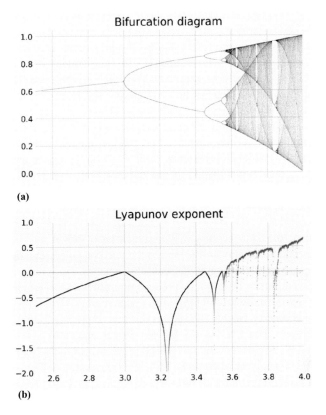

FIGURE 1.6 Chaotic dynamics and stability. (a) Bifurcation diagram and (b) Lyapunov exponent for the logistic map, Equation (1.4). Plot demonstrates evolution of different initial conditions as a function of parameter r. If the Lyapunov exponent is positive, then the system is chaotic. (Based on Strogatz 2015.)

flows, capturing both physics of large-scale motions and the statistics of small-scale variables. In this context, the importance of information-theoretical framework of dynamical systems cannot be overestimated.

1.3.1 INTRODUCTION

Nonlinear science gives us many interesting examples of turbulent phenomena and their behavior. Thus, an attempt to predict the long-time behavior of turbulent flow leaded to the concept of complex (nonlinear) dynamical systems which often exhibit several features such as chaos, bifurcations, coherent structures, similarity, self-organization, and feedback. We will consider some of them.

The dynamical systems theory, DST, (also known as nonlinear dynamics or chaos theory) is the branch of mathematics, dealing with time-varying nonlinear ordinary differential equations and iterative mappings (Birkhoff 1927; Katok and Hasselblatt

1995; Arnold 1998). The DST describes and predicts temporal evolution of complex systems (i.e., systems having a large number of interacting components) and also can be used for analysis of chaotic fields, processes, and signals. However, the application of DST to geophysical fluid dynamics is still a challenging task.

An idea to bring a dynamical system perspective to turbulence studies was suggested first in the 1940s independently by Landau and Hopf (Hopf 1948; Landau and Lifshitz 1987). They considered turbulence as an oscillating flow with a set of incommensurable quasi-periodical modes. As the Reynolds number increases, the oscillatory flow becomes more unstable and chaotic due to Andronov–Hopf bifurcations and finally reaches consistency with the Kolmogorov's theory of turbulence. However, the Landau–Hopf theory has not been justified by experiments with hydrodynamic turbulence. Later it was predicted that fluid turbulence is not oscillating flow but described by strange attractors – a main concept of the chaos theory.

In the end of the 20th century, CFD studies have shown that time-dependent Navier–Stokes equations may correspond to the universal hierarchy of different levels of *dynamic complexity* (simply speaking, dynamic complexity occurs when "obvious interventions or interactions produce non-obvious consequences").

Turbulent flows usually exhibit long-term dynamic transformations and very complex changing geometry. Such a situation is difficult to predict and/or describe using analytical methods. The DST of the Navier–Stokes equations, as a whole, offers several important insights into the turbulence problem, particularly related to stability and topological analyses of complex turbulent flows. Among them, we emphasize the following principle features of turbulent dynamical systems: 1) chaos, 2) coherent structures, 3) fractal, and 4) self-organization.

1.3.2 CHAOS

In common sense, the word *chaos* means "a state of disorder." Although there is still no universally accepted mathematical definition of chaos, this is an important attribute of dynamical systems. Lorenz (1993, page 204) defines chaos as "*The property that characterizes a dynamical system in which most orbits exhibit sensitive dependence; full chaos.*" More properties of chaos are also known as the following:

- No state of system repeats.
- Sensitivity to initial conditions (a property noted by Poincaré and Birkhoff).
- A small fluctuation of the input yields drastic changes in the output.
- Complex shape in phase space (fractal).
- Nonlinear response.
- Unpredictability of long-term behavior (butterfly effect).
- Chaos does not mean randomness (what is called "deterministic chaos").

Chaos can be described by mathematical models for many natural systems, e.g., physical, chemical, biological, and social, which demonstrate spatiotemporal evolution, unstable properties, irregularities, and seemingly unpredictable behavior. To "measure" chaos, two main characteristics – the Lyapunov exponents and the

Kolmogorov–Sinai Entropy (Sinai 1977), are used. The common feature of these measures is the sensitive dependence on initial conditions; specifically, the state of chaos is defined as the behavior of a dynamical system with positive topological entropy (Ott 2002). To illustrate such behavior, consider several examples (Figure 1.7).

The logistic map (introduced by Robert May in 1976) is a 1D discrete-time dynamical system with an unexpected degree of complexity. Historically it has been one of the most important and paradigmatic systems during the early days of research on deterministic chaos. The logistic map is defined by the following equation:

$$x_{n+1} = rx_n(1-x_n), 0 < x_n < 1, \ 0 < r < 4, \text{ with } n = 1,2,3..., \tag{1.4}$$

where x_n is the population of n-th generation and *r* is the growth rate. The logistic map (known also as quadratic map) is capable of very complicated behavior (Figure 1.6a). The logistic map illustrates chaotic dynamics of the system, i.e., the transition from regularity to chaos – process known also as "the route to chaos."

The Lorenz system is a classical model of transition to turbulence (introduced by Lorenz 1963). It is also a simple mathematical representation of dissipative dynamical system, exhibiting chaotic motions. The system is given by three nonlinear first-order ordinary differential equations:

$$\frac{dx}{dt} = -\sigma x + \sigma y, \quad \frac{dy}{dt} = -xz + rx - y, \quad \frac{dz}{dt} = xy - bz, \tag{1.5}$$

where x, y, and z represent positions and b, σ, and r, are constants, which determine the behavior of the system, and t is time. The Lorenz attractor is a set of chaotic solutions of the system. It is demonstrated that sensitive dependence on initial conditions is the essence of chaos known as the "butterfly effect" (Figure 1.7a). Equations (1.5) have been studied intensively (Sparrow 1982).

The Smale horseshoe map (introduced by Smale 1967) is the prototypical map possessing a chaotic invariant set. It is constructed as an infinite sequence of smooth stretch and fold operations and represents the building block of chaotic attractors (Figure 1.7b). The horseshoe map f is a diffeomorphism defined from a region S of the plane into itself. The horseshoe map is one-to-one, which means that an inverse f^{-1} exists when restricted to the image of S under f. The analytical feature of the horseshoe is *hyperbolicity*, which is characterized by the presence of expanding and contracting directions in the phase space of dynamical system (e.g., Lichtenberg and Lieberman 2010). Thus, the horseshoe map creates topological complexity via the squeeze/stretch and folding of a single square. Horseshoe map displays properties typical for *deterministic chaos* – the appearance of chaotic motions in purely deterministic dynamical systems. It is a simple model which reproduces the most complex (symbolic) dynamics found in nature.

Arnold's cat map is a chaotic 2D map from the torus into itself (Arnold and Avez 1968). It is given by transformation Tx = Ax, where A is a matrix with integer entries and T defines a map the torus. The image of Arnold's cat map can be written by the following equation:

$$\begin{bmatrix} x_{n+1} \\ y_{n+1} \end{bmatrix} = \begin{bmatrix} 1 & c \\ d & cd+1 \end{bmatrix} \begin{bmatrix} x_n \\ y_n \end{bmatrix} \mod(N), \tag{1.6}$$

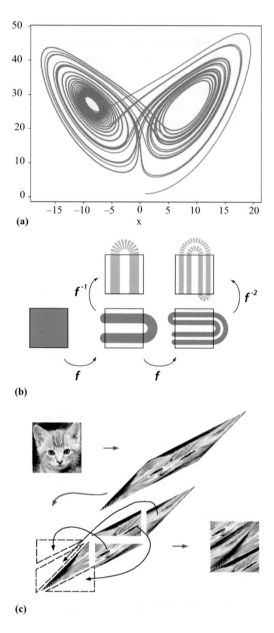

FIGURE 1.7   Examples of chaotic systems. (a) the Lorenz attractor, (b) the Smale horseshoe map, (c) the Arnold's cat map, and the logistic map (Figure 1.6a).

where c and d are positive integers and the modulo mod(N) denotes modular arithmetic; N is called the *modulus* of the congruence (e.g., modulus mod(1) takes the unit square into itself). Arnold had illustrated the map using a cat. It belongs to a class of dynamical systems which can be understood completely. They are extremely "chaotic" (Figure 1.7c). Arnold's cat map transformation is classical image encryption algorithm; it can be generalized to the 3D case as well. There are many other examples of chaotic dynamical systems, e.g., van der Pol oscillator, the Henon map, Baker's map, Rössler attractor, exponential map, cellular automata, etc. (see our references).

Could these and other mathematical models of chaos approach real world turbulence? Perhaps, this question is still open but it does not sound like science fiction: *"Chaos is a creator of information – another apparent paradox"* (Gleick 1987/2011).Our current understanding is that chaos can be related to a 2D system with a simple structure (i.e., restricted by a few degree of freedom), whereas turbulence is a spatiotemporal and high-dimensional system – phenomenon with many degree of freedom and irregular motion. This creates a huge complexity of turbulence relative to chaos. Last but not least, it is believed that chaos and turbulence both represent kind of dynamical system where the state appears unpredictable and unrepeatable.

In terms of the Navier–Stokes equations, the signature of chaos is no more than "sensitive or rough dependence on initial data." This fundamental property defines the relationship between *deterministic chaos* and turbulence in mathematical sense; however, today it may be difficult to say how the abstraction theory of chaos will make progress in natural science of turbulence...

1.3.3 COHERENT STRUCTURES

Since the 1960s, it has become clear that shear turbulence is not necessarily a random field of velocity (or vorticity) but may involve repetitive distinctive features with quasi-regular properties. *"No turbulent flow is completely without disorder, nor is there any completely without structure."* Later investigations have shown that many kinds of turbulent flows exhibit multiscale spatial organization and coherent motions. In particular, there is an enormous amount of data revealing the coherent structures in the turbulent boundary layer that has been collected over the past several decades. The data came both from many sources – laboratory observations, the Navier–Stokes numerical simulations, and theoretical considerations as well (see early reviews Hussain 1986; Robinson 1991 and books: Métais and Lesieur 1991; Ashworth et al. 1996; Hunt et al. 2000; Scott 2003; Shats and Punzmann 2006; Holmes et al. 2012; Tardu 2014; Tur and Yanovsky 2017).

Robinson (1991) defines coherent motions (or coherent structures) as

a three-dimensional region of the flow over which at least one fundamental flow variable (velocity component, density, temperature, etc.) exhibits significant correlation with itself or with another variable over a range of space and/or time that is significantly larger than the smallest local scales of the flow.

There are other definitions of coherent structures but we think that this one could be the most adequate in context with geophysical turbulence.

A coherent structure is characterized by a high level of space and phase correlations of vorticity, so-called "the coherent vorticity." Roughly speaking, they are recurrent patterns in turbulent flows. Coherent geophysical structures may occur in shear flow, jets, wakes, or boundary layers in the form of streamwise, hairpin, horseshoe vortex, filaments, and others types. These structures are well recognizable in the flow-visualization pictures and usually associated with "eddies" (although a coherent structure is not an eddy). In hydrodynamics, coherent structures mainly appear in the form of vortex streets (an example is a von Kármán vortex street behind a cylinder). Topologically, a coherent structure consists of many coexisting and non-overlapped elements (as saying "eddies"), which are not superimposed and have own spatial domains and boundaries.

Thus, distinguished properties of coherent structures can be specified as the following; they

- Exhibit the most evident geometrical features.
- Accumulate vorticity that retain for longer time.
- Include hairpin/horseshoe/ring-type vortices (or eddies).
- Produce energetically dominant regions.
- Have distinguished statistical properties.
- Consist of mixed vortexes/eddies elements.
- Exhibit spatiotemporal complexity and nonlinearities.
- Dominate at very high Reynolds number.
- Influence on mean flow.
- Demonstrate structural self-similarity.

There are more detailed characterizations of coherent structures that can exist in complex nonlinear media – gas, plasma, planetary boundary layer (PBL), the Cosmic web, optical media, etc. (e.g., Reguera et al. 2001).

The formation of coherent structures often is due to an instability of one kind or another depending on initial conditions (e.g., in laminar free shear flow coherent structures can occur due to the initial Kelvin—Helmholtz instability, which produces rolling-up vortex sheet). As well known, initial conditions play a key role in the development of turbulence and thereby define the origin and evolution of coherent structures. Interaction between coherent structures is an important mechanism of their triggering (known as the *bursting process*, e.g., burst-sweep event) or decay. Sometime a process of *regeneration* is observed – the occurrence of a new coherent structure after decay old one. Individual structures can split into substructures or merge with other coherent structures. Visually, coherent structures can be specified as quasi-deterministic low-dimensional dynamical systems of complex geometry and different scales.

There are various instrumental techniques for detection of coherent structures from experimental and/or numerical data (Holmes et al. 2012). The best know evidence is still *flow visualization*. Because individual coherent structures are 2D or 3D by definition, dominant energetically, and exhibit *morphological peculiarities*, their quantitative analysis requires the use of both spectral-statistical and pattern recognition techniques. The most popular and well-developed method is the *Karhunen–Loève* (KL) decomposition allowing the extraction of those dominant "modes" or empirical eigenfunctions that carry the greatest kinetic energy on average (that is a key feature of coherent structures). On the other hand, KL decomposition provides evaluation of a sequence of subspaces of increasing dimension, in which the Navier–Stokes equations can be approximated using the Galerkin projection method that yields relatively simple sets of ordinary differential equations needed for numerical modeling.

There is also an idea about the relevance of coherent structures to strange attractors because of the concentration of turbulent kinetic energy; however, it is difficult to obtain clear analytical or numerical evidence. Much more easily to explore coherent structures (and attractors) using a statistical approach or *fractal* concept (see Section 1.3.4).

Nowadays, a problem of identification and extraction of coherent structures from overall flow is considered as a subject of applied mathematics. In fluid mechanics, two main approaches – *Eulerian* and *Lagrangian*, are used for specification of turbulent flow. Correspondingly, these approaches are applicable to coherent structures as well. In practice, however, it is commonly believed that Lagrangian tools are more preferable than Eulerian approach (simply because of a float structure of observed or generated data).

Recent developments (Haller 2015; Serra and Haller 2016) show that turbulent coherent structures can be studied in four main contexts: 1) Eulerian coherent vortices (ECV) which are spatial patterns in the instantaneous velocity field, 2) Lagrangian coherent structures (LCS) which are persistent spatial patterns in tracer distributions, 3) Objective Eulerian coherent structures (OECS) which are instantaneous limits of LCS, and 4) Exact coherent states (ECS) which are persistent temporal patterns in the velocity field. ECV are well studied in fluid dynamics but have no universally detection algorithm. In contrast, LCS, OECS, and ECS have various detection capabilities and their application to complex dynamical systems requires developments of mathematical and numerical models (Haller 2015; Serra and Haller 2016). Several computer examples of LCS are shown in Figure 1.8.

From mathematical prospective, LCS by definition is a skeleton of the repelling material surfaces representing a map of trajectories in a dynamical system. LCS characterize a local structure of unsteady flow through the *Finite-Time Lyapunov Exponent* (FTLE) ridges, which are spatial derivatives of the scalar FTLE field. FTLE-based techniques were developed to extract LCS form overall flow, particularly for identification of complex flow structure and visualization of the velocity field (Haller and Yuan 2000; Haller and Sapsis 2011).

Among many types of natural coherent structures, LCS are of a great interest in geophysics and astrophysics (e.g., Bennett 2006; Venditti et al. 2013; Prants et al. 2017). Geophysical LCS can be detected and recognized using remote sensing observations (Chapter 5).

1.3.4 (Multi)Fractal

The word fractal was coined by Benoit Mandelbrot in 1975 from the Latin *fractus* and means "broken" or "fractured." Mandelbrot devised a mathematical way of

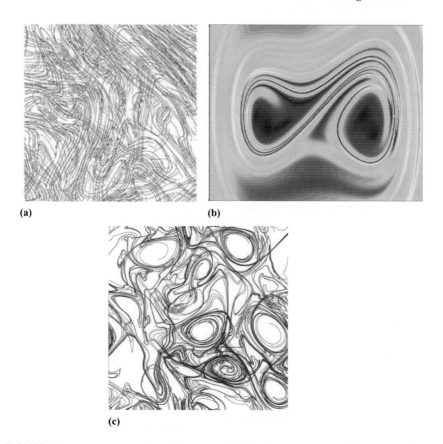

(a)   (b)

(c)

FIGURE 1.8   Lagrangian coherent structures LCS. (a) Hairpin vortices, (b) cat's eye eddies, and (c) vortex boundaries. Computer modeling. (After Haller and Sapsis 2011; Haller 2015).

describing irregular shapes, patterns, and processes which cannot be addressed by classic geometry. Fractals can be easily found in nature – such as coastlines, clouds, mountain ranges, trees, leaves, show flakes, crystals, lightning, sea shells, etc.

Today, fractal is used as a common criterion in the chaos theory and DST. The interest of physicists to this problem was stimulated by the pioneering work of Benoit Mandelbrot and his notable book "The Fractal Geometry of Nature" (1983). He has remarked that *"turbulence involves many fractal facets"* and claimed that *"Fractals... make it possible to apply the technique of self-similarity to the geometry of turbulence"* (Mandelbrot 1983, pages 97 and 102). Indeed, over decades, it was shown that turbulence is the source of many fractal, self-similar, and nonlinear scaling properties. Topological (geometrical) features of turbulent flow – chaotic patterns, coherent structures, and attractors– can be explored using a concept of *fractal dimension*.

Leaving aside mathematical details and rigor, we just outline the definition of fractal as geometrical object with non-integer dimension. Mandelbrot (1983, page 15) has stated that *"A fractal is by definition is a set for which the Hausdorff–Besicovitch dimension strictly exceeds the topological dimension."* The Hausdorff dimension is direct measure of roughness (or lack of smoothness) of the surface embedded in a 3D Euclidean space. It was first introduced by mathematician Felix Hausdorff in 1918. The significant advances were made by Abram Samoilovitch Besicovitch and this dimension is also commonly referred to as the *Hausdorff–Besicovitch dimension*.

From a geometrical point of view, the strange attractor is, as a rule, a fractal with the Hausdorff dimension (Mandelbrot 1983). We say "strange attractor" meaning that a system has chaotic and unpredictable behavior or the unfamiliar (irregular) geometric structure. In this regard, the Hausdorff dimension is a quantitative characteristic of complexity and chaotization of a motion; it is also a criterion to distinguish ergodic and non-ergodic properties of nonlinear dynamical systems (see, e.g., Schroeder 1991). Fractal geometry and mathematical details are discussed in excellent books (Feder 1988; Edgar 2008; Falconer 2014). Let's consider briefly basic characteristics of fractals and their applications in turbulence science.

Fractal dimension. Among a number of known definitions of the dimension, we define and understand the fractal dimension as *an index characterizing the complexity, similarity, and variability of objects, signals, processes, or phenomena*. It means that fractal dimension is universal statistical parameter of the state for any (non)linear chaotic dynamical system with many degrees of freedom. In certain sense, this definition exactly corresponds to current remote sensing problems and challenges.

The definition of the Hausdorff measure and dimension seem quite technical (Falconer 2014); calculations of this dimension for a real dataset are difficult; therefore, in practice, the box-counting method is commonly used. The *box-counting dimension* (also known as *Minkowski–Bouligand dimension* or *Minkowski dimension* or in some applications simply "fractal dimension") is given by

$$D = -\lim_{r \to 0} \frac{\log N(r)}{\log(r)} \text{ or } D = \lim_{r \to 0} \frac{\log N(r)}{\log(1/r)}, \tag{1.7}$$

where N(r) is the minimum number of cubes with the size r whose completely covers the volume r^n in n-dimensional space. The definition (1.7) means that $N(r) \sim r^{-D}$ for small r and the limit means that the dimension D is finite. The fractal dimension D cannot exceed the total Euclidean n-dimensional space and can be as integer as well non-integer (fractional). The dimension D as defined by (1.7) is really called a "capacity" and was first defined by Kolmogorov in 1958.

The box-counting dimension is computed using log/log diagram by plotting log(N) vs. log(1/r) for different cube sizes r and then the slope of a linear regression is defined. The value of the slope is the fractal dimension D (computational method is shown in Figure 1.9). As for real or numerical experiments the limit $r \to 0$ in (1.7) is just a graphic realization of the range of scales r where the relation (1.7) is still valid.

Note that the Hausdorff dimension in general is not equal to the box-counting dimension in mathematical sense by definition (Falconer 2014). Both dimensions can be equivalent for strictly self-similar sets such as classical fractals. Meanwhile, the box-counting method is simple, suitable, and useful for fractal analysis of many natural objects, scaling processes, and quasi-stochastic data sets (Mandelbrot 1983). Box-counting method is a very practicable in remote sensing data analysis.

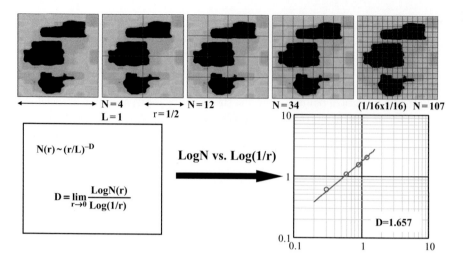

FIGURE 1.9 Computing box-counting fractal dimension. Schematic illustration.

Alternatively, in the case of a single object of complex geometry (e.g., atmospheric cloud, turbulent spot, or vortex), the fractal dimension D_s can be calculated using so-called *perimeter-area relation*

$$P \propto \left(\sqrt{A}\right)^{D_s},\qquad(1.8)$$

where P and A are measured perimeter and area of a single object with a boundary (Mandelbrot 1983). This particular relation is widely used for fractal analysis of individual objects visible in the images. For example, both fractal algorithms (1.7) and (1.8) have been tested and applied for geometrical analysis of foam/whitecap coverage from ocean aerial photography (Raizer et al. 1994).

Classical fractals. Most famous examples of classical fractals are the Cantor set, the Sierpinski Triangle, the Koch Snowflake, The Dragon Curve, the Julia set, and the Mandelbrot set (Figure 1.10). A given class of geometrical fractals is constructed using an equation that undergoes recursions. The shape is divided on many parts by the same scaling rule, and each part resembles the shape of the whole. This property is related to *self-similarity*. Thus, a fractal is a geometrical object that is similar to itself on all scales (such a fractal is known as "scaling fractal;" there are also "non-scaling" fractals).

A famous example is the Julia set (named after the French mathematician Gaston Julia who investigated this set in 1918), given by the family of complex quadratic polynomials:

$$f_c(c) = z^2 + c,\qquad(1.9)$$

where c is a complex parameter. By definition, the Julia set is the closure of the repelling periodic points. The Julia set is a 2D map over the complex plane; it exhibits self-symmetry, real axes symmetry, and rotational symmetry (depending on complex constant c). The Julia set associated with the complex function (1.9) may be depicted

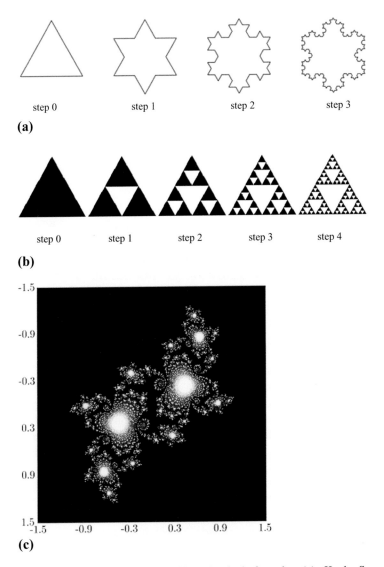

(a)

step 0   step 1   step 2   step 3

(b)

step 0   step 1   step 2   step 3   step 4

(c)

FIGURE 1.10 Geometrically self-similar classical fractals. (a) Koch Snowflake, $D = \log 4/\log 3 \approx 1.261$, (b) Sierpinski Triangle, $D = \log 3/\log 2 \approx 1.585$, and (c) the Mandelbrot or Julia set (inside every Mandelbrot set is an infinite number of Julia sets). The boundary of the Mandelbrot set has Hausdorff dimension $D = 2$.

using the following algorithm (Scientific Programming with Python. Internet https://scipython.com/book/chapter-7-matplotlib/problems/p72/the-julia-set/):

1. For each point, z_0, the complex plane such that real part $-1.5 \leq \mathrm{Re}\,[z_0] \leq 1.5$ and imaginary part $-1.5 \leq \mathrm{Im}\,[z_0] \leq 1.5$, iterate according to $z_{n+1} = z_n^2 + c$.

2. Color the pixel in an image (as shown in Figure 1.10c) corresponding to this region of complex plane according to the number of iterations for $|z|$ to exceed some critical value $|z|_{max}$ (or black if this does not happen before a certain maximum of iterations n_{max}).
3. Compute and plot the Julia set for complex $c = -0.1 + 0.65\iota$ using $|z|_{max} = 10$ and $n_{max} = 500$. For a Julia set with $c \ll 1$, the capacity dimension is

$$D = 1 + \frac{|c|^2}{4\ln 2} + O\left(|c|^3\right) \text{ (Ruelle 1995)}.$$

For the Mandelbrot set, the map is computed with all z-values equal to 0. Both the Julia set and the Mandelbrot set algorithms are available in MATLAB (e.g., Lynch 2004).

In general, a fractal has the following features (Falconer 2014):

- It has a fine structure at arbitrarily small scales,
- It is too irregular to be described in traditional Euclidean space,
- It is self-similar (at least approximately or stochastically),
- It has exhibited scale invariance,
- It has a Hausdorff dimension which is greater than its topological dimension,
- It has a simple and recursive definition.

Two categories – deterministic and stochastic (or random) fractals – are distinguished. Deterministic fractals (e.g., Julia set and Mandelbrot set) are generating using exact scaling transformations by the same rules at each iteration. Random fractals are created using random scaling transformations and applied for modeling and characterization of natural objects and stochastic processes. The basic random fractal model is a generalization of the Brownian motion, called fractional Brownian motion (Falconer 2014).

Self-similarity. This is a typical property of fractals. Self-similarity means that a whole or a small part of the system looks and behaves the same, or more specifically, that the function is the same under a discrete subset of dilations. A self-similar geometrical object is exactly or approximately similar to a part of itself. Mathematically, self-similarity can be defined more precisely. Any function $f(x, t)$ is self-similar (with self-similarity exponent γ) if it verifies that $f(\lambda x, t) = \lambda^\gamma f(x, t)$, $\lambda > 0$ is the scaling factor (dilations). According to this definition, three following types of self-similarity are distinguished:

- **Exact self-similarity** – it is the strongest type of self-similarity; the fractal appears identical at all scales. Fractal, defined by iterated function systems often display exact self-similarity (example is Koch Snowflake).
- **Quasi-self-similarity** – it is incomplete form of self-similarity; the fractal appears approximately (but not exactly) identical at different scales. The copies of the entire fractal can be distorted and degenerate forms. Fractals defined by recurrence relations are usually quasi-self-similar but not exactly self-similar (example is Mandelbrot set).

- **Statistical self-similarity** – it is the weakest type of self-similarity; fractal is generated by stochastic process and has numerical and statistical measures which are preserved across scales. Random fractals are statistically self-similar, but neither exactly nor quasi-self-similar (examples are random Koch curve and fractional Brownian motion). Statistical self-similarity is a property of many natural objects and processes.

The definition of statistical self-similarity can be formulated as the following. A stochastic process $\{X(t), t \geq 0\}$ is said to be self-similar if its finite dimensional distributions satisfy the scaling relation

$$X(\lambda t) \overset{d}{=} \lambda^H X(t), \tag{1.10}$$

where $\overset{d}{=}$ denotes equality of probability distributions and H refers to the Hurst exponent (self-similarity parameter). In practice, the Hurst exponent is a statistical measure used to classify *randomness* of time series or signal.

An important class of statistical self-similar random processes is known as "$1/f$ process," where f is frequency. It is generally defined as a process with power spectra $S(f) \propto f^{-\beta}$ with spectral exponent $\beta = 2H + 1$. The properties of the Hurst exponent are summarized as the following: $0 < H < 1$; $H = 0.5$ for a Brownian motion (random walk); $H > 0.5$ for a persistent (long-term memory, correlated) process; $H < 0.5$ for an anti-persistent (short-term memory, anti-correlated) process. Furthermore, the Hurst exponent is related to a fractal dimension of a one-dimensional time series $D = 2 - H$ or $2D = 5 - \beta$ ($1 < D < 2$) at $1 < \beta < 3$. For a self-affine surface in n-dimensional space $D = n + 1 - H$ (Mandelbrot 1983).

Scaling. In general sense, scaling is the expression of a physical law in terms of equations using powers of dimensionless quantities. The importance of scaling can be understood better with this quotation:

> *...scaling laws never appear by accident. They always manifest a property of a phenomenon of basic importance, "self-similar" intermediate asymptotic behavior: the phenomenon, so to speak, repeats itself on changing scales. This behavior should be discovered, if it exists, and its absence should also be recognized. The discovery of scaling laws very often allows an increase, sometimes even a drastic change, in the understanding of not only a single phenomenon but a wide branch of science. The history of science of the last two centuries knows many such examples.*

> (Barenblatt 2003, Preface, page xiii)

Two forms of scaling are considered, namely: first, obtained through a dimensional analysis and second, related to the manifestation of self-similarity. The first form of scaling is a very powerful concept in physics because it allows the analysis of a phenomenon on a vastly different scale than it is observed in reality. The theory and applications of dimensional analysis and scaling in physics, mechanics, and engineering are discussed in classical textbooks (Sedov 1993; Barenblatt 1996, 2003). The second form of scaling (in which we are focusing on) is closely related to *fractalization*, particularly to self-similarity of complex objects (fractals). More specifically, a fractal involves hierarchy of different scales and exhibits scaling within the certain (limited) spatial and/or temporal domains. Considering, e.g., the Koch curve in Figure 1.10a, the scale is decreased by 1/3 every iteration from 1/3, to 1/9 and 1/27, the detail increases four times from 4 to 16 and 64, with fractal dimension $D = \log(4)/\log(3) = 1.26$. This indicates that we can see far more small things than large ones if we apply scaling process.

Mathematically, scaling can be defined as a finite transformation that enlarges or diminishes details of the system. Technically, two key scaling approaches – similarity-based scaling and dynamic model-based scaling, are often used in natural science. In geophysical applications, scaling of the system should be defined as accurate as possible in order to model and/or predict behavior and dynamics of the system with highest probability. This makes multiple scaling one from the fundamental concept in geosciences and remote sensing (Quattrochi et al. 2017). Scale invariance and/or scaling remain a major problem in phase transition and turbulence (for more detail see, e.g., Lesne and Laguës 2012).

In nature, many fractal systems have a far more complex scaling relation than simple fractals, usually involving a range of scales that can depend on their location within the system. In this case, single fractal measure is not enough to describe self-similarity of entire system (the simplest example is the Cantor set with some measure distributed on it). Systems having scale-dependent interrelated fractal structure are known as *multifractals* and can be represented in terms of intertwined fractal subsets with different scaling exponents.

Multifractal Formalism. Here we refer to textbooks (Frisch 1995; Mandelbrot 1999; Harte 2001; Ott 2002; Falconer 2014). A multifractal system is a generalization of a fractal system in which a single exponent (the fractal dimension) is not enough to describe its dynamics; instead, a continuous spectrum of exponents is needed (Harte 2001). This concept, known also as *multifractal analysis*, initially was introduced by Mandelbrot (1999) in context with cascade models of turbulence.

In 1985, Parisi and Frisch (see Frisch 1995) introduced a *multifractal formalism* (MF) to study the behavior of velocity increments in fully developed turbulence, considering the Navier–Stokes equations and turbulent intermittency. Further developments have led to an essential modernization of the MF theory and the creation of an efficient mathematical framework for statistical analysis of complex nonlinear dynamical systems including turbulence. Multifractal analysis consists of determining local behavior of the system at different scales. Instead of a single fractal dimension (or capacity), defined from the relationship $N(r) \propto r^{-D_0}$ for $r \to 0$ with D_0 = const, formula (1.7), in terms of box-counting measure, the multifractal scaling of the system is characterized by $N(r) \propto r^{-f(\alpha)}$ for $r \to 0$, where $N(r)$ is the number of boxes of length r required to cover the system in a small range of singularity strength from α to $\alpha + d\alpha$. Power exponent $f(\alpha)$ is called the *multifractal spectrum* and describes the statistical distribution of α. This is the main characteristics of multifractals. The singularity spectrum is defined as

$$f(\alpha) = \lim_{\delta \to 0} \lim_{r \to 0} \frac{\log\left[N(r,\alpha+\delta) - N(r,\alpha-\delta)\right]}{\log 1/r} \tag{1.11}$$

and the generalized dimension is given by

$$D_q = \frac{1}{q-1} \lim_{r \to 0} \frac{\log \sum_{k=1}^{N} (p_k)^q}{\log r}, \tag{1.12}$$

where $p_k(r) \propto r\alpha_k$ is a probability measure versus singularity strength α. The generalized dimension D_q is computed as a function of the order of the probability moment q. In order to evaluate D_q from experimental data, first we need to determine the set of values p_k by using a suitable normalization and then to create a log-log plot. The generalized fractal dimension D_q and the singularity spectrum $f(\alpha)$ are connected via the Legendre transformation (Bohr et al. 1998):

$$(q-1)D_q = \min_{\alpha} \left[\alpha q - f(\alpha) \right]. \tag{1.13}$$

The MF now is defined by a set of equations:

$$f(\alpha) = q\alpha(q) - (q-1)D_q, \tag{1.14}$$

$$\alpha(q) = q \frac{d}{dq}(q-1)D_q, \tag{1.15}$$

$$\tau(q) = \left[q\alpha(q) - f(\alpha(q)) \right], \tag{1.16}$$

$$\frac{d}{dq}\tau(q) = \alpha(q), \tag{1.17}$$

$$\frac{d}{d\alpha}f(\alpha) = q(\alpha). \tag{1.18}$$

Multifractal measure is a graph $f(\alpha)$ versus α having the shape of an inverted parabola (Figure 1.11). The singularity spectrum $f(\alpha)$ has some general properties: 1) the maximum value of $f(\alpha)$ occurs at $q = 0$ and $f(\alpha) = D_0$ which is the box-counting dimension or the Hausdorff dimension, 2) $f(D_1) = D_1$, and 3) the line joining the origin to the point on the $f(\alpha)$ curve where $\alpha = D_1$ is tangent to the curve (Ott 2002). Roughly speaking, singularity spectrum $f(\alpha)$ characterizes the original measure and its width indicates overall variability of the system. MF (1.14)–(1.18) allows computing a spectrum or a set of fractal dimensions of dynamic system at different spatial or temporal domains. At present, MF is a powerful tool in multiresolution data/image analysis including remotely sensed observations as well; however, the study of turbulence in natural world is still of great interest and importance.

Applications to turbulence. In fluid dynamics, turbulent flow is characterized by irregular (chaotic) movement of particles of the fluid. Turbulent flow is a nonlinear dynamic system with an extremely large number of degrees of freedom. Unlike laminar flow, mixing, intermittent, and irregularity in turbulent flow are very high;

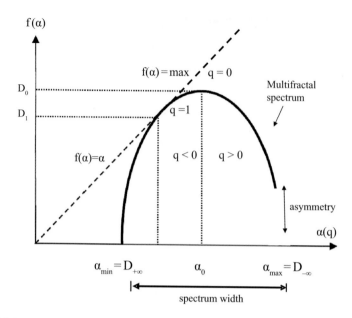

FIGURE 1.11   Multifractal formalism MF. Typical multifractal spectrum diagram.

therefore, the behavior of turbulent flow is complicated to predict and describe mathematically.

Because turbulence exhibits stochastic multiple scale structure and dynamics, many believe that turbulence (i.e., velocity field) has to be fractal or multifractal. From an observational perspective, this statement does not seem quite realistic because there is no convincing experimental evidence on it; nevertheless, there are a number of popular and applicable multifractal models, which are used for statistical analysis and simulation of turbulent features. Most advanced techniques operate with multifractal spectrum $f(\alpha)$, computed for time series or imaging data. Actually, two basic multifractal concepts are considered: 1) based on the self-similar solutions of the Navier–Stokes equations and 2) based on phenomenological multifractal approach.

The first concept presumes the scale invariance (or self-similarity) of unsteady incompressible 2D or 3D Navier–Stokes equations for isotropic turbulence that can be considered in terms of MF. The Navier–Stokes equations (back to Section 1.2.3) are formally invariant under the scaling transformation

$$\frac{\partial \mathbf{u}}{\partial t} + (\mathbf{u} \cdot \nabla)\mathbf{u} = -\frac{1}{\rho}\nabla p + \nu\nabla^2\mathbf{u} + \mathbf{F}(\mathbf{x},t), \quad \nabla \cdot \mathbf{u} = 0, \quad (1.19)$$

$$\mathbf{x} \to \lambda\mathbf{x}, \ \mathbf{u} \to \lambda^h\mathbf{u}, \ t \to \lambda^{1-h}t, \ \nu \to \lambda^{h+1}\nu, \ \rho \to \lambda^{h'}\rho, \ \varepsilon \to \lambda^{3h-1}\varepsilon \quad (1.20)$$

with $\lambda > 0$; h is an arbitrary scaling exponent. This arbitrariness allows the possibility of multiple scaling, hence, the solutions can, in principle, be multifractal (Schertzer and Lovejoy 1991). Under the assumption that energy dissipation ε is homogeneous in space and time, it follows that the scaling exponent is nor arbitrary but it should be fixed to h = h' = 1/3 that corresponds to the K41 theory. In general, self-similar solutions of (1.19) with (1.20) can be obtained using numerical simulations. There are some approximate solutions of the unsteady incompressible Navier–Stokes equations for fully developed turbulence ($v \to 0$) at low dimension. Recently, the wavelet-based MF has been developed and applied to turbulence (Massopust 2016). For more information on these and related works, we refer the interested reader to special literature.

The second, phenomenological concept, considers multifractal measure using the structure functions. Lagrangian structure functions of higher order are given by

$$S_n\left(r\right) = \left\langle \left[\left(\mathbf{u}\left(\mathbf{x}+\mathbf{r}\right) - \mathbf{u}\left(\mathbf{x}\right)\right)\mathbf{r}/r \right]^n \right\rangle \propto C_n r^{\varsigma(n)}, \qquad (1.21)$$

where ς(n) is a single scaling exponent. The average <...> is defined over the ensemble of particle trajectories evolving in the flow. The second order structure function $S_2(r) \propto C_2 \varepsilon^{2/3} r^{2/3}$ with scaling exponent ς(n) = n/3 corresponds to the Kolmogorov scaling law, with no corrections. In this case, the scaling exponent is related to the fractal dimension D via ς(n) = n/3 + (3 − D)(1 − n/3)(Frisch 1995). Specializing to n = 2, the power-law energy spectrum of 3D turbulence in the inertial range satisfies:

$$E\left(k\right) \approx C_\ell \varepsilon^{2/3} k^{-5/3} \left(k_\ell \right)^{-(3-D)/3}, \qquad (1.22)$$

where ℓ is the length-scale of the largest eddies in the Kolmogorov inertial interval. The K41 theory is recovered when D = 3, otherwise, at 1 < D < 3, the energy spectrum (1.24) becomes steeper than the Kolmogorov's $E(k) \propto k^{-5/3}$ spectrum. Note that (1.22) is identical to so-called β-model of turbulence (Frisch 1995).

Multifractal methods and models for characterization of dynamical turbulence are summarized in Figure 1.12. Some of them are interrelated and/or intersect by definition. It should be noted that the (multi)fractal measures, introduced here use phenomenological self-similar approach (but not solutions of the Navier–Stokes equations) that are believed to imitate turbulence dynamics, at least qualitatively. Unfortunately, direct experiments supported MF for turbulence are very limited and mostly conducted in laboratory wind tunnels (Harte 2001; Ott 2002).

In real-world observations, multifractal turbulent structures (of 2D or 3D geometry) can be detected and investigated using high-resolution imagery and topological processing (in early experiments, it was just flow visualization). The extracted (visible) trajectories, boundaries, patterns, surfaces, and/or volumes are analyzed using fractal-based methods as well as the Lyapunov characteristic exponent in the case of dynamical systems. Multifractal measures reflect scaling and dynamics of turbulent flow; therefore, an efficient criteria will be variations of the fractal dimension and/or multifractal spectrum, registered at different stages of turbulent flow. Indeed, turbulent coherent structures – strange attractors, have various attributes and/or configurations such as fractal-like vortexes, clusters, spots, intrusions, and multiphase flows. As a whole, detected *fractal signatures* can be perceived as

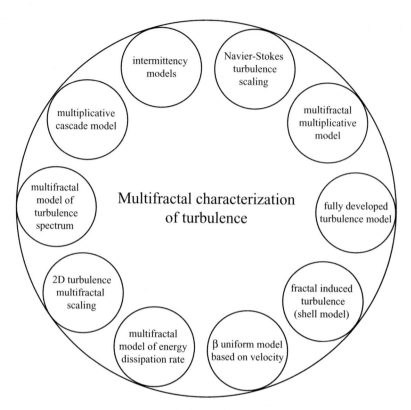

FIGURE 1.12   Multifractal methods and models for characterization of turbulence.

indicators of complex turbulent processes, fields, or distinct stochastic patterns in dynamical environment.

Geophysical applications. A number of geophysical applications have demonstrated the relevance and advantages of (multi)fractal analysis in studying a large class of stochastic natural phenomena and dynamical systems (e.g., Schertzer and Lovejoy 1991). In particular, these developments provided a broad enough framework to unify complex problems related to extreme events, variability, structural transformations, nonlinear interactions, and multiplicative cascades– the phenomena that have been identified with the term *nonlinear variability*. Fractal-based approach including MF is of a primary interest in remote sensing of multiscale dynamical phenomena/events in Earth's environments (specifically occurring in the atmosphere and oceans).

Fractals and multifractals already have been employed in a number of disciplines of nonlinear dynamics and geosciences (Dimri 2000; Lovejoy and Schertzer 2007) including weather and climate (Lovejoy and Schertzer 2013), ecology and aquatic science (Seuront 2010), hydrology (Bernardara et al. 2007), natural hazard (Cello and Malamud 2006), geology and seismology (Turcotte 1997), surfaces and scattering (Franceschetti and Riccio 2007), wave breaking field in sea (Raizer et al. 1994; Sharkov 2007), laser glint (Shaw and Churnside 1997), microwave and optical detection of ocean features (Raizer 2017, 2019). Today, (multi)fractals are also addressed in computer vision (Farmer 2014) and pattern generation (Banerjee et al. 2020). Other interesting but problematic area of (multi)fractal analysis is the macro financial environment and the stock market behavior – the problem, known as "market turbulence" (Mandelbrot and Hudson 2004).

Speaking on more simple things, not surprisingly that fractal viewpoint, created by Mandelbrot to explain natural complexity has found many applications outside the traditional fields of physics and mathematics. Nowadays, fractal geometry is not only an area of scientific discovery but a form of intellectual entertainment including computer art, design, abstract painting, and other visual artwork styles.

1.3.5 Self-Organization

Self-organizing systems were first studied in the 1950s and 1960s in thermodynamics and cybernetics. Ilya Prigogine generalized these data and established a new scientific view on the problem (Nicolis and Prigogine 1977; Klimontovich 1991). According to accepted definition, self-organization is a process of the formation of ordered structures in a non-equilibrium dynamical system. In classical statistical physics, this subject is related to study of out-of-equilibrium dynamics, i.e., dynamical phase transition.

At present, such *far-from-equilibrium* dynamical systems are called *self-organized criticality* (SOC) systems. The idea of SOC has inspired physicists for more than 25 years. This is a widely discussed topic with many examples and applications which is a part of a new discipline known as *synergetics* – the creation of "order from chaos" (or "the order in chaos").

The SOC has been introduced first in the paper (Bak et al. 1987; see also book Bak 1996) using the example of a sandpile.

Dynamical systems with extended spatial degrees of freedom naturally evolve into self-organized critical structures of states which are barely stable. Flicker noise, or frequency 1/f noise, can be identified with the dynamics of the critical state. This picture also yields insight into the origin of fractal objects.

(Bak et al. 1987)

SOC is considered today to be one of the mechanisms by which complexity arises in nature. In original model, "criticality" means that a system does not need to be tuned by external parameters, since an "instability threshold" is established by common physical conditions throughout a system. Mathematically, any SOC system exhibits a power law distribution of event sizes, as well as rich intermittent dynamics in time that is associated with the power-law scaling inherent in a wide variety of natural and artificial processes. This provides close relationship between SOC and (multi)fractal concepts in terms of dynamical data analysis.

SOC is constituted by out-of-equilibrium systems driven with constant rate and made of many interactive components, which possess the following fundamental properties (Sornette 2006):

- A highly nonlinear behavior, namely, essentially a threshold response.
- A very slow driving rate.

- A globally stationary regime, characterized by stationary statistical properties.
- Power distributions of event sizes and fractal geometrical properties (including long range correlations).

SOC typically observed in slowly driven non-equilibrium systems with many degrees of freedom and strongly nonlinear dynamics. SOC concept formulates conditions in which open systems self-organize, e.g., by means of critical threshold, boundary dissipation, scale separation, or metastability. Chaotic system tends to move forward from initial conditions to a critical state and structurizes itself without external control. Figure 1.13 illustrates schematically SOC concept, adapted to the case of a chaotic attractor.

Self-organization phenomena/events can be found in the in animate world: plasma, formation of cloud streets, convection instability, Taylor–Couette flow, roll patterns, hexagonal patterns (Bénard cells), turbulent transition, defects, crystallization, eco-systems, planetary systems, galaxies, etc. (Jensen 1998; Hergarten 2002; Solé and Bascompte 2006; Pruessner 2012; Marov and Kolesnichenko 2013). SOC systems can also be observed in the Earth's atmosphere in the form of atmospheric cascades (Selvam 2017). SOC models are used in mathematical systems such as cellular automata; it is an example of the related concept of *emergence*.

In turbulent flow, SOC structures appear spontaneously due to stochastic fluctuations, secondary instabilities, and strong nonlinear interactions. At some critical moment, turbulence becomes an ensemble of semi-organized highly correlated motions exhibiting self-similar properties. This event involves so-called "turbulent fragmentation process" or fragmentation cascades that finally may result to turbulence collapse. Under certain conditions, the emerging complex SOC structures (e.g., turbulent coherent structures) and relatively stable self-similar cell patterns can be

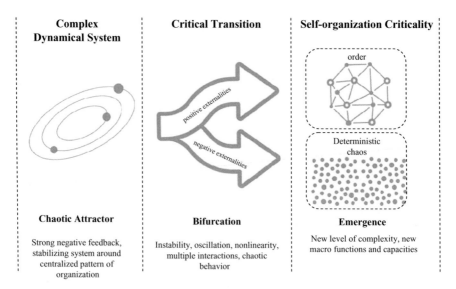

FIGURE 1.13 Self-organized criticality SOC concept. (Based on System Innovation.)

manifested as a fractal. From a mathematical perspective, the links between SOC and fractal geometry are numerous – many SOC systems can be described by fractal geometry when described in appropriate phase space. Any image of an SOC system (e.g., remotely sensed imagery of cloud dynamics in the atmosphere) is likely to be fractal, and the long-term dynamic evolution of an SOC system is likely to be fractal as well.

In the case of fully developed turbulence, the process of SOC and the formation of self-similar fractal (coherent) structures can be described using *random multiplicative models* introduced and extensively studied by the Russian school (Frisch 1995). Indeed, the theory provides a possible interpretation of the intermittency phenomenon which issues self-organization in turbulence as well. Other theories involve either the fractional Fokker–Planck equation or the Ginzburg–Landau model (their discussion is beyond the scope of this section). An appropriate numerical analysis of SOC can also be done in terms of the energy spectrum (McComb 2014).

Evolution of the energy of self-organization state can be described using the turbulent kinetic energy spectrum $E(k,t)$ in the form

$$\frac{\partial E(k,t)}{\partial t} = T(k,t) - 2\nu k^2 E(k,t) + W(k,t), \qquad (1.23)$$

where $T(k,t)$ is the transfer spectrum describing the energy redistribution due to interactions, k is the wavenumber, ν is the kinematic viscosity, and t is time. The second term $2\nu k^2 E(k,t)$ describes the energy dissipation due to viscosity and the last term $W(k,t) = 4\pi k^2 \langle \mathbf{u}(-\mathbf{k},t) \cdot \mathbf{f}(\mathbf{k},t) \rangle$ is the spectrum of the work by stirring forces, $\langle \mathbf{u}(\mathbf{k},t) \rangle$ is the averaged instantaneous velocity field. The asymptotic solution for the energy spectrum after self-organization $E(t) = E_\infty = \frac{\varepsilon_w}{2\nu} = const$ ($\varepsilon_w = \int W(k)dk$ is the energy injection rate) can be obtained at $\varepsilon_w = \varepsilon$ (where ε denotes the dissipation rate) and $t \rightarrow \infty$ (McComb et al. 2015). Equation (1.23) describes the time variation of the wavenumber distribution of turbulent energy; it is based on the von Karman–Howarth equation (see, e.g., Monin and Yaglom 1975).

There are some efforts to obtain 2D Navier–Stokes self-organized solutions for transitional shear flows using numerical simultions. Here we recall very laconic statement: "*...nonequilibrium, nonlinearity, and the dissipativity of a medium are fundamentally important for the formation of such* (i.e., self-organized) *structures*" (Belotserkovskii 2009). In this connection, it will be essential to incorporate SOC into turbulence scenario. There are two options to complete that: first is to incorporate SOC into the Navier–Stokes initial conditions and second is to integrate SOC into multifractal models. For example, SOC can be considered as a topological measure of the turbulent/non-turbulent interface in a turbulent boundary layer which may exhibit self-organized motion resulting to spontaneous emergence of self-organized patterns and their bifurcations.

Note on the literature. Here is a list of additional literature which provides more insights in dynamical systems and turbulence available for future reading. Introduction to chaos is in excellent textbook (Hirsch et al. 2004).

Dynamical systems, chaotic dynamics, stability, attractors, and turbulence are considered in books (Gaponov-Grekhov and Rabinovich 1992; Moon 1992; Ruelle 1995; Robinson 1998; Branover et al. 1999; Ott 2002, Devaney 2003; Mandelbrot 2004; Lichtenberg and Lieberman 2010; Shivamoggi 2014; Elhadj 2019).

Instabilities and chaos are discussed in the book (Glendinning 1994). Complex systems are described in books (Mitchell 2009; Majda 2016).

Physics of chaos, chaotic systems, and instabilities are considered in books (Zaslavsky et al. 1991; Bohr et al. 1998; Zaslavsky 2007). Chaotic theory of turbulence is considered in book (Klimontovich 1991).

1.4 COMPUTATIONAL FLUID DYNAMICS

Computational fluid dynamics (CFD) is the numerical analysis of fluid flow, heat transfer, and related phenomena. CFD solvers contain a set of algorithms used for modeling and simulation of the flow and its evolution. The foundation of CFD is built on the Navier–Stokes equations, the set of partial differential equations that describe fluid flow. With CFD, the area of interest is subdivided into a large number of cells or control volumes. In each of these cells, the Navier–Stokes partial differential equations can be rewritten as algebraic equations and then solved numerically, iteratively, that yields a complete picture of the flow throughout the domain.

CFD techniques date back to the early 1970s, and the first commercial CFD software became available in the early 1980s. Since then, CFD have come a long way increasing its capability of providing accurate numerical analysis of various flow regimes and geometries. CFD models have been developed for physical phenomena such as turbulence, multiphase flow, chemical reactions, and radiative heat transfer.

Nowadays, CFD has become an integral part of the applied science and engineering dealing with aerodynamics, hydrodynamics, magnetohydrodynamics (Biskamp 2003), and geophysics where turbulence is important factor. CFD are beneficial and cost-efficient way in the development of novel airspace technologies and applications as well. In particular, CFD become an exceptional artwork in fluid mechanics and turbulence research providing 1D – 3D simulations of complex hydrodynamic flows (such as, e.g, submarine wake), involving large numbers of input parameters, various geometries, and mixed boundary conditions.

1.4.1 Introduction

Most geophysical turbulent flows in our environment are too complex and diversified to be resolved by analytic calculations. In this case, the problems must be solved by CFD methods. Anderson (1995, pages 24–25) defines CFD as "the art of replacing the integrals or the partial derivatives (as the case may be) in these equations by discretized algebraic forms, which in turn are solved to obtain numbers for the flow field values at discrete points in time and/or space. The end product of CFD is indeed a collection of numbers, in contrast to a closed-form analytical solution."

Hence, the cornerstone of CFD is the fundamental governing equations of fluid dynamics – the continuity, momentum, and energy equations. According to this definition, CFD should be able for investigating any turbulent flow and predict its behavior in space and time that is otherwise impossible to measure or calculate purely theoretically. Obviously, methods of CFD have unique capabilities in geosciences and remote sensing; therefore, here is a brief survey of CFD models and techniques.

CFD have relatively short history. Mathematical schemes and techniques for numerical solutions of the basic equations of fluid dynamics were created in the 1950s, first realistic simulations of 2D turbulent flows were performed in the late 1960s, and numerical modeling of 3D flows was started in the late 1980s. Since that time, computer revolution and software developments have triggered CFD technology dramatically. Nowadays, CFD is everyday tool in scientific research, engineering design, optimization, and analysis of fluid mechanics problems.

The CFD simulation process consists of three primary steps (and many intermediate operations) which are the following:

1. **Pre-Processing**. This step defines the flow model and identifies the fluid domain of interest. The domain of interest is divided into smaller segments known as mesh generation step.
2. **Solver**. This step provides numerical solutions of a set of equations related to the problem. The computational method varies depending on the task, computer capability, and specifics of software codes.
3. **Post-Processing**. This step performs analysis of the results using different methods of data representation such as graphical illustrations (streamlines, plots, curves, etc.), flow visualizations, and datasets. This step may also provide statistical treatment of computer data, their further assessment and application.

CFD describe a turbulent flow through the numerical solutions of unsteady 2D and/or 3D Navier–Stokes equations. To provide this process, CFD software implements *discretization* of the governing equations using different mathematical methods; the most known are the finite difference method or the finite volume method. The fluid equations are replaced by discrete approximations at grid points that must be close enough so that the solution is independent of the grid point spacing. An important (automatic) procedure is the creation of the relevant computational domain (or computational grid) which includes flow mesh and initial and boundary conditions. To simulate flow dynamics, the computational domain has to be at least several times larger than the largest turbulent eddies. Therefore, comprehensive CFD research of turbulence requires powerful computers and large computing time. Hence, it is necessary to reduce significantly the cost of computer simulations. Such a reduction can be obtained using a concept of scale separation (e.g., Sagaut et al. 2013).

CFD have many advantages over experiments which are 1) relatively low cost of numerical research in comparison with laboratory studies, 2) better visualization and deeper insight into results, 3) testing various turbulence stages with variable set of parameters, 4) practically unlimited volume of information, 5) cyclical development and reproduction capability, and 6) scale-invariant model implementation.

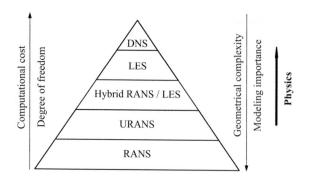

FIGURE 1.14 The hierarchy of CFD methods for turbulence modeling. (Based on Sagaut et al. 2013.)

TABLE 1.3
Comparison of CFD Methods in Turbulence Modeling

Method	Grid Size	Number of Time Steps	Reynolds Number Dependence and Typical Range	Model Description
DNS	10^{16}	$10^{7.7}$	Strong 10^2–10^8	Compressible flow, Navier–Stokes solver in 3D for complex geometry
LES	$10^{11.5}$	$10^{6.7}$	Weak 10^2–10^5	Incompressible flow due to low-pass filtering of Navier–Stokes equations
RANS/ URANS	$10^8 - 10^7$	$10^{3.5}$–10^4	Weak 10^2–10^5	Steady and unsteady flow, time-averaged equations of fluid motion, Reynolds decomposition

CFD offer state-of-the art analysis of a variety of turbulent flows through the numerical solutions of governing equations of fluid dynamics including 2D and/or 3D incompressible time-dependent Navier–Stokes equations with mixed boundary conditions. The approach and performance of the computations are defined by the chosen CFD method; the most common are the following:

- Direct numerical simulation (DNS).
- Reynolds-averaged Navier–Stokes (RANS) models.
- Large eddy simulation (LES).
- Hybrid RANS/LES method.

Figure 1.14 shows hierarchy of the listed methods in terms of their complexity and major capabilities for turbulence modeling, e.g., level of resolved physics. Computer capabilities of these methods can be understood from Table 1.3. In some methods, grid size aims at numerical accuracy; however it does not mean that grid refinement will add any new physics to the solution. Because the application of CFD to turbulence research is very important, let's consider these methods in more detail.

1.4.2 DIRECT NUMERICAL SIMULATIONS (DNS)

Direct numerical simulations (DNS) are the most straightforward approaches to the solution of unsteady turbulent flows since the governing equations are discretized directly and solved numerically. DNS can obtain an accurate 2D and 3D time-dependent description of the flow completely without resorting to any model assumptions. Unfortunately, turbulent flows encountered in engineering applications exhibit such a wide range of excited length and time scales (shock wave, boundary, and free shear layers) at high Reynolds number (Re $\approx 10^5 - 10^8$) that full-scale DNS are still beyond our capabilities (Sagaut et al. 2013).

In DNS approaches, the unsteady incompressible Navier–Stokes equations are solved directly using highly accurate numerical schemes on a very fine spatial grid, thereby all characteristic length and time scales are resolved. The numerical algorithm includes spectral method, pseudo-spectral method, and finite difference method. Briefly, the evolving flow field of a scalar variable (e.g., velocity) can be approximated by a finite dimensional expansion in terms of the basis functions

$$u(x, t) \approx u_N(x, t) = \sum_{n=0}^{N} a_n(t) \phi_n(x), \qquad (1.24)$$

where $a_n(t)$ represents the time-dependent expansion coefficients and $\phi_n(x)$ are the basis functions that are chosen to best represent the flow (e.g., Jiang and Lai 2009).

DNS use all spatial scales from the Kolmogorov microscale $\eta = (v^3/\varepsilon)^{1/4}$ to the integral length scale of flow L (which is associated with the motions containing most of the kinetic energy). In the 3D DNS, the minimum number of points (or cells) in a computational grid is approximately $N \sim \mathrm{Re}^{9/4} = \mathrm{Re}^{2.25}$ with numerical resolution $L/\eta \sim \mathrm{Re}^{3/4}$ (the result is based on the Kolmogorov K41 scaling, see Frisch 1995; Pope 2000). The number N is known also as the *number of degrees of freedom* involved in a turbulent flow (Frisch 1995). Typically, at Re = $10^4 - 10^6$, the required computational grid consists of $N \sim 10^9 - 10^{12}$ discrete points or cells that will take a large amount of computing time to solve equations of fluid dynamics. This and other estimates (e.g., Pope 2000; Zikanov 2010; Andersson et al. 2012) illustrate the difficulty of DNS at very high values of the Reynolds number.

DNS have the following advantages: 1) provide detailed investigation of dynamical and structural properties of turbulent flow with good visualization, 2) generate a large volume of data, 3) enable to compute flow at variable set of parameters within specified range, and 4) the closure problem is circumvented. The disadvantages are very high computational cost (requires supercomputer) and necessity to use fast and high-order numerical schemes and algorithms.

DNS are an efficient tool for many practical applications and fundamental fluid dynamics research. The basic result is a complete picture of the evolution of a time-dependent flow fields of velocity $\mathbf{u}(\mathbf{x},t)$ and pressure $p(\mathbf{x},t)$. Visualization of simulated flow (and its details) provides a better understanding of physics of the phenomenon including the mechanisms of turbulence production and dissipation. However, the computational cost of DNS is very high and it increases as the cube of the Reynolds number $t_{comp} \propto Re^3(Sc)^2$, for gases $Sc \sim 1$, for liquids like water $Sc \sim 10^3$, and for very viscous liquids $Sc \sim 10^6$ (Andersson et al. 2012). There are two other approaches that have been commonly used to reduce computational cost of flow modeling; they are considered below.

1.4.3 Reynolds-Averaged Navier–Stokes (RANS) Method

RANS method (known also as RANS equations) is the oldest numerical approach, which still remains a primary and popular method for practical analysis of turbulence. RANS equations govern the averaged flow quantities, with the whole range of the scales of turbulence being modeled. Its advantages are simplicity, low computational cost (relative to DNS and LES), a broad selection of additional turbulence models (see below), and significant experience in software application. The well-known drawbacks of RANS models are low level of physics-based data assessment, the inability to predict complex separated flows, inability to account turbulent fluctuations and model the mean flow with significant errors. Some disadvantages are often acceptable in engineering applications; however, some of them are serious issues in advanced scientific research.

RANS equations represent time-averaged equations of fluid motions, i.e., the Navier–Stokes equations (1.1)–(1.3). An idea was proposed by Osborne Reynolds in 1895 (Tennekes and Lumley 1972). In RANS equations, instantaneous quantity is decomposed into its time-averaged and fluctuating quantities. The derivation of the RANS equations is based on the *Reynolds decomposition*, in which a scalar quantity φ (velocity or pressure) is represented by the sum $\varphi = \overline{\varphi} + \varphi'$, where φ, $\overline{\varphi}$, and φ' ($\overline{\varphi'} = 0$) are the instantaneous, mean, and fluctuating terms, respectively. The averaging is defined as $\overline{\varphi} = \lim_{T \to \infty} \frac{1}{T} \int_0^T \varphi(t) dt$. Splitting each instantaneous quantity (velocity and pressure) into its mean and fluctuating components, substituting and averaging yield the RAMS equations (incompressible)

$$\frac{\partial \rho}{\partial t} + \frac{\partial \rho \overline{u}_i}{\partial x_i} = 0, \tag{1.25}$$

$$\frac{\partial(\rho \overline{u}_i)}{\partial t} + \frac{\partial(\rho \overline{u}_i \overline{u}_j)}{dx_j} = -\frac{\partial \overline{p}}{\partial x_i} + \frac{\partial}{\partial x_j}\left(\overline{\tau}_{ij} - \rho \overline{u'_i u'_j}\right) + F_i, \tag{1.26}$$

where $\overline{\tau}_{ij}$ is the viscous stress tensor (dependent on the coefficient of viscous) and $R_{i,j} = -\rho \overline{u'_i u'_j}$ is called the *Reynolds stress tensor*. It is a symmetric tensor with six independent components. Equations (1.25)–(1.26) operate with three components of vectors $\mathbf{x} = (x_1, x_2, x_3)$, $\mathbf{u} = (u_1, u_2, u_3)$; correspondingly, indexes $i, j = 1, 2, 3$ assign three space coordinates. The Reynolds decomposition and averaging introduce additional quantities $\rho \overline{u}_i \overline{u}_j$ and $\overline{u'_i u'_j}$ which are not known. The system of Equations

(1.25)–(1.26) now is an unclosed system; the number of unknown quantities (pressure, three velocity components, and six stresses) is larger than the number of the available equations (continuity and Navier–Stokes). This is termed as so-called turbulence "closure problem." In order to close RANS equations, the Reynolds stress tensor $R_{i,j}$ must be specified in some way. To resolve this problem, a closure model with additional equations is invoked. The choice of adequate RANS closure models for turbulence is difficult task. According to discussion (Doering and Gibbon 1995; Launder and Sandham 2002), in many cases, closure models often are semi-empirical approximations and/or "intuitive empiricism." Here we mention only two mainly used RANS closure models, namely:

1. **Eddy Viscosity Models** (via the Boussinesq hypothesis); Reynolds stresses are modeled using an eddy. This model is reasonable for simple turbulent shear flows: boundary layers, jets, mixing layers, channel flows, etc.
2. **Reynolds-Stress Models** (via transport equations for Reynolds stresses). This model is more advantageous in complex 3D turbulent flows with large streamline curvature and swirl, but the model is more complicated computationally than the first one.

Further discussion of closure models is beyond the scope of this book. Paper (Durbin 2018) provides detailed review of the problem.

RANS method usually involves additional turbulence models, which are classified in terms of transport equations. The most known are the following: 1) one-equation model, 2) Spalart –Allmaras model, 3) two-equation model, 4) standard k–ε model, 5) RNG k–ε model or Re-Normalisation Group (RNG) k–epsilon model, 6) realizable k–ε model, 7) standard k–ω model, 8) SST k–ω model or Menter's Shear Stress Transport k–omega model, 9) Reynolds stress model, 10) Low Reynolds number modifications, etc. There are many other more specific RANS-related turbulent models (about 200 in total, see our references). These models are used for calculations of viscosity, accounting swirling and/or rotation flows and/or complex near-wall flows, prediction of intermittency, as well as in the form of closure models. Steady RANS (sometimes referred to as SRANS) methods are routinely used for scientific and engineering applications; however, they do not predict unsteady flow. Nowadays, more attention has shifted to the modification referred to as Unsteady RANS (URANS). As a whole, the RANS/URANS methods are affordable numerical framework for turbulence modeling on a regular basis; however, RANS/URANS predictions may have large discrepancies due to the uncertainties in definition of the Reynolds stresses. This disadvantage is partially resolved using more relevant methods known as LES and hybrid RANS/LES.

1.4.4 Large-Eddy Simulation (LES)

LES was first proposed in 1963 by Smagorinsky for atmospheric flow prediction. LES is considered as an intermediate method between RANS and DNS. The key concept of LES is scale separation through filtering (Sagaut 2006). In LES, instantaneous large-scale eddies (grid-scale) are computed directly, while small-scale eddies (subgrid) are modeled. For this, a low-pass spatial filtering of the Navier–Stokes equations is applied. As a result, LES provides significant reduction of the computational cost relative to DNS by ignoring the small length scales, which are most computationally expensive to resolve.

A filter operation is defined by the convolution

$$\overline{\varphi}(x,t) = \iint_{\Omega} \varphi(x',t')G(x-x', t-t', \overline{\Delta})dt'dx', \qquad (1.27)$$

where Ω is the fluid domain, $G(\dots)$ is the filter convolution kernel (or filtering function), and $\overline{\Delta} = (\Delta x \cdot \Delta y \cdot \Delta z)^{1/3}$ is the filter size. In (1.27) various forms of the filtering function can be used; the most common are the box filter and the Gaussian filer.

The resulting LES equations for an incompressible fluid in Cartesian coordinates are

$$\frac{\partial u_i}{\partial t} + \frac{\partial \overline{u}_i \overline{u}_j}{\partial x_j} = -\frac{1}{\rho}\frac{\partial \overline{p}}{\partial x_i} - \frac{\partial \tau_{ij}}{\partial x_j} + \nu\frac{\partial^2 \overline{u}_i}{\partial x_j \partial x_j}, \quad \text{filtered Navier} - \text{Stokes equation} \quad (1.28)$$

$$\frac{\partial \overline{u}_i}{\partial x_i} = 0, \quad \text{filtered continuity equation} \qquad (1.29)$$

where $\tau_{ij} = \overline{u_i u_j} - \overline{u}_i\overline{u}_j$ denotes the *subgrid scale* (SGS) *stress tensor*. The SGS stress is a symmetric tensor, invariant to Galilean transformation (i.e., the SGS tensor should be the same for any frames moving with different velocities) and determines the dynamical coupling between large and small scales of turbulence. Thus, LES of turbulence is not only filtering approach but also SGS model.

There are a variety of SGS models available; they simulate the energy transfer between the resolve scales and the subgrid scales. SGS models usually adopt the Boussinesq hypothesis

$$\tau_{ij} - \frac{1}{3}\delta_{ij}\tau_{kk} = -2\nu_t\overline{S}_{ij}, \quad \overline{S}_{ij} = \frac{1}{2}\left(\frac{\partial \overline{u}_i}{\partial x_j} + \frac{\partial \overline{u}_j}{\partial x_i}\right) \qquad (1.30)$$

(δ_{ij} is the Kronecker delta). The most popular SGS model is referred to as the Smagorinsky–Lilly SGS model, where the eddy viscosity is defined as $\nu_t = (C_s\overline{\Delta})^2|\overline{S}|$ with $|\overline{S}| = \sqrt{2\overline{S}_{ij}\overline{S}_{ij}}$, and $|\overline{S}|$ is the characteristic filtered strain rate (also known as strain magnitude), and $C_K = 0.1 - 0.2$ is the Smagorinsky coefficient. Theoretical value $C_S \approx 0.18$ for a homogeneous isotropic flow was found by Lilly (e.g., Sagaut et al. 2013).The isotropic part of the SGS stresses τ_{kk} is not modeled but added to the filtered pressure term.

It is believed that LES has better potential to provide more accurate and more reliable results than RANS approach. In turbulence analysis, LES has considerable advantages over RANS; we mention only three major: 1) enable to compute unsteady flow, 2) resolve flow structure in detail, and 3) provide analysis of transitional separated flow.

The overall performance and accuracy of LES are defined mostly by SGS models. The problem of both RANS and LES is a verification and validation of the obtained results that may require the comparison of numerical data with test (experimental, theoreticall) data.

Although formally LES is intrinsically superior (can solve many more scales than RANS but not as many as DNS), sometimes LES yields results that are less accurate and less reliable that those by RANS. This makes RANS products more affordable, convenient, and popular for applications even in research areas. Moreover, RANS models have the benefit of longer history while experience in LES is limited. Choosing between RANS and LES is a trade-off between computational cost and complexity, on one hand, and potential accuracy, on the other hand. This problem can be partially overcome using hybrid methods. More about RANS and LES can be found in books (Pope 2000; Lesieur et al. 2005; Sagaut 2006).

1.4.5   Hybrid RANS/LES Method

This is relatively new approach which has been developed over the last decades to combine the essential benefits of two famous methods – low computational cost in RANS and higher accuracy in LES (Sagaut 2006; Sagaut et al. 2013). Hybrid RANS/LES method provides some sort of blend between the equations governing the Reynolds-averaged and spatially filtered Navier–Stokes equations. The idea is to simulate only the largest turbulent eddy structures that can be adequately resolved on a given mesh. The smallest remaining structures and the turbulence energy contained within them are then modeled using a subgrid scale model. This approach allows a more appropriate representation of the unsteady turbulent fluctuations than RANS alone, and it is computationally feasible for transitional flows in 3D domains.

In general, hybrid RANS/LES modeling framework includes

- Continuity equation for RANS.
- Filtered continuity equation for LES.
- Momentum equation for RANS.
- Momentum equation for LES.
- Kinetic energy equation for RANS.
- Kinetic energy equation for LES.
- Turbulent transport equation.
- SGS model.
- Turbulence model.
- Closure model.

Many researchers already demonstrate great capabilities of hybrid behavioral paradigm in modeling and simulation of complex unsteady turbulent flows up to $Re \sim 10^5 - 10^6$. A large variety of hybrid RANS/LES methods are currently in use such that there is the question of which hybrid RANS/LES method represents the optimal approach.

It is believed that hybrid RANS/LES methods are suitable for analysis of turbulent flow with dominated coherent structures and strong unsteady properties. According to the literature, hybrid RANS/LES methods are applied for prediction of 1) flow in boundary layer, 2) wall-bounded flow, 3) flow over solid boundary or body, 4) flow separation over and behind the obstacle, 5) swirling turbulent jet flow, 6) turbulent wake flow, 7) aeroacoustic flow, 8) unsteady cavitating flow, 9) flow induced by shock-wave boundary layer interactions, 10) combustion and reactive flows, and some other specific flows. Perhaps for this reason, new aeronautics and naval technologies have paid growing attention to the hybrid methods of turbulence modeling, encountered in advanced engineering and scientific applications. Detailed discussions and recent advantages in turbulence modeling using RANS, LES, and hybrid RANS/LES can be found in several books "*Progress in Hybrid RANS-LES Modelling*" (e.g., Hoarau et al. 2018, 2020); current publications are available in *Journal of Computational Physics*, *Theoretical and Computational Fluid Dynamics*, *Journal of Turbulence*, *Physics of Fluids*, and other scientific journals on fluid dynamics.

1.4.6 Closing Remark

CFD have relatively short history. Mathematical schemes and techniques for numerical solutions of the basic equations of fluid dynamics were created in the 1950s, first realistic simulations of 2D turbulent flows were performed in the late 1960s, and numerical modeling of 3D flows was started in the late 1980s. Since that

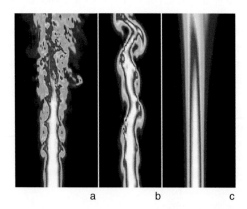

a   b   c

FIGURE 1.15   Comparison of different CFD methods for turbulence modeling. Prediction of turbulent jet: (a) DNS, (b) LES, and (c) RANS. LES requires less computational effort than DNS, while delivering more detail than the inexpensive RANS. (After Maries et al. 2012).

time, computer revolution and software developments have triggered CFD technology dramatically. Nowadays, CFD is everyday tool in scientific research, engineering design, optimization, and analysis. Examples (Figure 1.15) demonstrate capabilities of CFD in modeling and simulation of turbulent jet. There are many commercial CFD software packages (e.g., FLUENT, PHOENICS, FLOW3D, FIDAP) which are available for analysis of various problems of fluid dynamics and turbulence. User-friendly programs have become also accessible in open-sources and retail toolboxes. It sounds like you don't have to be an expert in CFD, you only need to understand fluid mechanics!

A large amount of literature on CFD already exists and the number of related publications grows dramatically. For the interested reader, we recommend great introductory textbooks (Anderson 1995; Cebeci 2004; Wilcox 2006; Schäfer 2006; Sagaut 2006; Rodriguez 2019) and several well written books (Biswas and Eswaran 2002; Belotserkovskii et al. 2005; Lesieur et al. 2005; Hoffman and Johnson 2007; Schiestel 2008; Belotserkovskii 2009; Jiang and Lai 2009; Zikanov 2010; Andersson et al. 2012; Cebeci 2013; Sagaut et al. 2013; Rebollo and Lewandowski 2014; Leschziner 2016; Volkov 2017; Jayanti 2018; Zhai 2020). Recent advances in computing turbulent flows are reviewed in paper (Argyropoulos and Markatos 2015). Useful information on CFD models and software products is also available at the website: http://www.cfd-online.com/.

1.5 CONCLUSIONS

In this chapter, we have outlined fundamental aspects of turbulence science. A number of key topics including the theory, modeling, and basic statements were briefly described and discussed. A review of the existing literature has shown that the greatest attention and efforts come from numerical studies and advantages, aiming to improve the understanding turbulence physics and elaborating practical solutions of the problem.

The core finding of this chapter is that there may be two different logics for turbulence research. One logic, compatible with theoretical (analytical) models and approximations, emphasizes further studies of statistical (spectral and energy) properties of turbulent flows using methods of fluid mechanics and phenomenological descriptions. The other logic, compatible with CFD, emphasizes continuous increase of computer science capabilities for turbulence modeling in order to generate novel results. First of all, this activity is concerned with numerical solutions of the time-dependent Navier–Stokes equations at high Reynolds numbers. More and more computer recourses and programs are required for sophisticated computations. An important issue is also the validation of CFD products against analytical solutions, different numerical codes, and experimental data. This problem involves statistical treatment of the obtained results, which may not be *identical or not reproductive* due to errors and uncertainties in CFD.

Do these efforts provide a progress? Of course, yes, but in most cases, there is a gap between numerical and experimental data. At the same time, measurements of natural turbulence are still limited and often remain without adequate theoretical analysis (e.g., most *in situ* observations of ocean turbulence were conducted in the 1970s and 1980s). Possible reasons of such a *discordance* are high cost, deficiency of modern measurement technology, and/or lack of expertise. There is also a common (by analogy with classical statistical mechanics) problem of the reconciliation of experimental time-averaged data (if any) and model ensemble-averaged data that may become a critical issue in the turbulence research due to strong spatiotemporal dynamics.

To expand the role of theory, obviously, it would be necessary to bring to meaningful consistency insights of the Navier–Stokes equations and statistical measures of turbulence. At present, the study of turbulence is very far from this point. The current approach stems from the fundamental works of Kolmogorov (1941a, b, 1962), who, by means of statistical analysis and hypotheses of similarity, developed the energy cascade theory and established the energy spectrum of homogeneous isotropic turbulence, known as the *"five-thirds law"* (which as many believe is an universal property). However, there is no rigorous mathematical proof of the Kolmogorov theory which would follow directly from the Navier–Stokes equations. The only case is the limit of infinite Reynolds number, Re $\to \infty$, (or the limit of *zero viscosity*, $v \to 0$) that corresponds to so-called "fully developed turbulence" – the state, which, perhaps, never exist in real world. To ensure that this is true, we are referring again to the documented pictures in famous van Dyke's *Album of Fluid Motion* (Van Dyke 1982) or to novel space images of the Planets.

In a view of geosciences and remote sensing, the turbulence measure is not a statistically *averaged quantity* or spectral power-law paradigm but rather, this is that we may call *"complex topological time-varying (extreme) 3D object."* It means that the problem should be studied in various spatiotemporal domains of high dimension but not only in the inertial region. An important subdivision of fluid dynamics, approaching this concept, is geophysical fluid dynamics where both global and local processes are populated with long-lived, large-scale, energetic turbulent structures such as eddies, vortices, waves, fronts, wakes, jets, and plumes. Some aspects of geophysical turbulence in the atmosphere and the oceans will be considered in Chapter 2.

REFERENCES

Anderson, J. D. 1995. *Computational Fluid Dynamics*. McGraw-Hill Int., New York.
Andersson, B., Andersson, R., Håkansson, L., Mortensen, M., Sudiyo, R., and van Wachem, B. 2012. *Computational Fluid Dynamics for Engineers*. Cambridge University Press, Cambridge, UK.
Aragón, J. L., Naumis, G. G., Bai, M., Torres, M., and Maini, P. K. 2008. Turbulent luminance in impassioned van Gogh paintings. *Journal of Mathematical Imaging and Vision*, 30(3):275–283. doi:10.1007/s10851-007-0055-0.
Argyropoulos, C. D. and Markatos, N. C. 2015. Recent advances on the numerical modelling of turbulent flows. *Applied Mathematical Modelling*, 39(2):693–732. doi:10.1016/j.apm.2014.07.001.
Arnold, L. 1998. *Random Dynamical Systems (Springer Monographs in Mathematics)*, Springer-Verlag, Berlin, Germany (corrected 2nd printing 2003).
Arnold, V. I. and Avez, A. 1968. *Ergodic Problems of Classical Mechanics*. Benjamin, New York.

Ashworth, P. J., Bennett, S. J., Best, J. L., and McLelland, S. J. (Eds.). 1996. *Coherent Flow Structures in Open Channels*. John Wiley & Sons, Chichester, UK.

Bailly, C. and Comte-Bellot, G. 2015. *Turbulence*. Springer International Publishing, Switzerland.

Bak, P. 1996. *How Nature Works: The Science of Self-Organized Criticality*. Springer Science, New York.

Bak, P., Tang, C., and Wiesenfeld, K. 1987. Self-organized criticality: An explanation of the $1/f$ noise. *Physical Review Letters*, 59(4):381–384. doi:10.1103/physrevlett.59.381.

Banerjee, S., Hassan, M. K., Mukherjee, S., and Gowrisankar, A. 2020. *Fractal Patterns in Nonlinear Dynamics and Applications*. CRC Press, Boca Raton, FL.

Barenblatt, G. I. 1996. *Self-similarity, and Intermediate Asymptotics: Dimensional Analysis and Intermediate Asymptotics (Cambridge Texts in Applied Mathematics Book 14)*. Cambridge University Press, Cambridge, UK.

Barenblatt, G. I. 2003. *Scaling: Cambridge Texts in Applied Mathematics*. Cambridge University Press, Cambridge, UK.

Batchelor, G. K. 1953. *The Theory of Homogeneous Turbulence (Cambridge Science Classics)*. Cambridge University Press, New York.

Batchelor, G. K. 1967. *An Introduction to Fluid Dynamics*. Cambridge University Press, Cambridge, UK.

Belotserkovskii, O. M. 2009. *Constructive Modeling of Structural Turbulence and Hydrodynamic Instabilities*. World Scientific Publishing, Singapore.

Belotserkovskii, O. M., Oparin, A. M., and Chechetkin, V. M. 2005. *Turbulence: New Approaches*. Cambridge International Science Publishing, Cambridge, UK.

Bennett, A. 2006. *Lagrangian Fluid Dynamics*. Cambridge University Press, Cambridge, UK.

Bernard, P. S. 2019. *Turbulent Fluid Flow*. John Wiley & Sons, Hoboken, NJ.

Bernard, P. S. and Wallace, J. M. 2002. *Turbulent Flow: Analysis, Measurement, and Prediction*. John Wiley & Sons, Hoboken, NJ.

Bernardara, P., Lang, M., Sauquet, E., Schertzer, D., and Tchiriguyskaia, I. 2007. *Multifractal Analysis in Hydrology: Application to Time Series*. Éditions Quæ, Versailles, Cedex.

Birkhoff, G. D. 1927. *Dynamical Systems*. Colloquium publications (American Mathematical Society), Providence, NY.

Birnir, B. 2013. *The Kolmogorov-Obukhov Theory of Turbulence: A Mathematical Theory of Turbulence (SpringerBriefs in Mathematics)*. Springer, New York, Dordrecht.

Biskamp, D. 2003. *Magnetohydrodynamic Turbulence*. Cambridge University Press, Cambridge, UK.

Biswas, G. and Eswaran, V. (Eds.). 2002. *Turbulent Flows: Fundamentals, Experiments and Modeling*. CRC Press, Boca Raton, FL.

Bohr, T., Jensen, M. H., Paladin, G. and Vulpiani, A. 1998. *Dynamical Systems Approach to Turbulence*. Cambridge University Press, Cambridge, UK.

Branover, H., Golbraikh, E., Eidelman, A., and Moiseev, S., 1999. *Turbulence and Structures: Chaos, Fluctuations, and Helical Self-Organization in Nature and Laboratory*. Academic Press, San Diego, CA.

Cardy, J., Falkovich, G., and Gawedzki, K. 2008. *Non-equilibrium Statistical Mechanics and Turbulence* (Eds. S. Nazarenko and O. V. Zaboronski). Cambridge University Press, Cambridge, UK.

Carlson, J., Jaffe, A., and Wiles, A. (Eds.). 2006. *The Millennium Prize Problems*. American Mathematical Society, Providence, RI.

Cebeci, T. 2004. *Turbulence Models and Their Application: Efficient Numerical Methods with Computer Programs*. Horizons Publishing – Springer, Long Beach, CA.

Cebeci, T. 2013. *Analysis of Turbulent Flows with Computer Programs*, 3rd edition. Butterworth-Heinemann, Oxford, UK.

Cello, G. and Malamud, B. D. (Eds.). 2006. *Fractal Analysis for Natural Hazards. Geological Society Special Publication 261*. Geological Society of London, London, UK.

Chandrasekhar, S. 1981. *Hydrodynamic and Hydromagnetic Stability*. Dover Publications, New York.

Chorin, A. J. 1994. *Vorticity and Turbulence*. Springer, New York.

Davidson, P. A. 2015. *Turbulence: An Introduction for Scientists and Engineers*, 2nd edition. Oxford University Press, Oxford, UK.

Davidson, P. A., Kaneda, Y., and Sreenivasan, K. R. (Eds.). 2013. *Ten Chapters in Turbulence*. Cambridge University Press, Cambridge, UK.

Deissler, R. 1998. *Turbulent Fluid Motion*. CRC Press, Boca Raton, FL.

Devaney, R. 2003. *An Introduction to Chaotic Dynamical Systems*, 2nd edition. Westview Press, Boulder, CO.

Dimri, V. P. (Ed.). 2000. *Application of Fractals in Earth Sciences*. A. A. Balkema Publishers, Rotterdam,The Netherlands.

Doering, C. R. and Gibbon, J. D. 1995. *Applied Analysis of the Navier–Stokes Equations*. Cambridge University Press, Cambridge, UK.

Drazin, P. G. 2002. *Introduction to Hydrodynamic Stability*. Cambridge University Press, Cambridge, UK.

Durbin, P. A. 2018. Some recent developments in turbulence closure modeling. *Annual Review of Fluid Mechanics*, 50(1):77–103. doi:10.1146/annurev-fluid-122316-045020.

Durbin, P. A. and Pettersson Reif, B. A. 2011. *Statistical Theory and Modeling for Turbulent Flows*, 2nd edition. John Wiley & Sons, Chichester, UK.

Eckert, M. 2019. *The Turbulence Problem: A Persistent Riddle in Historical Perspective*. Springer, Switzerland.

Edgar, G. 2008. *Measure, Topology, and Fractal Geometry (Undergraduate Texts in Mathematics)*, 2nd edition. Springer Science, New York.

Elhadj, Z. 2019. *Dynamical Systems: Theories and Applications*. CRC Press, Boca Raton, FL.

Falconer, K. 2014. *Fractal Geometry: Mathematical Foundations and Applications*, 3rd edition. John Wiley & Sons, Chichester, UK.

Farmer, M. E. 2014. *Application of Chaos and Fractals to Computer Vision*. Bentham Science Publishers, Sharjah, UAE.

Feder, J. 1988. *Fractals*. Springer Science, New York.

Foias C., Manley O.,Rosa R., and Temam, R. 2001. *Navier–Stokes Equations and Turbulence (Encyclopedia of Mathematics and its Applications)*. Cambridge University Press, Cambridge, UK.

Franceschetti, G. and Riccio, D. 2007. *Scattering, Natural Surfaces, and Fractals*. Academic Press – Elsevier, London, UK.

Frisch, U. 1995. *Turbulence: The Legacy of A. N. Kolmogorov*. Cambridge University Press, Cambridge, UK.

Frost, W. and Moulden, T. H. (Eds.). 1977. *Handbook of Turbulence: Volume 1 Fundamentals and Applications*. Plenum Press, New York.

Gaissinski, I. and Rovenski, V. 2018. *Modeling in Fluid Mechanics: Instabilities and Turbulence*. CRC Press, Boca Raton, FL.

Gaponov-Grekhov, A. V. and Rabinovich, M. I. 1992. *Nonlinearities in Action: Oscillations, Chaos, Order, Fractals (English by E. F. Hefter and N. Krivatkina)*. Springer-Verlag, Berlin.

Gleick, J. 1987. *Chaos: Making a New Science*. Viking Penguin Book, New York (Revised edition, Open Road Integrated Media, New York, 2011).

Glendinning, P. 1994. *Stability, Instability and Chaos: An Introduction to the Theory of Nonlinear Differential Equations*. Cambridge University Press, Cambridge, UK.

Haller, G. 2015. Lagrangian coherent structures. *Annual Review of Fluid Mechanics*, 47(1):137–162. doi:10.1146/annurev-fluid-010313-141322.

Haller, G. and Sapsis, T. 2011. Lagrangian coherent structures and the smallest finite-time Lyapunov exponent. *Chaos: An Interdisciplinary Journal of Nonlinear Science*, 21, 023115. doi:10.1063/1.3579597.

Haller, G. and Yuan, G. 2000. Lagrangian coherent structures and mixing in two-dimensional turbulence. *Physica D: Nonlinear Phenomena*, 147(3–4):352–370. doi:10.1016/s0167-2789(00)00142-1.

Harte, D. 2001. *Multifractals: Theory and Applications*. Chapman and Hall/CRC, Boca Raton, FL.

Heinz, S. 2003. *Statistical Mechanics of Turbulent Flows*. Springer, Berlin, Germany.

Hergarten, S. 2002. *Self-Organized Criticality in Earth Systems*. Springer-Verlag, Berlin, Heidelberg.

Hinze, J. O. 1959. *Turbulence. AnIntroduction to its Mechanism and Theory*. McGraw-Hill, New York.

Hirsch, M. W., Smale, S., and Devaney, R. L. 2004. *Differential Equations, Dynamical Systems, and an Introduction to Chaos (Pure and Applied Mathematics)*, 2nd edition. Academic Press – Elsevier, San Diego, CA.

Hoarau, Y., Peng, S.-H., Schwamborn, D., Revell, A., and Mockett, C. (Eds.). 2018. *Progress in Hybrid RANS-LES Modelling: Papers Contributed to the 6th Symposium on Hybrid RANS-LES Methods, 26–28 September 2016, Strasbourg, France*. Springer International Publishing, Switzerland.

Hoarau, Y., Peng, S.-H., Schwamborn, D., Revell, A., and Mockett, C. (Eds.). 2020. *Progress in Hybrid RANS-LES Modelling: Papers Contributed to the 7th Symposium on Hybrid RANS-LES Methods, 17–19 September 2018, Berlin, Germany*. Springer International Publishing, Switzerland.

Hoffman, J. and Johnson, C. 2007. *Computational Turbulent Incompressible Flow: Applied Mathematics: Body and Soul 4*. Springer-Verlag, Berlin, Heidelberg.

Holmes, P., Lumley, J. L., Berkooz, G., and Rowley, C. W. 2012. *Turbulence, Coherent Structures, Dynamical Systems and Symmetry*, 2nd edition. Cambridge University Press, Cambridge, UK.

Hopf, E. 1948. A mathematical example displaying features of turbulence. *Communications on Pure and Applied Mathematics*, 1(4):303–322. doi:10.1002/cpa.3160010401.

Hunt, J. C. R. and Vassilicos, J. C. (Eds.). 2000. *Turbulence Structure and Vortex Dynamics*. Cambridge University Press, Cambridge, UK.

Hussain, A. K. M. F. 1986. Coherent structures and turbulence. *Journal of Fluid Mechanics*, 173:303–356. doi:10.1017/s0022112086001192.

Jayanti, S. 2018. *Computational Fluid Dynamics for Engineers and Scientists*. Springer Science, Dordrecht, The Netherlands.

Jensen, H. J. 1998. *Self-Organized Criticality (Emergent Complex Behavior in Physical and Biological Systems)*. Cambridge University Press, Cambridge, UK.

Jiang, X. and Lai, C.-H. 2009. *Numerical Techniques for Direct and Large-Eddy Simulations*. CRC Press, Boca Raton, FL.

Kadomtsev, B. B. 1965. *Plasma Turbulence (translated from Russian by L. C. Ronson and edited by M. G. Rusbridge)*. Academic Press, London, New York.

Katok, A. and Hasselblatt, B. 1995. *Introduction to the Modern Theory of Dynamical Systems*. Cambridge University Press, Cambridge, UK.

Klimontovich, Yu. L. 1991. *Turbulent Motion and the Structure of Chaos: A New Approach to the Statistical Theory of Open Systems (translated from Russian by A. Dubroslavsky)*. Springer – Kluwer Academic Publishers, Dordrecht, The Netherlands.

Kollmann, W. 2019. *Navier–Stokes Turbulence: Theory and Analysis*. Springer, Switzerland.

Kolmogorov, A. N. 1941a. The local structure of turbulence in incompressible viscous fluid for very large Reynolds numbers. *Doklady Akademii Nauk SSSR*, 30:301–304 (reprinted in: *Proceedings of the Royal Society A: Mathematical, Physical and Engineering Sciences*, 434(1890):9–13. doi:10.1098/rspa.1991.0075).

Kolmogorov, A. N. 1941b. Dissipation of energy in the locally isotropic turbulence. *Doklady Akademii Nauk SSSR*, 32:16–18 (reprinted in: *Proceedings of the Royal Society A: Mathematical, Physical and Engineering Sciences*, 434 (1890):15–17. Available on the Internet https://www.jstor.org/stable/51981).

Kolmogorov, A. N. 1962. A refinement of previous hypotheses concerning the local structure of turbulence in a viscous incompressible fluid at high Reynolds number. *Journal of Fluid Mechanics*, 13(01):82–85. doi:10.1017/s0022112062000518.

Ladyzhenskaia, O. A. 1969. *The Mathematical Theory of Viscous Incompressible Flow* (*translated from Russian by R. A. Silverman and J. Chu*), 2nd edition, Gordon and Breach Science Publishers, New York.

Landau, L. D. and Lifshitz, E. M. 1987. *Fluid Mechanics. Course of Theoretical Physics, Volume 6 (translated from Russian by J. B. Sykes and W. H. Reid)*, 2nd edition. Pergamon Press, Oxford, UK.

Launder, B. E. and Sandham, N. D. (Eds.). 2002. *Closure Strategies for Turbulent and Transitional Flows*. Cambridge University Press, Cambridge, UK.

Lemarie-Rieusset, P. G. 2016. *The Navier–Stokes Problem in the 21st Century*. CRC Press, Boca Raton, FL.

Leray, J. 1934. Sur le mouvement d'un liquide visqueux emplissant l'espace. *Acta Mathematica (in French)*, 63(1):193–248. doi:10.1007/BF02547354. English translation. Available on the Internet https://arxiv.org/pdf/1604.02484.pdf.

Leschziner, M. 2016. *Statistical Turbulence Modelling for Fluid Dynamics - Demystified: An Introductory Text for Graduate Engineering Students*. Imperial College Press, London, UK.

Lesieur, M. 2008. *Turbulence in Fluids*, 4th edition. Springer, Dordrecht, The Netherlands.

Lesieur, M., Métais, O., and Comte, P. 2005. *Large-Eddy Simulations of Turbulence*. Cambridge University Press, New York.

Lesieur, M., Yaglom, A. and David, F. (Eds.). 2001. *New Trendsin Turbulence. Turbulence: nouveaux aspects. Les Houches Session LXXIV 31 July - 1 September 2000 (Les Houches - Ecoled'Ete de Physique Theorique)*. EDP Science – Springer, Paris, Berlin.

Lesne, A. and Laguës, M. 2012. *Scale Invariance: From Phase Transitions to Turbulence*. Springer-Verlag, Berlin, Heidelberg.

Lichtenberg, A. J. and Lieberman, M. A. 2010. *Regular and Chaotic Dynamics*, 2nd edition. Springer-Verlag, New York.

Lienhard, J. H. 1966. Synopsis of lift, drag and vortex frequency data for rigid circular cylinders. *Research Division Bulletin*. Washington State University, College of Engineering. Available on the Internet https://uh.edu/engines/vortexcylinders.pdf.

Loitsyanskii, L. G. 1966. *Mechanics of Liquids and Gases*. Pergamon Press, Oxford, UK (translated from Russian).

Lorenz, E. N. 1963. Deterministic nonperiodic flow. *Journal of the Atmospheric Sciences*, 20(2):130–141. doi:10.1175/1520-0469(1963)020<0130:dnf>2.0.

Lorenz, E. N. 1993. *The Essence of Chaos*. University of Washington Press, Seattle.

Lovejoy, S. and Schertzer, D. 2007. Scale, scaling and multifractals in geophysics: Twenty years on. In *Nonlinear Dynamics in Geosciences. Aegean Conferences* (Eds. A. A. Tsonis and J. B. Elsner). Springer Science, New York, pp. 311–337.

Lovejoy, S. and Schertzer, D. 2013. *The Weather and Climate: Emergent Laws and Multifractal Cascades*. Cambridge University Press, Cambridge, UK.

Lumley, J. L. and Yaglom, A. M. 2001. A century of turbulence. *Flow, Turbulence and Combustion*, 66(3):241–286. doi:10.1023/a:1012437421667.

Lyapunov, A. M. 1992. *General Problem of the Stability of Motion (translated and edited from Russian by A. T. Fuller)*. Taylor & Francis, London (original: Lyapunov, A. M. 1892. The general problem of the stability of motion. Kharkov Mathematical Society, Kharkov, in Russian).

Lynch, S. 2004. *Dynamical Systems with Applications using MATLAB®*. Springer Science, New York.

Majda, A. J. 2016. *Introduction to Turbulent Dynamical Systems in Complex Systems*. Springer, Switzerland.

Mandelbrot, B. B. 1983. *The Fractal Geometry of Nature*. W. H. Freeman, New York.

Mandelbrot, B. B. 1999. *Multifractals and 1/f Noise: Wild Self-Affinity in Physics*. Springer-Verlag, Berlin, Heidelberg.

Mandelbrot, B. B. 2004. *Fractals and Chaos: The Mandelbrot Set and Beyond*. Springer, New York.

Mandelbrot, B. B. and Hudson, R. L. 2004. *The (Mis)Behaviour of Markets: A Fractal Views of Risk Ruin and Reward*. Basic Books, New York.

Manneville, P. 2010. *Instabilities, Chaos and Turbulence (ICP Fluid Mechanics)*, 2nd edition. Imperial College Press, London.

Maries, A., Haque, A., Yilmaz, S. L., Nik, M. B., and Marai, G. E. 2012. Interactive exploration of stress tensors used in computational turbulent combustion. In *New Developments in the Visualization and Processing of Tensor Fields* (Eds. D. Laidlaw and A. Villanova). Springer-Verlag, Berlin, Heidelberg, Germany, pp. 137–156. Available on the Internet https://www.evl.uic.edu/documents/maries-2012-ieo.pdf.

Marov, M. Y. and Kolesnichenko, A. V. 2013. *Turbulence and Self-Organization: Modeling Astrophysical Objects*. Springer, New York.

Massopust, P. R. 2016. *Fractal Functions, Fractal Surfaces, and Wavelets*, 2nd edition. Elsevier – Academic Press, London, UK.

Mathieu, J. and Scott, J. 2000. *An Introduction to Turbulent Flow*. Cambridge University Press, Cambridge, UK.

McComb, W. D. 1990. *The Physics of Fluid Turbulence*. Clarendon Press – Oxford UniversityPress, Oxford, UK.

McComb, W. D. 2014. *Homogeneous, Isotropic Turbulence: Phenomenology, Renormalization and Statistical Closures*. Oxford University Press, Oxford, UK.

McComb, W. D., Linkmann, M. F., Berera, A., Yoffe, S. R., and Jankauskas, B. 2015. Self-organization and transition to turbulence in isotropic fluid motion driven by negative damping at low wavenumbers. *Journal of Physics A: Mathematical and Theoretical*, 48(25), 25FT01. doi:10.1088/1751-8113/48/25/25ft01.

Métais, O. and Lesieur, M. (Eds.). 1991. *Turbulence and Coherent Structures*. Springer, Dordrecht, The Netherlands.

Mitchell, M. 2009. *Complexity: A Guided Tour*. Oxford University Press, New York.

Moiseev, S. S., Pungin, V. G., and Oraevsky, V. N. 1999. *Nonlinear Instabilities in Plasmas and Hydrodynamics*. CRC Press, Boca Raton, FL.

Monin, A. S. 1978. On the nature of turbulence. *Soviet Physics Uspekhi*, 21(5):429–442. doi:10.1070/pu1978v021n05abeh005554.

Monin, A. S. and Yaglom, A. M. 1971. *Statistical Fluid Mechanics, Volume 1: Mechanics of Turbulence (English translation, edited by J. L. Lumley)*. The MIT Press, Cambridge, MA.

Monin, A. S. and Yaglom, A. M. 1975. *Statistical Fluid Mechanics, Volume 2: Mechanics of Turbulence (English translation, edited by J. L. Lumley)*. The MIT Press, Cambridge, MA.

Moon, F. C. 1992. *Chaotic and Fractal Dynamics: Introduction for Applied Scientists and Engineers*. John Wiley & Sons, New York.

Nicolis, G. and Prigogine, I. 1977. *Self-Organization in Nonequilibrium Systems: From Dissipative Structures to Order through Fluctuations*. John Wiley & Sons, New York.

Nieuwstadt, F. T. M., Westerweel, J., and Boersma, B. J. 2016. *Turbulence: Introduction to Theory and Applications of Turbulent Flows*. Springer, Switzerland.

Ott, E. 2002. *Chaos in Dynamical Systems*, 2nd edition. Cambridge University Press, Cambridge, UK.

Pikovsky, A. and Politi, A. 2016. *Lyapunov Exponents: A Tool to Explore Complex Dynamics*. Cambridge University Press, Cambridge, UK.

Piquet, J. 2001. *Turbulent Flows: Models and Physics*, 2nd printing. Springer, Berlin, Germany.

Prants, S. V., Uleysky, M. Yu., and Budyansky, M. V. 2017. *Lagrangian Oceanography: Large-scale Transport and Mixing in the Ocean (Physics of Earth and Space Environments)*. Springer International Publishing, Switzerland.

Pruessner, G. 2012. *Self-Organised Criticality: Theory, Models and Characterisation*. Cambridge University Press, Cambridge, UK.

Pope, S. B. 2000. *Turbulent Flows*. Cambridge University Press, Cambridge, UK.

Quattrochi, D. A., Wentz, E., Lam, N. S.-N., and Emerson, C. W. 2017. *Integrating Scale in Remote Sensing and GIS*. CRC Press, Boca Raton, FL.

Raizer, V. 2017. *Advances in Passive Microwave Remote Sensing of Oceans*. CRC Press, Boca Raton, FL.

Raizer, V. 2019. *Optical Remote Sensing of Ocean Hydrodynamics*. CRC Press, Boca Raton, FL.

Raizer, V. Y., Novikov, B. M., and Bocharova, T. Y. 1994. The geometrical and fractal properties of visible radiances associated with breaking waves in the ocean. *Annales Geophysicae*, 12:1229–1233.

Rebollo, T. C. and Lewandowski, R. 2014. *Mathematical and Numerical Foundations of Turbulence Models and Applications*. Birkhauser – Springer, New York.

Reguera, D., Bonilla, L. L., and Rubi, J. M. (Eds.). 2001. *Coherent Structures in Complex Systems*. Springer-Verlag, Berlin, Heidelberg.

Richardson, L. F. 1922. *Weather Prediction by Numerical Process*. Cambridge University Press, London (reprint second edition in 2007).

Richter, J. P. (Ed.). 1970. Plate 20 and Note 389. In: *The Notebooks of Leonardo Da Vinci*. Dover Publications, New York.

Robinson, C. 1998. *Dynamical Systems: Stability, Symbolic Dynamics, and Chaos*, 2nd edition. CRC Press, Boca Raton, FL.

Robinson, S. K. 1991. Coherent motions in the turbulent boundary layer. *Annual Review of Fluid Mechanics*, 23(1):601–639. doi:10.1146/annurev.fl.23.010191.003125.

Rodriguez, S. 2019. *Applied Computational Fluid Dynamics and Turbulence Modeling: Practical Tools, Tips and Techniques*. Springer, Switzerland.

Ruelle, D. (Ed.). 1995. *Turbulence, Strange Attractors, and Chaos*. World Scientific Publishing, Singapore.

Sagaut, P. 2006. *Large Eddy Simulation for Incompressible Flows: An Introduction*, 3rd edition. Springer-Verlag, Berlin, Heidelberg, Germany.

Sagaut, P., Deck, S., and Terracol, M. 2013. *Multiscale and Multiresolution Approaches in Turbulence - Les, Des and Hybrid Rans/Les Methods: Applications and Guidelines*, 2nd edition. Imperial College Press, London, UK.

Schäfer, M. 2006. *Computational Engineering - Introduction to Numerical Methods*. Springer-Verlag, Berlin, Heidelberg.

Schertzer, D. and Lovejoy, S. (Eds.). 1991. *Non-Linear Variability in Geophysics: Scaling and Fractals*. Kluwer Academic Publishers, Dordrecht, The Netherlands.

Schiestel, R. 2008. *Modeling and Simulation of Turbulent Flows*. ISTE – John Wiley & Sons, Hoboken, NJ.

Schroeder, M.1991. *Fractals, Chaos, Power Laws: Minutes from an Infinite Paradise*. Dover Publications, Minneola, New York.

Scott, A. 2003. *Nonlinear Science: Emergence and Dynamics of Coherent Structures*, 2nd edition. Oxford University Press, Oxford and New York.

Sedov, L. I. 1993. *Similarity and Dimensional Methods in Mechanics (translated from Russian by A. G. Volkovets)*, 10th edition. CRC Press, Boca Raton, FL.

Selvam, A. M. 2017. *Self-organized Criticality and Predictability in Atmospheric Flows: The Quantum World of Clouds and Rain*. Springer International Publishing, Switzerland.

Sengupta, T. K. 2012. *Instabilities of Flows and Transition to Turbulence*. CRC Press, Boca Raton, FL.

Serra, M. and Haller, G. 2016. Objective Eulerian coherent structures. *Chaos: An Interdisciplinary Journal of Nonlinear Science*, 26(5), 053110. doi:10.1063/1.4951720.

Seuront, L. 2010. *Fractals and Multifractals in Ecology and Aquatic Science*. CRC Press, Boca Raton, FL.

Shats, M. and Punzmann, H. (Eds.). 2006. *Turbulence and Coherent Structures in Fluids, Plasmas and Nonlinear Medium (Lecture Notes in Complex Systems)*. World Scientific Publishing, Singapore.

Shaw, J. A. and Churnside, J. H. 1997. Fractal laser glints from the ocean surface. *Journal of the Optical Society of America*, 14(5):1144–1150. doi:10.1364/josaa.14.001144.

Shivamoggi, B. K. 2014. *Nonlinear Dynamics and Chaotic Phenomena: An Introduction*, 2nd edition. Springer Science, Dordrecht, The Netherlands.

Sinai, I. G. 1977. *Introduction to Ergodic Theory (translated from Russian by V. Scheffer)*. Princeton University Press, Princeton, NJ.

Smale, S. 1967. Differentiable dynamical systems. *Bulletin of the American Mathematical Society*, 73(6):747–818. doi:10.1090/s0002-9904-1967-11798-1.

Smyth, W. D. and Carpenter, J. R. 2019. *Instability in Geophysical Flows*. Cambridge University Press, Cambridge, UK.

Solé, R. V. and Bascompte, J. 2006. *Self-Organization in Complex Ecosystems*. Princeton University Press, Princeton, NJ.

Sornette, D. 2006. *Critical Phenomena in Natural Sciences: Chaos, Fractals, Selforganization and Disorder: Concepts and Tools*, 2nd edition. Springer-Verlag, Berlin, Heidelberg.

Sparrow, C. 1982. *The Lorenz Equations: Bifurcations, Chaos, and Strange Attractors*. Springer-Verlag, New York.

Strogatz, S. H. 2015. *Nonlinear Dynamics and Chaos: With Applications to Physics, Biology, Chemistry, and Engineering*, 2nd edition. CRC Press, Boca Raton, FL.

Tardu, S. 2014. *Transport and Coherent Structures in Wall Turbulence*. ISTE – John Wiley & Sons, Hoboken, NJ.

Tartar, L. 2006. *An Introduction to Navier–Stokes Equation and Oceanography (Lecture Notes of the Unione Matematica Italiana)*. Springer, The Netherlands.

Tennekes, H. and Lumley, J. L. 1972. *A First Course in Turbulence*. The MIT Press, Cambridge, MA.

Ting, D. S-K. 2016. *Basics of Engineering Turbulence*. Elsevier – Academic Press, Amsterdam, The Netherlands.

Townsend, A. A. 1976. *The Structure of Turbulent Shear Flow*, 2nd edition. Cambridge University Press, New York.

Tritton, D. J. 1988. *Physical Fluid Dynamics (Oxford Science Publications)*, 2nd edition, Oxford University Press, Oxford, UK.

Tsai, T.-P. 2018. *Lectures on Navier–Stokes Equations. Graduate Studies in Mathematics, 192*. American Mathematical Society, Providence, RI.

Tsinober, A. 2009. *An Informal Conceptual Introduction to Turbulence: Second Edition of An Informal Introduction to Turbulence (Fluid Mechanics and Its Applications)*, 2nd edition. Springer, Dordrecht, The Netherlands.

Tur, A. and Yanovsky, V. 2017. *Coherent Vortex Structures in Fluids and Plasmas*. Springer International Publishing, Switzerland.

Turcotte, D. 1997. *Fractals and Chaos in Geology and Geophysics*, 2nd edition. Cambridge University Press, Cambridge, UK.

Van Dyke, M. 1982. *An Album of Fluid Motion*, 14th edition. The Parabolic Press, Stanford, CA.

Venditti, J. G., Best, J. L., Church, M., and Hardy, R. J. (Eds.). 2013. *Coherent Flow Structures at Earth's Surface*. Wiley – Blackwell, Chichester, UK.

Volkov, K. (Ed.). 2017. *Turbulence Modelling Approaches: Current State, Development Prospects, Applications*. In Tech, Croatia.

Voropayev, S. I. and Afanasyev, Y. D. 1994. *Vortex Structures in a Stratified Fluid: Order from Chaos*. Chapman & Hall/CRC Press, Boca Raton, FL.

White, F. M. 2006. *Viscous Fluid Flow*, 3rd edition. McGraw-Hill, New York.

Wilcox, D. C. 2006. *Turbulence Modeling for CFD*, 3rd edition. DCW Industries, La Canada, CA.

Yaglom, A. M. and Frisch, U. 2012. *Hydrodynamic Instability and Transition to Turbulence. 100 (Fluid Mechanics and Its Applications)*. Springer, Dordrecht, Heidelberg.

Zaslavsky, G. M. 2007. *The Physics of Chaos in Hamiltonian Systems*, 2nd edition. Imperial College Press, London.

Zaslavsky, G. M., Sagdeev, R. Z., Chernikov, A. A., and Usikov, D. A. 1991. *Weak Chaos and Quasi-Regular Patterns*. Cambridge University Press, New York.

Zhai, Z. 2020. *Computational Fluid Dynamics for Built and Natural Environments*. Springer Nature, Singapore.

Zikanov, O. 2010. *Essential Computational Fluid Dynamics*. John Wiley & Sons, Hoboken, NJ.

Zimin, V. D. and Frik, P. G. 1988. *Turbulent Convection*. Nauka, Moscow, USSR (in Russian: Зимин В. Д., Фрик П. Г. Турбулентная конвекция. Москва, Наука, 1988).

2 Geophysical Turbulence

The real is only one realization of the possible.

Ilya Prigogine

2.1 INTRODUCTION

In natural science, geophysical turbulence is a subtopic of Geophysical Fluid Dynamics (GFD). Textbooks and monographs (Stern 1975; Pedlosky 1979; Monin 1990; Salmon 1998; McWilliams 2006; Cushman-Roisin and Beckers 2011; Vallis 2017; Zeitlin 2018; Özsoy 2020) provide comprehensive knowledge of the subject matter. GFD deals with environmental motions (flows) in the Earth's mantle, oceans, atmosphere, and external plasmas including planetary boundary layer (PBL) and some astrophysical fluids (e.g., stars, galaxies), which are almost invariably turbulent.

According to Monin (1990), *"geophysical fluid dynamics can be defined as the fluid dynamics of natural flows of rotating baroclinic stratified fluids and gases."* Only large-scale and mesoscale turbulent motions (with horizontal scales L ≥ 10–100 km) fall within the scope of GFD. On submesoscales (L ~ 1–10 km) and microscales turbulent motions (L ≤ 0.1 km), it is just classical fluid dynamics with geophysical applications. For example, the effect of the small-scale turbulence in the stratified ocean and atmosphere is predominantly given by internal mixing processes. Table 2.1 gives typical scales of the motions in the Earth's atmosphere and ocean (Cushman-Roisin and Beckers 2011).

As a whole, GFD combines fluid mechanics and applied mathematics involving a comprehensive computational framework. Serious researchers have used DNS, LES, or RANS to provide analysis and prediction of turbulence and transport processes for different geophysical scenarios. One example is numerical modeling of stratified oceanic mixed layer, where interactions with a boundary (e.g., ocean's floor or bottom obstacle) give rise to turbulent flow with chaotic small-scale eddies of various size (Kantha and Clayson 2000).

Geophysical turbulence often occurs when vortices remain stable despite the existence of a strong turbulent environment. Vortices vary much more slowly, and the scales much larger, then the energy-containing scale of the turbulence. Jupiter's Great Red Spot is one of the most striking examples (Figure 2.1). Jupiter's turbulent beauty comes from its wild atmosphere full of storms, artfully swirling clouds and big bands of color wrapping around the gas giant.

Vivid environmental examples also include atmospheric jet streams and large-scale ocean currents such as the Gulf Stream and the Equatorial Undercurrent.

DOI: 10.1201/9781003217565-2

TABLE 2.1
Length, Velocity, and Time Scales in the Earth's Atmosphere and Oceans

Phenomenon	Length Scale	Velocity Scale	Time Scale
Atmosphere:			
Microturbulence	10–100 cm	5–50 cm/s	few seconds
Thunderstorms	few km	1–10 m/s	few hours
Sea breeze	5–10 km	1–10 m/s	6 hours
Tornado	10–500 m	30–100 m/s	10–60 minutes
Hurricane	300–500 km	30–60 m/s	Days or weeks
Mountain waves	10–100 km	1–20 m/s	Days
Weather patterns	100–5,000 km	1–50 m/s	Days or weeks
Prevailing winds	Global	5–50 m/s	Seasons to years
Climatic variations	Global	1–50 m/s	Decades and beyond
Ocean:			
Microturbulence	1–100 cm	1 -10 cm/s	10–100 seconds
Internal waves	1–20 km	0.05–0.5 m/s	Minutes to hours
Tides	Basin scale	1–100 m/s	Hours
Coastal upwelling	1–10 km	0.1–1 m/s	Several days
Fronts	1–20 km	0.5–5 m/s	Few days
Eddies	5 -100 km	0.1–1 m/s	Days or weeks
Major currents	50 -500 km	0.5–2 m/s	Weeks to seasons
Large scale gyres	Basin scale	0.01–0.1 m/s	Decades and beyond

(*Sources:* Cushman-Roisin and Beckers 2011).

FIGURE 2.1   Jupiter's Great Red Spot This latest image of Jupiter, taken by NASA's Hubble Space Telescope on August 25, 2020, was captured when the planet was 406 million miles from Earth. (Image from NASA).

Geophysical turbulence differs from the traditional turbulence inherent in hydrodynamics and aerodynamics by scales, energy, rotation, and stratification. One of the most common generation mechanisms of geophysical turbulence is energy transfer in stratified and rotating flow. Turbulence, influenced by stratification and rotation, is known as *geostrophic turbulence*. Geostrophic turbulence describes oceanic motion at mesoscales, from 50 to 500 km, and atmospheric motion at synoptic scales, from 500 km to 5,000 km, i.e. at the scale of our weather system, with meandering jets and long lived vortices.

Geostrophic turbulence is characterized by strongly nonlinear, rapidly rotating, stable stratified flow. This subject is relevant to the large-scale flow in the Earth's oceans and atmosphere. The theory of geostrophic (and quasi-geostrophic turbulence) turbulence represents an extension of the theory of 2D turbulence (Salmon 1998; Olbers et al. 2012). It describes the kinematic and dynamic properties of planetary waves, barotropic and baroclinic instabilities, dynamics of mesoscale eddies, and other large-scale turbulent motions.

An important role in the development of geophysical turbulence plays the Coriolis force and the centrifugal force. Large-scale vortex dynamo and transport processes (advection and diffusion) are also distinguishing attributes of geophysical turbulent flows. In the ocean and the atmosphere, geophysical turbulence is usually associated with multiscale nonlinear dynamics of water or air masses including mixing processes in boundary layers, circulations, currents, fronts, wave structures, bottom topography, eddy/vortex (quasi-horizontal) motions, wakes, and atmospheric jet streams.

Other interesting topic of GFD is the *wave–mean (flow) interactions*. According to (Bühler 2014), interactions with mean flows such as shear flows or vortices are generally a two-way process: the waves are affected by the mean flow, on the one hand, and the mean flow itself responds to the presence of the waves, on the other hand. The influence of waves on the mean flow has been well recognized in the near-shore circulation and around the surf zone. Wave–mean interactions lead to 3D turbulence that is difficult to predict even using DNS. Available today theory is based on the so-called Generalized Lagrangian Mean Equations (Bühler 2014). Although instantaneous and mean flows are clearly understood in the theory, the separation of geophysical flow into "mean flow," "waves," and "turbulence" is truly difficult to categorize because of stochastic wave–mean interactions. Thus, some essential aspects of GFD remain an open problem including those associated with wave-flow interactions, vortex formations, and turbulence generation.

This chapter is a brief overview of geophysical turbulence. We will consider and discuss only those phenomena (shown in Figure 2.2) which, in our opinion, are of practical interest in geosciences and remote sensing.

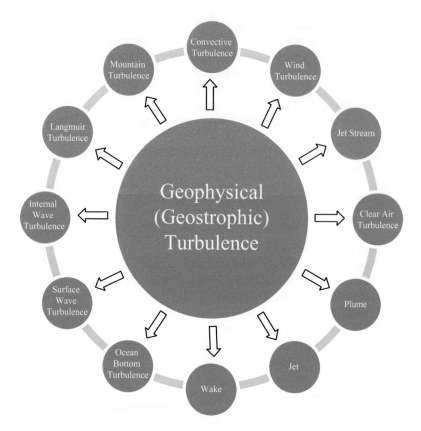

FIGURE 2.2   Basic classification of geophysical/geostrophic turbulence.

2.2   BASIC EQUATIONS

The theory of GFD is based on fundamental equations of fluid mechanics applied to a stratified fluid continuum, moving on a rotating sphere. There are many different combinations of physical conditions for geophysical turbulence. For instance, the evolution of the coupled ocean-atmosphere system can be described by the Boussinesq equation with effects of buoyancy and rotation (Vallis 2017).

A set of GFD governing equations can be written as following (McWilliams 2006):

$$\frac{\partial \rho}{\partial t} + \nabla \cdot (\rho \mathbf{u}) = 0, \qquad\qquad \text{continuity} \qquad\qquad (2.1)$$

$$\frac{d\mathbf{u}}{dt} + 2\Omega \times \mathbf{u} = \frac{1}{\rho}\nabla p + \nu\nabla^2\mathbf{u} + \mathbf{F}, \qquad \text{momentum} \qquad\qquad (2.2)$$

$$\frac{\partial(\rho\tau)}{\partial t} + \nabla \cdot (\rho\tau) = \rho S^{(\tau)}, \qquad\qquad \text{material tracer} \qquad\qquad (2.3)$$

$$\frac{d(\rho)}{dt} - \frac{1}{c^2}\frac{dp}{dt} = Q(\rho), \qquad\qquad \text{thermodynamic} \qquad\qquad (2.4)$$

$$\rho = \rho(p, T, \tau_n). \qquad\qquad\qquad \text{state} \qquad\qquad (2.5)$$

In these equations, p is pressure, ρ is density, T is temperature, τ_n is chemical tracer concentrations, τ denotes any material tracer concentration, Ω is the rotation rate of a rotating frame of reference ($2\Omega \times \mathbf{u}$ is the Coriolis force), $\mathbf{F}$ is the body force, $S^{(\tau)}$ stands for all the non-conservative sources and sinks of τ (e.g., boundary sources, molecular diffusion, reaction rate), Q is source term related to thermodynamic quantities, and c is speed of sound.

Equations (2.1)–(2.5) are deterministic equations. Because of stochastic spatiotemporal dynamics of the Earth's environment, understanding and modeling of geophysical turbulence require a statistical description of a turbulent flow with rotation and stratification. According to (Monin 1990), solutions of GFD equations can be found for *statistically steady turbulent flow* with time-independent mean. In real world, geophysical turbulence is caused by a number of internal dynamical factors associated with *nonstationary* (or *thermohydrodynamic)* conditions and/or time-dependent fluctuations of physical parameters (e.g., density or temperature). For example, thermohaline fine structure and vertical motions of water mass become key features of turbulent flow in the ocean. More complicated example – dynamics and structure of geophysical (or geostrophic) turbulence in thin atmospheric and oceanic boundary layers when both 2D and 3D turbulence can exist. In this case, 2D turbulence accompanies by a forward cascade, from small to large scales, whereas 3D turbulence accompanies by an inverse cascade, from large to small scales. These two cascades lead to different form of the energy spectrum of turbulence: in the inverse- and forward-cascade parts, $E(k) \propto k^{-5/3}$ and $E(k) \propto k^{-3}$, respectively (Baumert et al. 2005; Vallis 2017). Similar properties of 2D and 3D flow (2D $\Leftrightarrow$ 3D) are also inherent to homogeneous, isotropic, and statistically steady magnetohydrodynamic turbulence (Biskamp 2003).

There are several basic techniques applied for practical analysis and prediction of geophysical (geostrophic) turbulence; the most common are: 1) spectral and pseudo-spectral methods, 2) dimensional analysis and similarity theories, 3) dynamical system models, and 4) CFD modeling and simulation. These (and others) methods represent two different theoretical concepts related to statistical and dynamical models. In many cases, it is assumed that the large-scale "mean" turbulent flow (or current) can be separated from the small-scale components of the flow. The connection between these concepts (in terms of GFD) has been established for the case of *homogeneous isotropic turbulence* only when average or "mean" properties of an incompressible flow are considered. Otherwise, the model becomes too complicated in order to reproduce real data or observations and an additional information is needed – *"Never trust a fact, or a simulation, without a supporting interpretation"* (McWilliams 2006). Remote sensing provides abundant information to do this.

2.3   OCEAN (MARINE) TURBULENCE

The ocean is turbulent in the sense that motion, circulation, and mixing of the water masses occur continuously and randomly in the range of the scales from a few centimeters to thousands of kilometers. Thus, ocean turbulence is inherently 3D, multi-scale, and, as a whole unpredictable (in a broad sense) and can be referred to the category of stochastic hydrodynamic flows of variable spatiotemporal structure. Accordingly, ocean turbulence is considered from different points of view – physical oceanography, fluid dynamics, statistical mechanics, thermodynamics, nonlinear physics and fractal geometry, environmental and planetary sciences. For example, some oceanographers believe that only a small fraction (about 5%) of the ocean is fully "actively turbulent." The reason is mostly due to a shift in the accepted definition of turbulence among oceanographers (Gibson 1999).

There is also a concept of so-called "fossil turbulence" – the phenomenon, which often dominates in the ocean and can be interpreted as fossils of *primordial hydrodynamic states*. Fossil turbulence signatures in hydro-physical fields preserve information about previous turbulence events (Gibson 1988, 1999). This concept has certain advantages in the case studies, e.g., related to long-term evolution of submerged turbulent wake (i.e., the wake, which retains some properties after collapse) and/or the formation of stable slowly moving turbulent patterns (spots or intrusions). In such scenarios, various persistent "footprints" may memorize turbulence "history." However, the fossil turbulence is not easily measurable and/or recognizable by standard oceanographic methods.

Other important and perhaps the most basic concept is associated with so-called "shear turbulence." Typically it occurs in response to the various types of flow instability. In addition to mean shear instability as a source of turbulence, transient shear is essential to both the cascade and eventual dissipation of turbulence. Shear is also inherent in convective, stratified, boundary-layer, and geostrophic flows as well. Shear turbulence is anisotropic due to the preferred direction for the mean flow and it usually is inhomogeneous. However, in the case of small scales and large enough Reynolds numbers, shear turbulence will be equivalent to 3D homogeneous turbulence (in accordance with the Kolmogorov theory).

The connection between real-world observations and models of ocean turbulence is of great importance and interest for many applications including geosciences and remote sensing. In early days, more attention has been paid to *in situ* exploration of microstructure and statistical properties of turbulence in the world's oceans. Microstructure fluctuations in velocity, temperature, density, or any other hydrophysical field in the continuously moving ocean interior were found as sufficient evidence of turbulence.

Numerous observations and measurements conducted by towed and autonomous microstructure sensors across the world's ocean demonstrated the existence of a rich *taxonomy* (i.e., certain scientific classification) of turbulent features dependent on depth, stratification, and mixing processes in the upper ocean. In particular,

> turbulence near the sea surface is physically accessible but presents measurement challenges that, largely because of wave motion, have yet to be fully overcome. In contrast, the abyssal ocean offers the challenge of inaccessibility, huge pressure, and often patchy and relatively weak turbulence with small dissipation rates.

(Thorpe 2004)

In this review, he also illustrated some of the features related to ocean turbulence (Figure 2.3).

Nowadays, the study of mixing, turbulence, and transport and their role in ocean circulation is one of the major topics of Physical Oceanography (Talley et al. 2011; Prants et al. 2017). It is concerned with kinematics and dynamics, fluxes and stress, air-sea transfers, waves, tides, and flow over topography. It is well known that turbulent mixing in the near-surface boundary layer plays an important role in many air-sea exchange processes, and therefore, the problem of turbulence observations is of

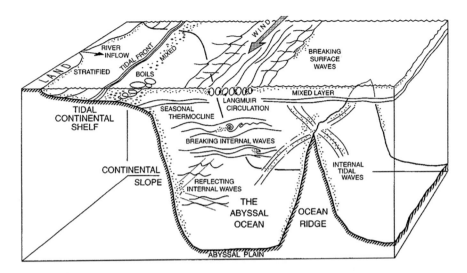

FIGURE 2.3 Sketch (not to scale) showing some of the features mentioned in text that are related to ocean turbulence. Regions of strong turbulence are stippled. (After Thorpe 2004).

a great interest in Satellite Oceanography. For example, sea surface turbulent fluxes, derived from remote sensing data, are useful for global energy and water flux research and applications.

To better understand these and other challenges in turbulence research, below we will discuss ocean turbulence phenomenology. The textbooks (Monin and Ozmidov 1985; Thorpe 2005, 2007) are classical references to this subject (see also reviews Gargett 1989; Thorpe 2004). Additionally, excellent books (Kitaigorodskii 1973; Phillips 1980; Monin and Krasitskii 1985; Apel 1987; Nihoul and Jamart 1989; Burchard 2002) provide valuable information about turbulence in the ocean. We also refer the interested reader to the *Encyclopedia of Ocean Science* (Steele et al. 2009), containing a large number of excellent thematic papers.

2.3.1 OCEAN VERTICAL STRUCTURE

Oceanographers divide the ocean into zones both vertically and horizontally. In particular, the ocean is divided by depth into five following zones (Figure 2.4):

1. **Epipelagic Zone** – the surface layer extends from the surface to 200 meters (656 feet). This layer is influenced by light producing heat and increasing biomass by enhancing photosynthesis.
2. **Mesopelagic Zone** – this zone locates below the epipelagic zone and extends from 200 meters (656 feet) to 1,000 meters (3,281 feet). The light that penetrates to this depth is extremely faint.
3. **Bathypelagic Zone** – this zone extends from 1,000 meters (3,281 feet) down to 4,000 meters (13,124 feet). It is referred to as the midnight zone or the dark zone.
4. **Abyssopelagic Zone** – this zone is also known as the abyssal zone or simply as the abyss. It extends from 4,000 meters (13,124 feet) to 6,000 meters (19,686 feet). The water temperature is near freezing, and there is no light at all.

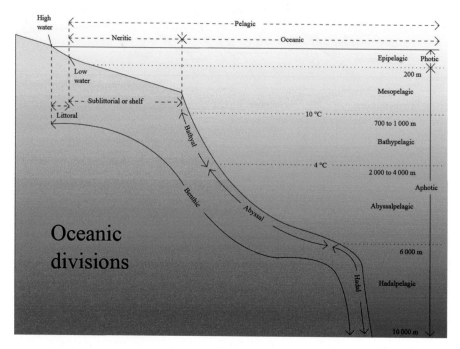

FIGURE 2.4   Five depth oceanic zones. (Wikimedia).

5. **Hadalpelagic Zone** – this zone extends from 6,000 meters (19,686 feet) down to the bottom. The temperature of the water here is just above freezing, and the ambient pressure is around 1000 times greater than at the water surface (approximately eight tons per 6.5 square cm). In spite of the pressure and temperature, life can still be found here. Hadalpelagic areas are mostly found in deep water trenches and canyons. The deepest point in the ocean is located in the Mariana Trench off the coast of Japan at 10,911 meters (35,797 feet).

Ocean five-zone depth classification is based on the differentiation by pressure, light penetration, temperature, oxygen, mineral nutrients, life forms, and ecology. Moreover, except at high latitudes, the ocean is divided into three horizontal depth zones based on density: the mixed layer, pycnocline, and deep layer. At high latitudes, the pycnocline and mixed layer are absent. Let's consider these layers in more detail (Figure 2.5).

1. **The upper ocean layer** of approximately 1,000–1,200 m deep, connects with the surface and represents the most dynamic and active part of the ocean. It is characterized by the intensive air-sea interaction, mixing (mechanical and convective), turbulence, heat, surface wave dynamics, and wave breaking. Upper ocean turbulence has a broad spectrum of time and spatial scales, forced by surface fluxes of buoyancy, momentum, and surface waves. The vertical structure of the upper ocean is primarily defined by the temperature and salinity,

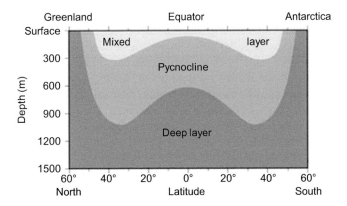

FIGURE 2.5   Three-layer model of ocean hydrodynamic structure.
(*Source:* American Meteorological Society).

which together control the water column's density stratification. Within the upper ocean layer, a number of distinct layers can be distinguished that are formed by different processes over different spatiotemporal scales. Those include a near-surface well-mixed layer and a seasonal thermocline, which is connected to the permanent thermocline or main pycnocline. Thus, it is reasonable to subdivide the upper ocean layer into two basic regions:

(a) **The near-surface boundary layer** (called also the upper ocean mixing layer or OBL) is the most active layer of 100–200 m deep. The mixed layer is the upper portion of the boundary layer where air-sea exchanges generate surface turbulence. Due to water mixing, this layer becomes vertically uniform by temperature and salinity, and thus density. Much of the turbulence kinetic energy (TKE) dissipation within the ocean takes place in the near-surface boundary layer. Large-scale coherent flow structures (known as Langmuir circulations) are an important part of turbulent boundary-layer processes and can be observed in the mixed layer. There are other flow coherent structures in the mixed layer known as "temperature ramps" or "microfronts." They appear due to generated in the shear flow turbulent eddies, which strain the mean temperature gradients and create spatial temperature nonuniformities.

(b) **The transition layer** of 200–1,200 m deep, known also as intermediate layer exists between the mixed layer and the deep ocean layer; it is characterized by the *pycnocline*, or *thermocline* or *halocline*. This region tends to be stratified and is often called the seasonal thermocline because its stratification varies with the seasons. Some *in situ* measurements indicated the enhancement of the turbulence (appearing turbulent "bursts") around well-stratified thermocline. The phenomenon is known as a "stratified turbulence." Turbulence in the transition layer is driven primarily by a combination of shear extending down below the mixed layer, penetrative convection, and breaking high-frequency internal waves.

2. **The deep ocean layer** is the lowest layer in the ocean, existing between the transition layer (thermocline) and the seabed (ocean bottom), at a depth below 1,200 m. Depths of 3,000–6,000 m are found over 74% of the ocean basins with 1% deeper. This layer has fairly constant cold temperatures. The mean temperature of water here is about 2°C (35.6°F) and salinity is 34–35 psu. Temperature, salinity, and density remain relatively constant below the main thermocline. Water temperature drop from the upper ocean to the deep ocean results from the large heat loss. In today's world, there are two principal places where deep waters form – the North Atlantic (called the North Atlantic Deep Water (NADW)) and Antarctica (called the Antarctic Bottom Water (ABW)), by sinking of dense water with a temperature less than 4°C (39.2°F) from the surface to great depth. From these regions, a cold deep water layer spreads over the entire ocean basins. Both NADW and ABW are important factors in the global climate system formation. Deep ocean circulations or currents occur under the influence of density differences caused by temperature and salinity changes. Density stratification can produce water mass transformation (mixing) in deep ocean basin. This mixing, occurring across *isopycnal surfaces*, is known as *diapycnal mixing*; it is described by the *diapycnal mixing coefficient*, $\kappa_v = \Gamma \dfrac{\varepsilon}{N^2}$, where the buoyancy $N = (g\rho^{-1}\partial\rho/\partial z)^{1/2}$ is a measure of stratification strength, ε is the TKE dissipation rate, and Γ is the mixing efficiency; in practice, $\Gamma = 0.2$ (Osborn 1980). The diapycnal mixing coefficient lies in the range of $\kappa_v \approx 10^{-4} - 10^{-5}\,\mathrm{m^2s^{-1}}$ and increases greatly to the bottom. Diapycnal mixing causes enhanced *turbulent* exchange (flux) across density surfaces. As a whole, the deep ocean layer is characterized by diapycnal mixing, fluxes of heat, salt, and dissolved gasses which create buoyancy fluxes across the ocean basin. Deep ocean waters also contain a lot of nutrients such as phosphates, nitrates, and carbonates.

3. **The bottom/benthic boundary layer (BBL)** of 5–60 m thick locates below 4,000 m (~ 13,000 feet) and greater immediately overlying the bottom of the ocean, no matter how deep. The BBL is the deepest, the darkest, and coldest part of the ocean. BBL is characterized by strong gradients of flow as well as high concentrations of dissolved and particulate matter. Bottom topography often plays an important role in the distribution of water masses and the location of currents. The BBL comprises the near-bottom layer of water, the sediment-water interface, and the top layer of sediment that is directly influenced by the overlying water. The BBL is usually considered to consist of 1) an outer Ekman layer, 2) a very thin (~ 10^{-3}m) viscous layer, and 3) transition layer called also the logarithmic layer.

It is assumed that the velocity profile u(z)in the BBL above the seafloor can be described by a law-of-the-wall, given by the von Kármán–Prandtl universal (i.e., Reynolds number independent) logarithmic law

$$\varphi = \frac{u}{u_*} = \frac{1}{\kappa}\ln\eta + C, \qquad \eta = \frac{u_*y}{v}, \tag{2.6}$$

or simply for unstratified oceanic BBL

$$\frac{u}{u_*} = \frac{1}{\kappa} \ln \frac{z}{z_0}, \tag{2.7}$$

In Equation (2.6), y is the distance from the wall, $\kappa = 0.4$ is the dimensionless von Kármán constant, u_* is the friction velocity, v is the kinematic viscosity, and the constant C is defined at wall-bounded shear flow at high Reynolds numbers. In Equation (2.7), z is height above the bottom, z_0 is the bottom roughness length. Stratified oceanic BBL requires the use of a *modified* law-of-the-wall, which accounts the wall shear stress and the pressure gradient. It also involves the buoyancy effect in the near-bottom lateral flow, defined by the Ozmidov scale $L_O = (\varepsilon/N^3)^{1/2}$ which is the outer scale of isotropic turbulence at stable stratification (Ozmidov 1965; Perlin et al. 2005).

More advanced approach which describes the Reynolds-number-dependent scaling law has been suggested by (Barenblatt 1993) and (Barenblatt and Prostokishin 1993) in the form

$$\varphi = \frac{u}{u_*} = \left(C_0 \ln Re + C_1\right) \eta^{\alpha(Re)}, \tag{2.8}$$

where the constants C_0, C_1, and α must be universal. Although the scaling law (2.8) has been tested for turbulent flow in pipes, it could also be acceptable for turbulence in oceanic BBL (and others wall-bounded turbulent flows) at relatively high Reynolds numbers. This question, however, is still open due to limited oceanographic data available.

The three-layer model (Figure 2.5) is appropriate to study hydrodynamic and thermodynamic properties of ocean environments. Because of the strong dependency of hydro-physical parameters in the depth, turbulence, existed or generated in each ocean layer, is specified by its *origin, scale, structure*, and *hydrodynamic stability* of fluid motions. In the stratified ocean, these factors play a key role in the *intermittent mixing (or convective mixing)* that represents unorganized stochastic process and produces turbulent vertical transport (turbulent fluxes). As a result, a turbulent water column in the upper ocean becomes sufficiently nonuniform with a possible appearance of the fine-stepped vertical structure –microdiscontinuities in the density, temperature, salinity, velocity, etc. (see Section 2.3.6). Meanwhile, the three-layer model (Figure 2.5) is valuable for satellite and aerospace observations with a view to the reasonable and beneficial use of oceanic data. In this context, several major categories of ocean turbulence will be considered in the next sections.

2.3.2 Surface Turbulence

Surface mixing and turbulent flows (or simply surface turbulence) are the key hydro-physical processes in the OBL. It is believed that winds generate turbulence in the upper ocean both through direct action of wind stress on the surface and through an indirect process involving dynamics of surface waves. The main natural causes of turbulence production are surface wave breaking, wave-wave interactions, wave-current interactions, modulation instabilities of different order, and impacts of atmospheric precipitation.

The wind stress is the shear stress exerted by the wind on the surface of large bodies of water. Wind stress is dominant external energy source producing mixing and turbulence at the air-sea interface. Indeed, turbulence intensity is a measure of the fluctuating component of the wind. According to credible description (Talley et al. 2011)

wind stress generates motion, which is strongest at the surface and decreases with depth, that is, with vertical shear in the velocity. These motions include waves that add turbulent energy to increase mixing, particularly if they break. For a surface layer that is initially stably stratified, sufficiently large wind stress will create turbulence that mixes and creates a substantially uniform density or mixed layer. This typically results in a discontinuity in properties at the mixed layer base.

Wind stress is defined as force per unit area exerted by the atmosphere on the ocean. The net wind stress can be expressed in terms of bulk properties as

$$\tau_w = \rho_a u_*^2 = \rho_a C_D u^2 = \rho_a \left[\frac{\kappa u}{\ln(z/z_0)} \right]^2, \qquad (2.9)$$

where τ_w is the total wind stress, ρ_a is air density ($1.22 \text{ kg} \cdot \text{m}^{-3}$), u_* is wind friction velocity, u is the corresponding wind speed (usually at 10 m above the surface), κ is the von Kármán constant, and C_D is dimensionless drag coefficient dependent on wind speed (e.g., $1.2 \cdot 10^{-3}$ at $4 < u < 11$ m/s). In Equation (2.9), the logarithmic velocity profile law for a neutrally stratified atmosphere is used (z_0 is the surface roughness length and z is the vertical coordinate over the surface). Semi-empirical formula $\tau_w = \rho_a C_D u^2$ in Equation (2.9) is based on a similarity hypothesis proposed by von Kármán for turbulent flow. Surface wind stress τ_w represents the drug force, produced by wind over the surface. Typical oceanic values of wind stress are $\tau_w \approx 0.1$ Pa (pascal unit). Turbulence appears at the air-sea interface due to instability which produces fluctuations of surface roughness and affects the value of the drug force. Because wind stress is a quadratic function of wind speed, gusty winds produce larger stresses than would steady winds of the same average speed. As a whole, wind-generated turbulence mechanically stirs the mixed layer resulting to the development of small-scale instabilities that can change stratification of the OBL.

Wind stress directly accelerates fluid parcels lying at the sea surface. It is assumed that wind stress is the principal means for forcing the ocean circulation, through a near-surface wind-driven frictional layer called the surface Ekman layer. Its typical thickness is about 30–50 m. Wind stress drives ocean circulation via frictional Ekman transport in the near-surface layer through the mass convergences (downwelling) and divergences (upwelling) in that layer. The convergences and divergences are directly related to the wind stress curl. Global wind stress and wind stress curl products are shown in Figure 2.6 (Talley et al. 2011).

Wave turbulence is non-equilibrium statistical mechanics of interacting nonlinear dispersive waves. Examples are waves on a fluid surface excited by winds or ships, and waves in plasma excited by electromagnetic waves. In oceanographic literature, the term "wave turbulence" is referred sometimes to the turbulence induced by *non-breaking* surface waves (also known as "non-breaking wave turbulence") that was first noted by Phillips in 1961 (Phillips 1980).

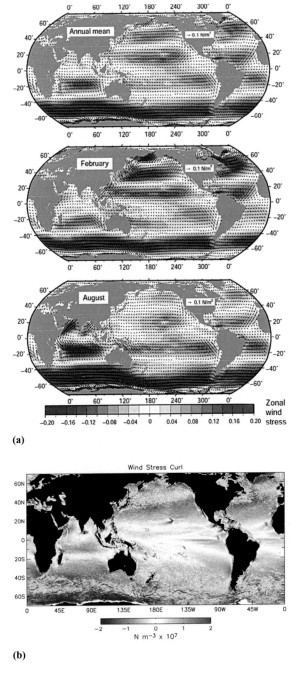

(a)

(b)

FIGURE 2.6   Global wind stress and wind stress curl. (After Talley et al. 2011, pages 143, 144, with permission from Elsevier).

Wave turbulence is the most abundant type of sea surface turbulence, occurring due to nonlinear cascades of wave energy through various scales, from where they are injected to where they are dissipated. Theoretical treatment (e.g., Nazarenko 2011) has revealed that wave turbulence is associated with strictly direct and inverse energy cascades. Respectively, similar to the Kolmogorov spectrum of hydrodynamic turbulence, these cascades are characterized by constant-flux states with power-law energy spectra $E_k(k) \propto Ak^{-q}$, A and q are constant. These spectra, known today as Kolmogorov–Zakharov (KZ) energy spectra (see books Zakharov et al. 1992; Zakharov 1998) can be obtained as exact analytical solutions of the wave kinetic equation (WKE); note that WKE is the wave analogue of the Boltzmann kinetic equations for particle interactions.

From a theoretical point of view, the simplest case is the turbulence of weakly interacting surface waves. The WKE that accounts for pumping, damping, wave propagation, and interaction is given by (Zakharov et al. 1992)

$$\frac{\partial n_k}{\partial t} = F_k - \gamma_k n_k + I_k^{(3,4)}, \tag{2.10}$$

where n_k is the spectral density of the wave action, γ_k is the decrement of linear damping and F_k describes pumping, and $I_k^{(3,4)}$ is the collision integral describing resonant three-wave ("3") or four-wave ("4") interactions. In Equation (2.10), the subscript "k" means dependence on wave vector.

For gravity surface waves in deep water with dispersion relation $\omega_k \equiv \omega(k) \approx (gk)^{1/2}$, ω is the wave frequency and k is the wavenumber, the collision integral $I_k^{(4)}$ is calculated using four-wave resonant conditions: $\omega(k_1) + \omega(k_2) = \omega(k_3) + \omega(k_4)$ and $k_1 + k_2 = k_3 + k_4$. Physics of surface wave-wave interactions involve the generation of side harmonics and parametric wave excitation effects that have been studied in early works (Phillips 1980; Craik 1985). After years, four-wave interaction phenomena were tested in the ocean remote sensing experiments using Russian side-looking aperture radar, SLAR "Toros" (Volyak et al. 1987; Raizer 2017), and IKONOS satellite imagery (Raizer 2019).

In common case of N-wave resonant interactions, the following conditions must be satisfied:

$$\omega(k_1) \pm \omega(k_2) \pm \cdots \pm \omega(k_N) = 0, \tag{2.11}$$

$$k_1 + k_2 + \cdots + k_N = 0, \tag{2.12}$$

where $\omega(k)$ is the frequency of the wave with wavenumber k = |k|. Signs plus or minus depend on the type of the N-wave interaction process. However, it seems *unlikely* that weak turbulence theory enables the description of such a multi-mode (or multi-wave) interaction process because the collision integral $I_k^{(N)}$ in Equation (2.10) cannot be simplified to approach an analytic solution. Perhaps, "N-wave turbulence" (if any) in nature can be discovered using empirical methods including remote sensing observations as well (Raizer 2019).

Meanwhile, weak turbulence theory is naturally applicable to spectral description of surface gravity and capillary waves on deep and shallow water. For example, the WKE self-similar solutions yield KZ spectra for capillary wave turbulence (with term $I_k^{(3)}$), $E_k \propto k^{-17/4}$, and for gravity wave turbulence (with term $I_k^{(4)}$), $E_k \propto k^{-7/2}$, which are both different from the Kolmogorov spectrum $E_k \propto k^{-5/3}$ (here $E_k = \omega_k n_k$, where n_k is the corresponding solution of the WKE).

A number of theoretical and laboratory studies of surface wave turbulence have been conducted over the last several decades (some data are available in book Shrira and Nazarenko 2013); however, experimental evidence of the predicted wave turbulence spectra and their variability in *open ocean* still remains the challenge. There also have been a few attempts of studying localized wave-turbulence interactions in the ocean (e.g., Veron et al. 2009). From this viewpoint, the flowing problems deserve further investigation: 1) the kinematics of wave turbulence, 2) separating turbulence and waves, 3) modulation of turbulence by surface waves, and 4) dynamics, structure, and scaling of wave-induced turbulent flows (usually associated with Langmuir turbulence).

Wave breaking turbulence – turbulence, generated by breaking surface waves, is a prominent natural phenomenon associated with great energy dissipation rate (about 80%). Wave breaking is one of the most important energy transfer mechanisms on the ocean surface. Wave breaking is known in fluid mechanics as a hydrodynamic process of wave transformation occurring due to loss of the wave system stability and/or equilibrium. The most prominent stage of wave breaking is the initial over-turning motion of the wave crest and forming a forward moving sheet of water which plunges down into the water in front causing splashes, air entrainment, and eddies (Banner and Peregrine 1993).

Breaking thereby is strongly nonlinear, transient, two-phase, highly dynamical, turbulent 3D free-space flow which rapidly changes the properties of the air-sea interface. Wave breaking turbulence is essentially 3D as well (some 1D or even 2D models and/or approximations may not be relevant for real-world ocean conditions; today CFD solvers provide more adequate numerical data).

Wave breaking is an important mechanism of turbulence production, mixing, fluxes, and hydrodynamic forces. Wave breaking dissipates energy through the entrainment of air bubbles into the flow and the generation of currents and turbulence. This enhances gas exchange between atmosphere and the oceans considerably. Active wave breaking event produces various disperse structures known as foam, whitecap, bubbles, spray, and the near-surface aerosols. The energy, dissipated due to wave breaking, and the amount of air entrainment are very sensitive to wind speed. This apparent wind speed dependence offers additional possibilities for the assessment of ocean turbulence data through remote sensing of foam/whitecap and bubble production under a variety of surface conditions.

In oceanography, four types of breaking surface waves are distinguished (Figure 2.7):

1. Spilling breakers – wave crests spill forward, creating foam and turbulent water flow.
2. Plunging breakers – wave crests form spectacular open curl; crests fall forward with considerable force.
3. Collapsing breakers – wave fronts form steep faces that collapse as waves move forward.
4. Surging breakers – long, relatively low waves whose front faces and crests remain relatively unbroken as waves slide up and down.

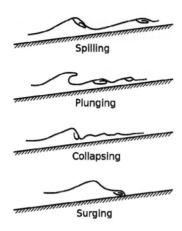

FIGURE 2.7   Classification of breaking wave types.

In the ocean, the scales of breaking waves may cover a very wide range from a few centimeters up to the length of the dominant gravity waves. A number of laboratory and field experiments have been performed to study geometry, kinematics, statistics, and temporal dynamics of breaking waves in deep and shallow waters. Some information is available in books (Bortkovskii 1987; Massel 2007; Babanin 2011). Key observation characteristics of breaking waves include: 1) statistics of wave crest lengths known as Phillips's $\Lambda(c)$-rate, c is the speed of breaking, 2) spectral energy dissipation due to wave breaking (see, e.g., Babanin 2011; Romero et al. 2012), 3) evolution of the wavenumber spectrum due to wave breaking 4) frequency of wave breaking, 5) scale of breaking waves, and 6) TKE dissipation due to breaking waves. For example, TKE dissipation rate per unit length of the crest of a breaking surface gravity wave is estimated as $\varepsilon_b = b\rho_0 c_b^5 / g$, where b is the breaking strength parameter which varies in the range $10^{-3} - 10^{-2}$, c_b is the speed of advance of the wave crest, and g is the gravitational acceleration (Thorpe 2007). Typical values of the TKE dissipation rate were found in the range $\varepsilon_b = 10^{-2} - 10^1$ with upper limit of $O(10^2)\text{W} \cdot \text{kg}^{-1} = \text{m}^2/\text{s}^3$. Field experiments (Sutherland and Melville 2015; Thomson et al. 2016) reveal depth scaling $\varepsilon_b(z) \propto (z/H_s)^{-\lambda_s}$ with tuning parameter $\lambda_s = 1 - 1.6$ dependent on sea state (z is the depth and H_s is the significant wave height). Thus, the decay of wave breaking turbulence with depth, defined by $\varepsilon_b(z)$, is stronger than the classic law-of-the-wall dependence $\varepsilon_{wl}(z) = u_*^3 / (\kappa z) \propto z^{-1}$. Near-surface turbulence observations show a mean dissipation enhancement of 1–2 orders of magnitude due to the effect of wave breaking (Steele et al. 2009).

The classification of basic types of breaking waves described above does not include the phenomenon known as *microbreaking* (or microscale wave breaking) which also produces surface turbulence of the smallest scale. Microbreaking is usually referred to breaking of very short wind waves (e.g., capillary waves with wavelengths < 2 cm) without air entrainment. Microbreaking may be relatively insignificant in the generation of turbulence within the OBL; however, it is believed that micro-breaking is a dominant mechanism in heat and gas transfer at low to moderate winds. Additionally, there is some evidence (based mostly on laboratory experiments) that microbreaking changes mean slopes and stability of short gravity waves that trigger their break and lead to enhancing surface turbulence. Furthermore, modulations of wave slopes due to microbreaking may cause the occurrence of *transient features* (such as parasitic ripples, bulge, and toe) resulting in considerable variations of (localized) surface roughness and turbulence structure; this "hardly visible" but important hydrodynamic effect is potentially detectable using a sophisticated remote (optical/IR, microwave) sensor.

Turbulence generated by breaking waves is very intermittent and coexists with turbulence generated by other sources such as shear stress, convection, internal waves, and Langmuir circulation. The depth of the turbulence injection layer due to wave breaking is estimated as $z_b = O(0.1H_s) \approx 0.2 - 1.0$ m; at high winds, z_b can be much higher than predicted. However, the detailed structure and scale of the turbulent flow, generated by breaking waves, are still not resolved conclusively. Little is known how wave (both non-breaking and breaking) turbulence interacts with ambient turbulence, and these complex interactions could be a subject of further investigations.

Wave breaking, foam, and whitecap coverage provide strong signatures in remote sensing of the ocean surface. This allows us to take significant advantages in discovering and better understanding wave breaking phenomena in the ocean. Nowadays, high-resolution airspace optical and microwave observations can be successfully used for investigations of spatial statistics and structural features of wave breaking fields and may on occasion be important in turbulence research as well. Photographs of non-breaking and breaking wave turbulence are shown in Figure 2.8.

2.3.3 NEAR-SURFACE TURBULENCE

The near-surface boundary layer (sometimes referred to the upper-ocean boundary layer, OBL) is the upper part of the ocean, approximately < 200 m in depth, that is dominantly (directly or indirectly) influenced by the surface fluxes such as heat,

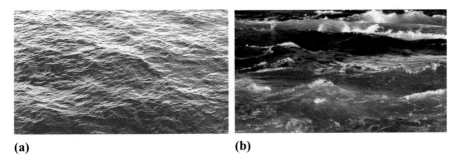

(a)   **(b)**

FIGURE 2.8 Photographs of (a) non-breaking and (b) breaking wave turbulence.

moisture, momentum, wind stress, and surface waves. Vertical structure of the OBL also includes the mixed layer and the upper part of the pycnocline. Local mixing is predominantly forced by wind forces and the atmosphere-ocean interaction. Exchange of mass, momentum, heat, and other water properties through the atmosphere-ocean interface are mediated by turbulence (Kraus and Businger 1994). Much of the mixing and TKE dissipation within the ocean takes place in the OBL. The most important processes, associated with OBL turbulence, are the following (Fedorov and Ginsburg 1992; Thorpe 2005; Steele et al. 2009; Soloviev and Lukas 2014).

Wind and buoyancy forces (Talley et al. 2011; Steele et al. 2009; Chassignet et al. 2012) are preliminary factors driving natural turbulence in the upper ocean. The near-surface layer of depth ~ 30–40 m is forced directly by the atmosphere through wind action and buoyancy (heat and freshwater) exchange. Wind- and buoyancy-generated turbulence causes the surface water to be well mixed and vertically uniform in temperature, salinity, and density.

Wind forcing, heating and cooling, and rainfall and evaporation have a profound impact on the distribution of mass and momentum in the near-surface mixed layer and ocean interior as well. Wind forcing can also set up oceanic currents and cause changes in the mixed layer temperature and salinity through horizontal and vertical advection.

On the other hand, regular meteorological and environmental factors such as wind speed, surface physical properties, and wave-breaking activity affect air-sea fluxes and turbulence production significantly. Because wind speed over the ocean exhibits great variability in space and time, the quantitative definition of wind force (like the Beaufort wind force scale) through global wind measurements is not free from difficulties and the most promising technique today could be based on remote sensing capabilities. To emphasize this option, Figure 2.9 illustrates an example of wind-retrieval map delivered from satellite radar data. A new algorithm (Monaldo et al. 2014) with 500-m sampling grid is used to generate a mean wind speed field from SAR measurements.

Buoyancy effects in the fluid have been first investigated by Turner (1973). In the ocean, the surface buoyancy or buoyant force (which is equal to the weight of the water displacement) is the sum of air-sea heat and freshwater fluxes. Buoyancy forces are responsible for developing the ocean's stratification, including pycnocline, thermocline, and halocline. As a result, buoyancy forcing changes the density of seawater.

Buoyancy effects can either force or damp turbulence. Forcing occurs in the case of unstable density stratification, i.e., when heavy fluid overlies light fluid and damping tends in the case of stable stratification, i.e., when light fluid overlies heavier fluid. A fluid parcel oscillates vertically with buoyancy or Brunt–Väisälä frequency

$N = \sqrt{-\dfrac{g}{\rho}\dfrac{d\rho}{dz}}$, where g is the acceleration of gravity, ρ is the average potential density,

a function of position and depth, and z is vertical coordinate (at stable water column $N^2 > 0$, at unstable water column $N^2 < 0$ and density variations enhance turbulence). Typical value of the buoyancy frequency for the seasonal thermocline is $N \sim 10^{-2}\,s^{-1}$. As mentioned above, the buoyancy flux and its variability directly influence on the

FIGURE 2.9 Radarsat-1 SAR wind speed retrieval off the U.S. East Coast (including Maryland and Delaware) on 10:58:02 UTC 21 Jan 2001. This image shows winds blowing offshore toward the southeast. The wind barbs represent the National Centers for Environmental Prediction's Climate Forecast Reanalysis wind speed and direction for reference. (After Monaldo et al. 2014).

atmosphere-ocean exchange that also effects on the near-surface turbulence. The buoyancy flux B_0 can be expressed as a sum of contributions from different fluxes (Talley et al. 2011)

$$Q_0 = Q_{sw} + Q_{lw} + Q_{lat} + Q_{sen}, \qquad (2.13)$$

$$B_0 = -\left(\frac{g\alpha}{\rho c_p}\right) \cdot Q_0 + g\beta(E - P)S_0, \qquad (2.14)$$

where c_p is specific heat capacity of water ($\sim 4 \cdot 10^3$ J $\cdot$ kg^{-1} $\cdot$ K^{-1} for seawater, K denotes Kelvin), S_0 is sea surface salinity, α is the effective thermal expansion coefficient, β is the effective haline contraction coefficient, and E and P are rates of evaporation and precipitation. Flux B_0 has units m^{2}s^{-3} and can be interpreted as the buoyant production of TKE. In Equation (2.13), the net surface heat flux entering the ocean Q_0 includes solar (shortwave) radiation Q_{sw}, net infrared (long-wave) radiation Q_{lw}, latent heat flux due to evaporation Q_{lat}, and sensible heat flux due to air and water having different surface temperatures Q_{sen}.

A negative buoyancy flux ($B_0 < 0$), due to either surface warming or precipitation, tends to make the ocean surface more buoyant and stable. Conversely, a positive buoyancy flux ($B_0 > 0$), due to either surface cooling or evaporation, tends to make

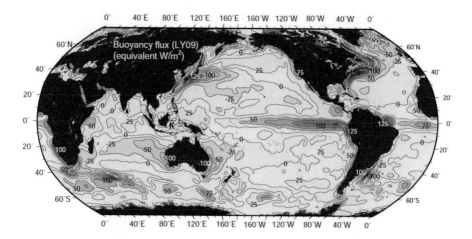

FIGURE 2.10 Global air-sea buoyancy flux map. Positive values indicate that the ocean is becoming less dense. Contour interval is 25 W/m². (After Talley et al. 2011, page 141, with permission from Elsevier).

the ocean surface less buoyant. As the water column loses buoyancy, it can become convectively unstable with heavy water lying over lighter water (Talley et al. 2011).

The buoyancy flux is also an important attribute of the synoptic weather system. Figure 2.10 shows a global air-sea buoyancy flux map. According to the description (Talley et al. 2011)

> *buoyancy loss (density gain) is most vigorous in subtropical western boundary current separation regions, where heat loss is large. The other region of large buoyancy loss, due to heat loss, is the subpolar North Atlantic and Nordic Seas. The associated northward transport of heat and hence buoyancy in the Atlantic is related to the meridional overturning circulation. Buoyancy gain is largest in the tropics, particularly over the cool surface waters in the eastern equatorial Pacific.*

Scaling laws are intimately connected to dimensional analysis and similarity theory and describe the scale invariance in many natural phenomena. In particular, scaling laws are widely used for characterizing the state of turbulent flow in the oceanic and atmospheric PBL. There are several fundamental dimensionless quantities acceptable in oceanography which are known as the following:

1. Kolmogorov scale $\eta = L_K = (\nu^3/\varepsilon)^{1/4}$, typically $6 \cdot 10^{-5} - 10^{-2}$m.
2. The Monin–Obukhov length scale $L_{MO} = -u_*^3 / (\kappa B_0)$, where B_0 is the buoyancy flux. A physical interpretation of L_{MO} is given by the Monin–Obukhov similarity theory (1954). Typically $L_{MO} = 1 - 10^2$m.
3. Ozmidov length scale $L_O = (\varepsilon/N^3)^{1/2}$, typically 1–10 m.
4. Corrsin length scale $L_C = (\varepsilon/S^3)^{1/2}$, where S is the background shear ($S = du/dz$).
5. Buoyancy scale $L_b = (B_0/N^3)^{1/2}$.
6. Buoyancy Reynolds number $Re_b = (L_O/L_K)^{4/3} = \varepsilon/(\nu N^2)$, typically 10–100.
7. Richardson number $Ri = (N/S)^2$.

These (and some others) scaling relationships and their combinations provide *empirical* framework for qualitative classification of turbulent motions associated with stability, buoyancy, and shear effects. In particular, transitions from shear-dominated to buoyancy-affected, and then to buoyancy-dominated regimes within the mixing OBL, can be examined well using various outer-scale ratios that allow the researchers to interpret experimental (and sometimes model) data in a unique manner. Because buoyancy and shear are the most important causes of turbulence production, the length scales, after all, determine the criteria of "turbulent" state in the world's oceans.

Nowadays, many experimental and model data, related to ocean turbulence, are tested using various scaling relationships which also can be applied to optimize numerical studies. In fact, scaling laws may predict different levels of stratification and stability and even regimes of turbulence; however, they do not provide a *posterior prediction of the behavior* of turbulence in the first place. The most valuable results can be obtained using CFD (e.g., LES or RANS) capabilities. In turbulence research, "scaling versus equations" is still a controversial problem.

Convection by definition is a physical process of either the mass transfer or the heat transfer due to bulk movement of molecules within fluids such as gases and liquids. The term "convection" comes from the Latin "*convectiō*" which means to carry/bring together. In fluids, convective heat and mass transfer takes place through both diffusion (the random, Brownian motion of individual particles of the fluid) and advection (in which matter or heat is transported by the larger-scale motion of currents in the fluid). In the context of heat and mass transfer, the term "convection" is used to refer to the sum of advective and diffusive transfer. Turner (1973) defines continuous convective motions arising from isolated sources as *plumes* and the buoyant blobs produced by ephemeral sources as *thermals*.

There are two major types of convection – *free convection* or natural convection, in which the fluid motion occurs by natural means such as buoyancy from density changes, and *forced convection*, in which fluid motion is generated by an external source in order to increase the heat transfer.

Many geophysical and astrophysical regime thermal convection processes are extremely turbulent. The turbulence is characterized by a dimensionless parameter called the Rayleigh number, which can be thought of as a ratio of the buoyant and viscous forces acting within the fluid. For example, a criterion for natural double-diffusion convection can be defined in terms of the thermal Rayleigh number,

$Ra_T = \dfrac{\alpha g \Delta T h^3}{\nu k_T}$, and saline Rayleigh number, $Ra_S = \dfrac{\beta g \Delta S h^3}{\nu k_S}$. Here α and β are the

thermal expansion and salinity contraction coefficients, k_T and k_S are molecular diffusivities of heat and salt, respectively (see Section 2.3.6), ν is the kinematic viscosity, g is the gravitational acceleration, h is the characteristic length-scale of convection (e.g., thickness of convective layer), and ΔT and ΔS are the corresponding positive temperature and salinity difference between the top and bottom of the layer. Values of several physical parameters of water are given in Table 2.2.

Generally, the condition from the onset of natural thermal convection is characterized by the critical Rayleigh number Ra_{cr} for a given thermodynamic system. The value of the critical Rayleigh number depends on the particular geometry and boundary conditions. For convection between parallel plates, the value Ra_{cr} lies between

TABLE 2.2

Physical Properties of Fresh Water (at 10°C) and Air (at 15°C) at Standard Atmospheric Pressure

Physical Property	Notation	Water	Air
Thermal expansion coefficient	α	$2.6 \times 10^{-4}/°C$	$3.5 \times 10^{-3}/°C$
Kinematic viscosity	ν	1.3×10^{-6} m²/s	1.5×10^{-5} m²/s
Thermal diffusivity	κ_T	1.4×10^{-7} m²/s	2.2×10^{-5} m²/s

657 (free boundaries) and 1708 (rigid or *no-slip* boundaries). At Ra > Ra$_{cr}$, the system becomes unstable and natural convection takes place. Thermal convection is estimated to be on the order of Ra $\sim 10^{20}$ in the ocean and convection is turbulent. In order to accurately simulate these scales of turbulence, it is vital to experiment at equally great parameters.

According to review (Soloviev and Klinger 2001), free convection in the ocean

> *is one of the key processes driving mixed layer turbulence, though mechanical stirring driven by wind stress and other processes is also important. Therefore, understanding convection is crucial to understanding the mixed layer as well as property fluxes between the ocean and the atmosphere.*

Because density is a function of temperature and salinity, it is possible to observe *thermal convection* due to the vertical temperature gradient, *haline convection* due to the vertical salinity gradient, or *thermohaline convection* due to their combination. More specifically, there are several types of natural convection known as 1) penetrative convection, 2) double diffusive convection, 3) diffusive convection, 4) patchy convection, 5) cellular convection or *Rayleigh—Bénard* convection, and 6) **chimney convection or convective chimney**.

The most interesting and important type of natural convection is the Rayleigh–Bénard convection (RBC). This type of convection occurs in a relatively thin water layer (~ 1 mm) or the first 2 kilometers of the Earth's atmosphere. RBC is the buoyancy-driven flow of a fluid heated from below and cooled from above. In a plane horizontal layer, the fluid notion develops a regular pattern of convection cells known as Bénard cells (or Rayleigh–Bénard cells). The heat increases the fluid's temperature, so reducing its density and resulting in the buoyancy forces that drive convection. Only an unstable situation gives rise to generation of Bénard cells (Figure 2.11).

The mechanisms that determine the fluid dynamics are: 1) thermal diffusion due to the temperature gradient in the layer, 2) buoyancy-driven convection due to temperature differences which affect buoyancy, and 3) inner friction that resists the deformation of the fluid parcels. The interplay of these three mechanisms produces "updraft plumes" which then turn into a regular hexagonal pattern (Bénard cells). The theory of RBC is based on the well-known dimensionless Boussinesq equations. Book (Koschmieder 1993) provides excellent description and more details.

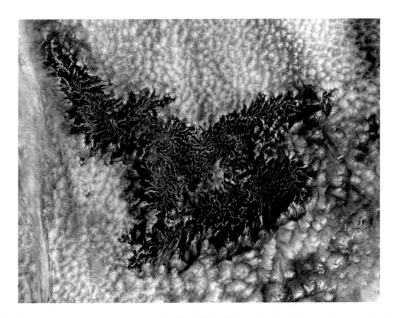

FIGURE 2.11 Natural convection: the Rayleigh–Bénard cells. Open-cell and closed-cell clouds off California, Pacific Ocean. True-color MODIS image from August 16, 2002. (Image from NASA).

Convection plays an important role in the movement of deep ocean water contributing to ocean currents. This phenomenon usually refers to *open ocean convection* and/or deep (ocean) convection. The mean value of TKE dissipation rate in the convective layer is estimated as $<\varepsilon> \approx 0.72B_0$, B_0 is the surface buoyancy flux, and can reach $3.2 \cdot 10^{-7}\,W \cdot kg^{-1}$ (Thorpe 2007). Deep convection occurs only in a few special locations around the world: Greenland Sea, Labrador Sea, Mediterranean Sea, Weddell Sea, Ross Sea, and Japan (or East) Sea (for more detail, see books Talley et al. 2011; Siedler et al. 2013).

Langmuir circulation (or Langmuir turbulence) is another transient response to impulsive wind forcing (Langmuir 1938). Irvine Langmuir, the eminent American chemist and physicist, was the first who recognized the dynamical significance of the phenomenon. Langmuir circulation (LC) is a system of horizontal coherent counter-rotating vortices that develop in the ocean, sea, or lake surface layer with moderate to strong winds blowing the surface. Movement in LC having longitudinal, literal, and vertical current components is substantially 3D.

In fluid dynamics and oceanography, Langmuir turbulence is a turbulent flow consisting of coherent Langmuir circulation structures (cells) that exist and evolve over a range of spatial and temporal scales. These structures arise through an interaction between the ocean surface waves and the currents. Large-scale Langmuir cells may subduct smaller cells, rotors, or other vortices generated by breakers, enhancing subsurface turbulence and dispersion, although this effect is presently unquantified (Thorpe 2005). Langmuir turbulence occurs when a surface boundary layer is forced

FIGURE 2.12   Langmuir circulation in ocean.

by wind in the presence of surface waves. It is also suggested that Langmuir turbulence may be more effective than shear-driven turbulence in deepening the mixed layer.

In oceanographic literature, Langmuir turbulence and its significance in inducing diapycnal mixing in the upper ocean is commonly described by the turbulent Langmuir number, $La_t = (u_*/u_{s0})^{1/2}$, where u_* is the surface friction velocity in the water and u_{s0} is the surface Stokes drift. Langmuir turbulence becomes a dominant process when $La_t \leq 0.3 - 0.5$, whereas shear-driven turbulence is dominant for significantly higher La_t (for more detail, see McWilliams et al. 1997).

LCs are visually evident as numerous long parallel lines or streaks of flotsam ("windrows") that are mostly aligned with the wind (Figure 2.12). The streaks are formed by the convergence caused by helical vortices with a typical depth and horizontal spacing of 4–6 m and 10–50 m, but they can range up to several hundred meters horizontal separation and up to two to three times the mixed layer depth. Alternate cells rotate in opposite directions, causing convergence and divergence between alternate pairs of cells. The cells can be many kilometers long. LC generally occurs only for wind speeds greater than 3 m/sec and appears within a few tens of minutes of wind onset.

LC is a well-established important mechanism providing a link between the wind-wave field and the development of the mixed layer. In practice, LC motions are turbulent. The common theory of LC referred to as the Craik–Leibovich theory (Leibovich 1983) is now utilized and supported by computer simulations and new field measurements (see book Steele et al. 2009 for further discussions).

Kelvin–Helmholtz (KH) instability is known in oceanography as the dynamic disturbances, occurring in a stably stratified shear flow. The instability is characterized by distinctive finite-amplitude billows (Figure 2.13). A classical case is two fluids of different densities moving with various speeds when small disturbance (e.g., wave) produces the KH instability which generates billows – a kind of organized structure. The billows intensify the deepening of the diurnal mixed layer. Ultimately, the billows overturn and generate small-scale turbulence. Billows may be triggered also on an interface between warm and cold water when the warm water moves to the right faster than the cold water. For example, wind forces can cause the development

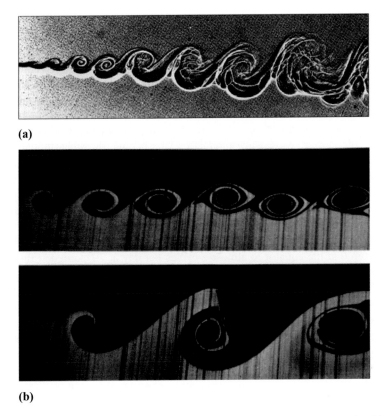

(a)

(b)

FIGURE 2.13   Kelvin–Helmholtz (KH) instability. The instability is characterized by distinctive finite-amplitude billows. (a) Flow structure (After Brown and Roshko 1974, 2012) and (b) KH instability of superposed streams

(*Source:* van Dyke 1982, photo # 146).

of shear KH-type instability and the origin of horizontal eddies which influence on the temperature profiles of the near surface boundary layer. This process can cause the layer transition from warm to cold conditions, creating both KH instabilities and thermal-dynamical instabilities.

The common theory predicts the onset of instability and transition to turbulent flow when the destabilizing influence of the velocity shear overcomes the stabilizing effect of the buoyancy force. The KH instability can be described with the Taylor–Goldstein equation (e.g., LeBlond and Mysak 1978). The stage of transition following the formation of billows on a thin interface depends on the Richardson number, $Ri = N^2/(du/dz)^2$, u is a representative flow speed and z is depth. Instability can occur only at small values $Ri < 1/4$ and the flow disturbances leading to billows are growing; as a result, turbulent mixing will generally occur. At large values $Ri > 1/4$ turbulent mixing across the stratification is suppressed.

There are numerous observations of KH billows in the near-surface layer of the ocean (Thorpe 2005). Turbulent mixing induced by KH instability is essential aspect of ocean-atmosphere dynamics (Vallis 2017). Moreover, KH instability is an essential mechanism in the development of turbulence in the stratified interior of the ocean, particularly associated with onset of internal wave turbulence (see Section 2.3.5).

Near-inertial waves (NIWs) in upper ocean or near-inertial motions or oscillations are a type of mechanical waves, existing in rotating fluids. NIWs have been observed for over 80 years throughout the world ocean. NIWs travel over hundreds of kilometers toward the equator because of decreasing inertial frequency with latitude. They can be generated by numerous mechanisms including wind forcing events like sporadic storms, wave-wave interaction (the parametric subharmonic instability), or the interaction between topography and geostrophic currents.

NIWs have a frequency ω close to the Coriolis frequency f_c and can be recognized by a prominent peak (called "inertial frequency") in the energy spectra of surface horizontal drifter velocities (Figure 2.14). NIWs exist only in the frequency range $f_c < \omega < N$, i.e., between the Coriolis frequency (lower boundary), f_c, and the buoyancy frequency (upper boundary), N. The inertial frequency depends on latitude, going from 0 at the equator to almost 2 cycles per day at 70° latitude (Talley et al. 2011). In most oceanic conditions, about half of the total internal kinetic energy (IKE) is concentrated in the near-inertial frequency band of the spectrum thereby freely propagating NIWs are restricted to frequencies higher than the local f_c.

In the upper ocean, NIWs are generally assumed to be the result of wind forcing (or wind-driven shear). The wind-generated NIWs (their group is also known as internal swell) are weakly dispersive waves with large horizontal and small vertical scales which propagate as much as thousand kilometers equatorward. Fluctuations in wind speed and direction result in persistent oscillations at near-inertial frequencies. Such oscillations are observed almost everywhere in the upper ocean and dominate

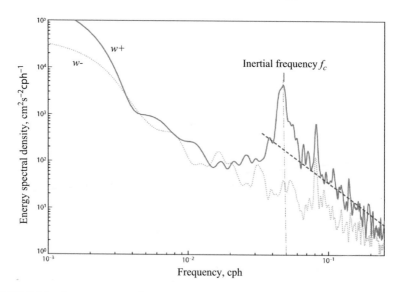

FIGURE 2.14 Energy spectra of surface horizontal drifter velocities of near-inertial waves. The solid blue line ($w+$) is clockwise motion, and the dashed blue line ($w-$) is counterclockwise motion. (Adapted from Alford et al. 2016).

the horizontal velocity component of the internal wave field. Observations demonstrate the latitudinal and seasonal variability of the near-inertial oscillations in some ocean regions (Thorpe 2005; Steele et al. 2009; Talley et al. 2011).

NIWs are highly energetic and susceptible to instabilities and contain a significant fraction of the ocean's internal wave energy and shear. Oceanographers believe that NIWs are a significant source of energy leaking the turbulent diapycnal mixing in the ocean interior; however, it is not entirely clear what fraction of the TKE is emitted to the near-surface mixed layer. It is assumed that averaged over large space and time scales TKE dissipation rate attributable to NIWs can be estimated as $<\varepsilon_{NIW}> \approx$ $7 \cdot 10^{-10} < (N/N_0)^2 < 10^{-8}\,W \cdot kg^{-1}$, $N_0 = 0.0052\,s^{-1}$ is a reference buoyancy frequency (Gregg 1989). More detailed descriptions of NIWs and the corresponding references can be found in book (Siedler et al. 2013).

As a whole, the variation of the OBL structure and subsurface instabilities play a crucial role in the coupled ocean-atmosphere dynamics and the generation of the near-surface turbulence (including also two-phase turbulent flows). Mixing and turbulence in the upper ocean ultimately effect on the air-sea fluxes of momentum, heat and mass, and global ocean circulation as well.

2.3.4 Deep-Ocean Turbulence

Deep-ocean turbulence is associated with ocean convection, internal waves, and *thermocline*. The generation and breaking of internal (or baroclinic) tides upon sloping seafloors are a potential source of mixing in the deep-ocean (see Section 2.3.4). Convective turbulence occurs as a combination of the Rayleigh–Taylor instability and shear-induced Kelvin—Helmholtz instability (Thorpe 2004, 2005). Some results suggest that turbulence generated by internal wave breaking near the thermocline/pycnocline interface may boost overall (ambient) turbulence level in mixed layer and spread deeper. Turbulence in the pycnocline is often patchy in time and space, with intermittent bursts of turbulent mixing.

The BBL Mechanism. Turbulence production in BBL is sustained by shear resulting from the interaction of the flow with bottom topography (Section 2.3.7). In any case, the physical presence of boundaries (sea surface, seabed, or the pycnocline) will affect the form and scale of turbulence within the mixed layer. Therefore, there appears to be no reason to suppose that BBL turbulence differs in any fundamental way from that dominated by tidal flows in shallow water (Thrope 2005).

The Internal Wave Mechanism. The stratified ocean interior is permanently in motion especially driven by internal waves with typical amplitudes of several tens of meters. In this case, turbulence parameters – TKE dissipation rate ($\varepsilon \approx 10^{-7}$ – $10^{-5}\,m^2 s^{-3}$) and turbulent diffusivity ($k_z \approx 10^{-4}$ – $10^{-3}\,m^2 s^{-1}$) – are high in the lower 100 m above sloping topography and they are 100–1000 times larger than observed in the open ocean, away from boundaries (Thorpe 2005; Velarde et al. 2018). It was also found that the interplay between the large-scale internal waves and the small-scale motions near the buoyancy frequency acts to create both the diapycnal turbulent mixing and the isopycnal transport of the mixed waters into the ocean interior. Thus, most internal wave – turbulence transitions – is associated with nonlinear deformation of internal wave motions near local buoyancy frequencies. Internal wave generated turbulence in a deep ocean layer dominates over BBL turbulence production.

The Decay Mechanism. The turbulence decay known also as "eddies decay" in the absence of sources is referred to spatiotemporal evolution of deep-ocean turbulence. It becomes essential in the presence of the thermocline. In the stable stratified ocean, depending on the thermocline depth, turbulence occurs in form of patches of variable temporal and spatial (horizontal) structures. These patches may take form of turbulent spots spreading at local areas and rising to the surface. There are only a few "direct" measurements of the turbulence decay rate in terms of temporal evolution of the flow following the removal of sources; most investigations of the decay have been made in laboratory experiments (Baumert et al. 2005).

2.3.5 INTERNAL WAVES AND TURBULENCE

In this section, we consider some aspects of internal waves, turbulence, and mixing in the ocean. The subject is too vast to be treated in its entirety. We encourage the interested reader to explore myriad details available in the scientific literature. Excellent textbooks and monographs (Roberts 1975; Lighthill 1978; Apel 1987; Miropol'sky 2001; Thorpe 2007; Sutherland 2010; Massel 2015; Morozov 2018) provide relevant information about physics of internal waves including observations, theory, and applications.

By definition, ocean Internal waves (IWs) are a type of gravity waves that occur within the ocean interior. IWs are generated whenever a source of energy displaces fluid vertically in the presence of density stratification. Unlike surface waves, IWs can propagate both vertically and horizontally depending on density gradient and source. IWs manifest themselves by oscillations of the water layers on which they center (Figure 2.15) and the mean sea level is not affected at all. However, IWs can induce local currents that influence on surface roughness and turbulence. The roughness change is a prominent effect potentially detectable using airspace radar, optical imager, and passive microwave radiometer as well (Raizer 2017). Spatial modulations of surface roughness cause periodic-like variations in backscatter and light reflection signals, which "reproduce" in the images IW patterns (see Chapter 5). Parameters and scales of the IWs can also be estimated using remotely sensed data.

The various mechanisms of IW generation in the ocean are known. In general, they are associated either with bottom topography (bathometry), or with disturbances within the stratified fluid such as convection, flows, currents, and turbulence, or with moving (oscillating) solid bodies. Here are the most important mechanisms (see also Raizer 2019).

- **Centrifugal and gravitational forces** of moon, sun, and some other planets generate tides. The most significant of these is the principle lunar tide with approximately half-day period (called also "semi-diurnal lunar tide"). The ocean tide effectively is a shallow IW and barotropic.
- **Flow over bottom topography** generates disturbances which usually take place when the fluid moves over the obstacle located on the ocean floor. The obstacle may represent as an isolated single hill as well periodical hill of

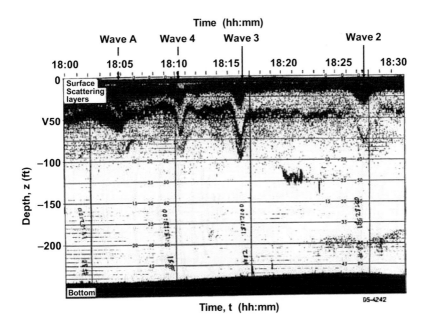

FIGURE 2.15 Large-scale internal waves generated on continental shelf. Original acoustic echo-sounder record of solitary wave displacements. (After Apel et al. 2007).

variable amplitude. The theory provides estimates of parameters of IWs depending on configuration (amplitude and period) of the obstacle.

- **Tidal flow over bottom topography** generates IWs (known as *lee waves*) when oscillating barotropic flow of a density-stratified fluid moves over underwater hill called "sill." This generation mechanism is referred to as "baroclinic conversion" because it transforms the energy of barotropic tides (in which the entire ocean column oscillates) to baroclinic IWs. Through other processes, the large-scale IWs convert the energy to smaller scales, ultimately dissipating through mixing. For tidally generated IWs, the effects of Coriolis forces are crucial.

- **Wind forcing and/or atmospheric instability** can generate IWs due to air-sea interactions that may trigger oscillations of the inertial current at the mixed upper layer of the ocean. Such IWs are observed mostly in continental shelf under the influence of fluctuations of atmospheric pressure.

- **Generation by surface** (also known as *spontaneous creation*) is associated with resonant coupling of high-frequency gravity surface waves which can produce a relatively low-frequency internal wave (Brekhovskikh and Goncharov 1994). This is a somewhat forgotten but potentially efficient process, depending on the stratification and swell characteristics.

- **Ekman layer instability** mechanism (mostly related to atmospheric IWs) supposes that instability of turbulent flow in stratified Ekman boundary layer can also generate IWs in the ocean.

- **Oscillating body** in a uniformly stratified fluid creates a cross-shaped pattern of IWs. The structure of the wave beams generated by a vibrating cylinder is a superposition of plane waves having different spatial structure but identical frequencies. The IW pattern is sometimes referred to as a *St. Andrew Cross* (Sutherland 2010).

- **Collapsing submerged turbulent wakes** can be efficient generators of IWs due to turbulent mixing and density fluctuations. These wave-type disturbances can propagate as in horizontal as well in vertical directions. The internal stratified layer may transmit the disturbances from the depth to the surface causing the changes in surface roughness. This event is potentially detectable by the optical sensor under appropriate conditions (Raizer 2019).

Much of IW research has been organized around spectral characteristics of IWs. Garrett and Munk (1972) synthesized existing observations into a kinematic model spectrum that describes the observed distribution of IW energy in spatial (wavenumber) and frequency space. This spectrum is known as the Garret–Munk (GM) spectrum. The GM spectrum is essentially a semi-empirical model which provides a framework for the study of diverse ocean spectra. The GM spectrum assumes horizontal anisotropy and vertical isopycnal displacement and takes the general form (e.g., LeBlond and Mysak 1978, page 529)

$$S_{IW}\left(k_h,\omega\right) \propto \mu^{-1}\omega^{-p+1}\left(\omega^2 - f_c^2\right)^{-1/2}, \quad k_h \leq \mu \quad \text{or} \tag{2.15}$$

$$S_{IW}\left(k_h,\omega\right)\mu\omega^{-p}, \quad \omega^2 \gg f_c^2, \quad k_h < \mu \tag{2.16}$$

where $k_h = \left(k_1^2 + k_2^2\right)^{1/2}$ is the horizontal component of the wave vector (k_1 and k_2 are the Cartesian components of the wave vector), ω is the frequency ($f_c < \omega < N$), μ is a cut-off wavenumber, f_c is the Coriolis frequency, and N is the buoyancy frequency. It is assumed that parameters p and q both lie between 5/3 and 2. There are a number of modifications of the GM spectrum which take into account saturated IW sub-ranges and the presence of tidal peaks often found in real observations (Levine 2002).

In most conceptual spectral models, IWs and turbulence are directly coupled. Thorpe (2007, page 54) has explained:

> *In practice in the stratified ocean, both turbulence and internal waves may affect the energy spectrum. Although, it is sometimes possible to use frequency to distinguish between internal waves with frequencies ($\omega < N$) and turbulence at higher frequencies, both the waves and the turbulence are transient and may change their nature in time. The internal wave field, for example, may at times be dominated by the passage of groups of waves generated near the sea surface that pass through the sampling region, breaking and periodically generating or intensifying turbulence.*

The relation between IWs and turbulence is complex and an important aspect of physical oceanography. IWs produce turbulence, e.g., by breaking, and vice versa, ambient or induced turbulence can generate IWs. IWs radiate turbulent patch which downwards into the pycnocline from the bottom of the near surface mixed layer. Regions of mixing caused by breaking IWs can be similarly localized within the larger field of oscillatory motion induced by the waves (Thorpe 2007).

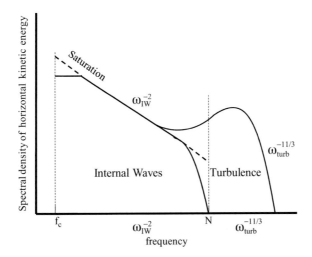

FIGURE 2.16   Schematic illustration of separating "wave" and "turbulence" parts from the spectrum of internal waves IWs.

Spectral properties of IWs are important for understanding of their spatial structure. Entire wavenumber ocean energy spectra (or slope spectra) are often used for statistical characterization of both IWs and turbulence; for this goal, *in situ* measurements of velocity, temperature, and/or density variations over the depth are conducted (Klymak and Moum 2007a, 2007b). The calculated spectra are divided into several subranges. The most general division is between large-scale motions related to IWs and small-scale motions related to turbulence. Therewith high-frequency (i.e., high-wavenumber) part of the spectrum is characterized through simple power law. Figure 2.16 clarifies this statement schematically. Entire spectrum of IWs and turbulence depend on regional characteristics of the ocean, and therefore, they may not be always identical. Strong nonlinear IWs and solitons, generated on the continental shelf, as a rule, have multimode structure which can be identified using spectral filtering. In principle, this may help to separate "wave" and "turbulence" components and reveal IW-turbulence transition region as well. The technique, however, requires detailed oceanographic observations.

The experimental results indicate that two basic hydrodynamic scenarios of the coupling "IWs – turbulence" are distinguished: 1) IWs produce turbulence, i.e., "IWs → turbulence," and 2) turbulence generates IWs, i.e., "turbulence → IWs." Mechanisms, effects, and signatures are quite different.

In the first scenario, "IWs → turbulence," the process results in localized mixing. Breaking IWs are suggested to be the dominant source of turbulence in the ocean interior. Thorpe (2004) describes mechanism of turbulence generation by IWs as the following:

In brief, internal waves propagate through the density-stratified ocean, carrying energy from their generation source to deeper or shallower regions. They break, producing mixing and vertical diffusion and dissipating energy…through their local enhancement of shear, which leads to an instability called Kelvin-Helmholtz instability…. Turbulence

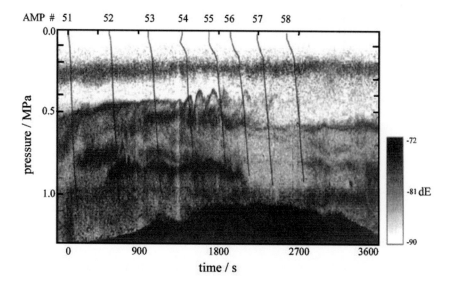

FIGURE 2.17 An example of Kelvin–Helmholtz billows produced by breaking of internal waves. An acoustic image showing pressure versus time of the acoustic scattering strength (1 MPa =100 m). The black lines mark trajectories of a free-fall microstructure probe (AMP) used to sample turbulence in and around the billows. (After Thorpe 2004).

is reached via a transition that involves a series of stages in which wave-like undula-tions "roll-up," overturning the density field in "billows" where the fluid becomes stati-cally unstable and small-scale vortical motions occur. (An example of billows detected acoustically is shown in Figure 2.17).

More complicated but common mechanism is based on the relationship between small-scale (or micro-scale) turbulence, IWs, and fine structure of hydro-physical fields (Monin and Ozmidov 1985). An important role in the development of small-scale turbulence plays dynamics of internal tides (Marchuk and Kagan 1989; Vlasenko et al. 2005). In the first scenatio, "IWs → turbulence," turbulence also dominates the diapycnal redistribution of suspended materials.

In the second scenario, "turbulence → IWs," submerged stratified turbulent flows (e.g., wake) emit IWs in the near field. Sometimes, wake-emitted IWs may be mani-fested through multiple surface modulation mechanisms such as 1) hydrodynamic modulation of surface capillary-gravity waves – event well known as surface alter-nating "slicks" and "rip currents," 2) spatial modulation of the molecular surface film consisting of seawater surfactants and/or organic compounds, 3) temperature modu-lation of thermal cool-skin sublayer, and 4) spatial modulation of surface wave breaking and foam coverage fields at low and/or high winds. An important "turbu-lence → IWs" process is the generation of turbulent spot which can propagate in both horizontal and vertical direction (see also Section 2.5). Under certain condition, surface manifestations of IWs are detectable using passive microwave remote sensing techniques (Raizer 2017).

2.3.6 DOUBLE-DIFFUSION AND TURBULENCE

Double-diffusive convection (DDC) is the buoyancy-driven flow with fluid density depending on two scalar components having different rates of molecular diffusion (diffusivities). In oceanography DDC is commonly referred to as thermohaline convection. It is assumed that the mechanism of forming thermohaline structure favorable for DDC is associated with the redistribution of the potential energy between the saline and thermal components of mesoscale and macroscale stratification (Fedorov 1978).

Many researchers (see books Turner 1973; Fedorov 1978; Brandt and Fernando 1995; Huang 2009; Radko 2013 and also reviews Schmitt 1994, 2003) have found that DDC can occur in a stratified fluid where the vertical gradient of either temperature or salinity is destabilizing. The process of DDC is characterized by the coefficient of molecular conductivity of heat k_T and by the coefficient of molecular diffusion of salt k_s. For sea water, typical values of these coefficients are $k_T \approx 1.4 \cdot 10^{-3} \, \text{cm}^2/\text{s}$ and $k_s \approx 1.3 \cdot 10^{-5} \, \text{cm}^2/\text{s}$ and $k_T/k_s \approx 100$. Thus, heat defuses approximately 100 times faster than salt. This leads to double-diffusive instability (or *thermohaline convective instability*) which plays a crucial role in mixing, turbulence, and transport processes.

The dynamics of unbounded double-diffusive systems are controlled by several key parameters: the Prandtl number $Pr = v/k_T$, the Schmidt number $Sc = v/k_s$, the diffusivity ratio $\tau = k_S/k_T$ (or the Lewis number $Le = k_T/k_s = Sc/Pr$), and the density ratio R_ρ which is defined as

$$R_\rho = \frac{\alpha \bar{T}_z}{\beta \bar{S}_z}, \qquad (2.17)$$

where $\alpha = -\dfrac{1}{\rho_0}\left(\dfrac{\partial \rho}{\partial T}\right)$ and $\beta = \dfrac{1}{\rho_0}\left(\dfrac{\partial \rho}{\partial S}\right)$ are the thermal expansion and haline (salinity) contraction coefficients, $\dfrac{\rho - \rho_0}{\rho_0} \approx -\alpha\left(T - T_0\right) + \beta\left(S - S_0\right)$ is the equation of state in the linear form, ρ is density, and ρ_0, T_0, S_0 are reference density, temperature, and salinity, respectively, $\bar{T}_z \equiv \dfrac{\partial \bar{T}}{\partial z}$ and $\bar{S}_z \equiv \dfrac{\partial \bar{S}}{\partial z}$ are mean vertical gradients of temperature (T) and salinity (S), and z is the vertical coordinate (it is taken to be positive upward). The ratio R_ρ describes the relative contributions of temperature and salinity to the density stratification (assuming a linear equation of state).

Double diffusive convection is possible only if one of the thermohaline components (temperature or salinity) gives hydrostatically unstable contribution into density stratification provided that the whole density stratification is hydrostatically stable.

FIGURE 2.18 Laboratory experiment on fingering convection. Salt fingers are produced by setting up a stable temperature gradient and pouring salt solution on top. (After Huppert and Turner 1981; Radko 2013)

Depending on mean temperature and salinity stratification, there are two types of DDC referred to as "salt fingers" and the "diffusive convection." Fingering due to double-diffusive stability/instability in water is shown in Figure 2.18 (laboratory experiment by Huppert and Turner 1981). According to Equation (2.17), there are the following criteria for the development of different types of DDC (Figure 2.19):

- If salinity contribution is hydrostatically unstable ($\bar{T}_z > 0$, $\bar{S}_z > 0$, $\bar{\rho}_z < 0$, $R_\rho > 1$), the salt finger convection is possible.
- If temperature contribution is hydrostatically unstable ($\bar{T}_z < 0$, $\bar{S}_z < 0$, $\bar{\rho}_z < 0$, $0 < R_\rho < 1$), the diffusive convection is possible.
- If both temperature and salinity contributions are hydrostatically stable ($\bar{T}_z > 0$, $\bar{S}_z < 0$, $\bar{\rho}_z < 0$, $R_\rho < 0$), there is no possibility for DDC to be developed.
- In the case of neutral stratification, obviously, $R_\rho = 1$.

Generally, salt finger convection is considered active when $1 < R_\rho < 2$, and diffusive convection is considered active when $0.5 < R_\rho < 1$.

The most prominent signature of DDC in the ocean is the thermohaline staircase, i.e., a sequence of mixed layers which are extremely uniform both in temperature and salinity. Thermohaline staircases, observed in the ocean are ranging from tens to hundreds of meters thick, separated by steep-gradient interfaces, which are typically several meters thick. Staircases are found where large-scale variations of temperature T and salinity S both increase upward in a manner that favors vigorous salt fingering. Thermohaline staircases are distinguished by their extreme stability in space and time.

An understanding of how turbulence affects staircase formation and diffusive fluxes becomes increasingly relevant. Some oceanographic observations (e.g., Shibley and Timmermans 2019) indicate that there is a subtle relationship between

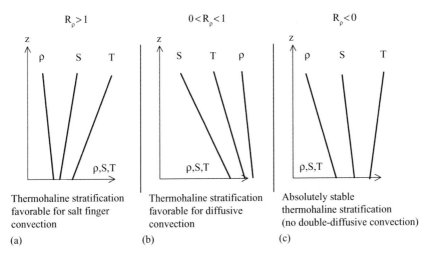

FIGURE 2.19 Types of thermohaline convection. (a) Thermohaline stratification favorable for salt finger convection and (b) Thermohaline stratification favorable diffusive convection. (c) Absolutely stable thermohaline stratification (no double-diffusive convection). (After Raizer 2017).

shear-driven turbulence and DDC. In particular, interaction of double diffusion and shear-driven turbulence may lead to several effects: 1) enhancement of ambient turbulence that tends to destroy a thermohaline staircase structure, 2) tendency to mix temperature and salinity at equal rates providing relatively stable fine-scale structure of temperature-salinity fields, and 3) turbulence redistribution and generation of thermohaline intrusions. The intrusion is driven by density flux divergence due to larger density flux by salt fingers than that due to diffusive convection. The typical vertical thickness of the intrusion is 10–100 m (Fedorov 1978). Dynamics of thermohaline intrusions can be explained by similarity theory (Barenblatt 1978a, 1978b). Thus, in the ocean regions, occupied by double-diffusive instabilities and/or salt-fingers, turbulent events disrupt the thermohaline microstructure. The phenomenon plays a significant role in the duration of submarine wakes depending on ocean stratification.

Significant uncertainties still exist in the formulation of double-diffusive flux-gradient laws (Radko 2013). The effects of background shear, intermittent turbulence, and IWs on the salt-finger transport are typically not taken into account and can be substantial (Radko 2013). Statistics of double-diffusive microstructure, particularly thermohaline staircases, require more attention. Moreover, the generation of salt fingers in the near-surface layer under the influence of external factors could also be a subject of study. All listed problems are of considerable interest and importance in ocean remote sensing and advanced applications.

2.3.7 Bottom Turbulence

Bottom turbulence or the near-bottom turbulence is turbulence existing in the BBL. This turbulence is generated by a number of internal processes which are 1) interaction of current shear with the bottom, 2) density stratification leading to mixing and turbulence entrainment, 3) breaking internal waves, generated over topography, known as lee waves, and 4) tides over rough topography. Turbulence can spread out from BBL along isopycnals (i.e., constant-density surfaces) into the ocean interior. The averaged TKE dissipation rate is defined by the mean velocity of flow $\bar{u}$ associated with the near-bottom Reynolds number $Re_{nb} \approx 10^3 - 10^5$ (see also Section 2.3.1). At distances, z, from seabed rigid boundary, the TKE dissipation rate generally follows the law-of-the-wall; it is defined as $\varepsilon = u_*^3 / \kappa z$ according to Equation (2.7). Typical values of the TKE dissipation rate found for bottom turbulence are $\varepsilon = 10^{-7} - 10^{-10} W \cdot kg^{-1}$.

Bottom turbulence is a determinant factor in many problems such as bottom friction and energy dissipation in water circulation, sedimentation, bottom erosion, recycling of nutrients, trapping and release of pollutants, etc. Bottom turbulence also plays an important role in the formation of currents over seabed and the damping of the motions within the basin. More information about BBL, bottom turbulence, and sediment transport from the seafloor can be found in books (Nihoul 1977; Nielsen 1992; Boudreau and Jorgensen 2001; Thorpe 2005).

2.4 ATMOSPHERIC TURBULENCE

Turbulence is a key topic of atmospheric science. Atmospheric motions are always turbulent. *Atmospheric turbulence is irregular fluctuations occurring in atmospheric air flow. These fluctuations are random and continuously changing and are superimposed on the mean motion of the air* (American Meteorological Society 2018). Atmospheric turbulence spans a huge range of scales: a time scale of less than 1 s to typically 1 h and length scales from 1 mm up to the order of 1,000 km. Atmospheric turbulence is generated by vertical wind shear and convection (temperature gradient) in the lower part of the troposphere, known as the *atmospheric boundary layer*, ABL (Stull 1988; Garratt 1992; Kaimal and Finnigan 1994; Wallace and Hobbs 2006; Tampieri 2017). The ABL depth varies in space and time, ranging from tens of meters in strongly statically stable situations, to several kilometers in convective conditions over deserts. Over sea, the ABL depth is typically a few hundred meters and rather constant on the time scale of a day. Turbulence in the ABL transfers momentum, heat, moisture, carbon dioxide, and other trace constituents between the Earth's surface and the atmosphere. It also impacts on air pollution and the erosion of soil. ABL is the most important level of the atmosphere, representing an essential element of GFD/PBL.

ABL is usually classified into three types: neutral, convective, and stable, based on atmospheric stability (buoyancy effects) and production of turbulence by wind shear. Turbulence in the stable boundary layer (SBL) is generated by shear and destroyed by negative buoyancy and viscosity. The strength of turbulence in the SBL is much weaker and it is much shallower in comparison to the neutral and convective boundary layers. Structure of the ABL is shown schematically in Figure 2.20.

The properties of atmospheric turbulence have been studied extensively over the past 100 years; today it is still a major practical importance in meteorology, weather, hydrology, aviation, communication and remote sensing imagery. Considerable research attention includes the following problems: 1) fluctuations of velocity fields, 2) stability and anisotropy, 3) turbulent transport, 4) action of wind shear and/or buoyancy forces, 5) stratification and intermittency, 6) dynamics of coherent

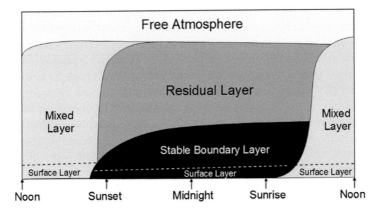

FIGURE 2.20   Structure of the atmospheric boundary layer (ABL). A typical daily cycle of the ABL in fair weather. (Based on Stull 1988).

structures, and 7) formation and mechanisms of turbulent fluxes. These environmental processes and factors affect assessments of atmospheric turbulence from measurements as well as deteriorate predictions of turbulence models.

2.4.1 Types and Classification of Atmospheric Turbulence

Atmospheric turbulence is an important hazard problem in the aviation industry and meteorology; it is known as *aviation turbulence* (e.g., Pandharinath 2014; Gultepe and Feltz 2020). Therefore, the scientific understanding and detailed specification of turbulent events, occurring in the air is of critical importance for aircraft safety. The commonly used classification includes the following types of atmospheric turbulence:

1. **Convective (thermal) turbulence** – basic type. Thermal turbulence is caused by solar heating of the surface, which in turn heats the lower atmosphere resulting in uneven convective currents, which lead to turbulence. Atmospheric convection is produced either by instable stratification of the atmosphere or nonuniform heating of the underlying surface. Both forms of convection – cellular (e.g., Rayleigh–Bénard convection) and random convection can exist. Thus, thermal turbulence is associated with unstable, neutral, or even weakly stable stratification (Vinnichenkoet al.1980).
2. **Mechanical (frictional) turbulence** – basic type. Turbulence is caused by air (shear) flow over the landscape, tall structures, terrain such as mountains, hills, bluffs, or buildings. The normal horizontal air flow is disturbed and transformed into a complicated pattern of eddies and other irregular air movements. This turbulence can be defined as small-scale, short-term, random, and frequent changes to the velocity of air.
3. **Mountain wave turbulence** – basic type. Turbulence is generated by *mountain waves* (often referred also to as lee waves). The flow over elevated terrain produces cloud formations and turbulent airflow patterns. It is responsible for

some of the most violent turbulence that is encountered away from thunderstorms. In particular, deformation of air flow by mountain barriers can produce wavelike vertical disturbances known as lee waves and *rotor turbulence*.

4. **Temperature inversion and turbulence**. Temperature inversion – a deviation (reversal) from the normal temperature profile with altitude. This occurs when a layer of warm air develops on top of a layer of cooler air. Inversion can cause turbulence at the boundary between the inversion layer and the surrounding atmosphere. The greater the positive temperature difference, the more intense the inversion, and the more stable, the lower atmosphere.

5. **Frontal turbulence**. Turbulence is caused by lifting of warm air, a frontal surface leading to instability, or the abrupt wind shift between the warm and cold air masses. The most severe cases of frontal turbulence are generally associated with fast-moving cold fronts.

6. **Thunderstorm turbulence**. Turbulence which occurs within developing convective clouds and thunderstorms, in the vicinity of the thunderstorm tops and wakes, in downbursts, and in gust fronts.

7. **Wind shear**. Wind shear is the change in wind direction and/or wind speed over a specific horizontal or vertical distance. If this shear is sufficiently strong within relatively thin layers of the free atmosphere, then the layers can be overturned by the development of waves which grow and finally break down into turbulent flow. Vertical wind shear is linked to KH instability.

8. **Clear air turbulence** (CAT). CAT is erratic air flow that occurs in the free atmosphere away from any visible convective activity or thunderstorm (see also Section 2.4.5). CAT generally occurs in a zone of high horizontal and vertical wind shear around the jet streak. CAT is also associated with a characteristic cirrus clouds pattern known as billows. The billows are an indication of breakdown into turbulent flow in the form of KH instability.

9. **Wake (vortex) turbulence**. Wake turbulence is a disturbance in the atmosphere that forms behind an obstacle (e.g., aircraft) as it passes through the air. This turbulence is characterized by a counter-rotating vortex pair (circular swirling of air). It includes various components, the most important in which are wingtip vortices and jet wash (see Section 2.5).

This classification is based mostly on descriptive analysis of turbulence observations at different atmospheric conditions. Turbulence is also separated into four levels of intensity: 1) light, 2) moderate, 3) severe, and 4) extreme, which are used in aviation industry. These categories and their threshold deviations are described in Table 2.3.

The development of atmospheric turbulence is associated not only with dynamics of air flow but also with thermal structure of the atmosphere. Vertical distribution of temperature is the most common criterion used in defining atmospheric regions. Averaging atmospheric temperatures over all latitudes and across an entire year gives us the average vertical temperature profile that is known as a standard atmosphere. Thermal atmospheric nomenclature includes four distinct layers – the troposphere, stratosphere, mesosphere, and thermosphere (Figure 2.21). A further region at about 500 km above the Earth's surface is called the exosphere. Atmospheric layers are

TABLE 2.3

Four Categories of Turbulence Intensity in Aviation

Intensity	Aircraft Reaction	Reaction Inside Aircraft
Light	Turbulence that momentarily causes slight, erratic changes in altitude and/or attitude (pitch, roll, yaw). Light Chop.	Occupants may feel a slight strain against belts or shoulder straps. Unsecured objects may be displaced slightly. Food service may be conducted and little or no difficulty is encountered in walking.
Moderate	Turbulence that is similar to Light Turbulence but of greater intensity. Changes in altitude and/or attitude occur but the aircraft remains in positive control at all times. It usually causes variation in indicated speed. Moderate Chop.	Occupants feel definite strains against seat belts or shoulder straps. Unsecured objects are dislodged. Food service and walking are difficult.
Serve	Turbulence that causes large, abrupt changes in altitude and/or attitude. It usually causes large variations in indicated airspeed. Aircraft may be momentarily out of control.	Occupants are forced violently against seat belts or shoulder straps. Unsecured objects are tossed about. Food service and walking are impossible.
Extreme	Turbulence in which the aircraft is violently tossed about and is practically impossible to control. It may cause structural damage.	

(*Source:* Aeronautical Information Manual 2020, chapter 7-1-21, TBL 7-1-9).

characterized by variations in temperature resulting primarily from the absorption of solar radiation; visible light at the surface, near ultraviolet radiation in the middle atmosphere, and far ultraviolet radiation in the upper atmosphere.

Atmospheric motions occur over a huge range of scales (Table 2.4; developed by Orlanski 1975). The meteorologically significant motions are the subject of *dynamical meteorology*. Subrange related to turbulence is usually associated with spatiotemporal variability of PBL/ABL; from this viewpoint, atmospheric turbulence can be divided into three following categories:

1. **Large-scale or macro-scale turbulence** with time interval over 1 h, length scale $\sim 10^5 - 10^6$ m and low-frequency fluctuations of velocity $10^{-3} - 10^{-1}$ Hz that contain most of the energy in the turbulence field.
2. **Intermediate-scale, local or mesoscale turbulence** with time of the order of a few minutes, length scale $\sim 10^3 - 10^5$ m and the mid-frequency velocity fluctuation range $10^{-1} - 10^1$ Hz. This range corresponds to the *inertial subrange*, where turbulence energy tends to decrease as the frequency increases according to the Kolmogorov's 5/3 power law.
3. **Small-scale or microscale turbulence** with time interval of the order of seconds, length scale $\sim 10^{-3} - 10^3$ m and the high-frequency velocity fluctuations $\sim 10^2 - 10^3$ Hz that corresponds to portion of the turbulence spectrum, where turbulence energy is usually dissipated.

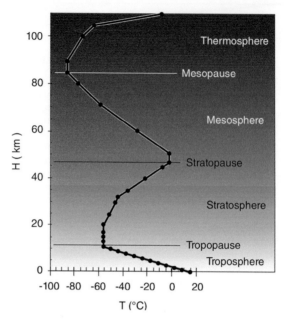

FIGURE 2.21 Nomenclature of thermal atmospheric layers. Layers in the Standard Atmosphere. (After Stull 2017).

TABLE 2.4

Scales of Horizontal Motions in the Atmosphere

Size	Scale	Name
20,000 km	macro α	planetary scale
2,000 km	macro β	synoptic scale
200 km	meso α	
20 km	meso β	mesoscale
2 km	meso γ	
200 m	micro α	boundary-layer turbulence
20 m	micro β	surface-layer turbulence
2 m	micro γ	inertial subrange turbulence
200 mm		
20 mm	micro δ	fine-scale turbulence
2 mm		
0.2 µm	viscous	dissipation subrange
	molecular	mean free path
0.002 µm		molecular

(*Sources:* Orlanski 1975).

This classification defines scaling properties of turbulent atmosphere as a stochastic dynamical system. Vertical turbulent motions characterize stability of the atmosphere that directly correlate to different types of weather systems and their severity.

Horizontal turbulent motions produce the *vortical* patterns and play a key role in the formation of inner and outer shear layers, vortexes, and coherent structures.

Scaling concept is also convenient for remote sensing purposes due to apparent sensitivity of electromagnetic wave propagation to atmospheric conditions and variations (profiles) of the atmospheric parameters – temperature, humidity, pressure, density, wind speed, liquid water content, ozone concentration, etc. Experiences show that both *ensemble-averaged* and *time-averaged* atmospheric data are applicable for analysis of the spatiotemporal field fluctuations. Finely, scaling provides the standard description of the statistical properties of homogeneous and isotropic turbulence in terms of the structure functions, spectral, and (multi) fractal characteristics.

2.4.2 LARGE-SCALE TURBULENCE

Large-scale atmospheric motions with typical horizontal scales approximately between 10 and 1,000 km exhibit various mesoscale phenomena such as thunderstorms, *cellular convection*, squall lines, supercells, complex terrain flows, inertiagravity waves, mountain waves, density currents, land/sea breezes, heat island circulations, jet streams, coherent structures/instabilities, and fronts. They are studied comprehensively using *in situ* observations, remote sensing, and both idealized and highly realistic mathematical models; this subject is referred to as *mesoscale meteorology* which is a branch of dynamical meteorology.

Many of large-scale atmospheric phenomena are potential sources of *quasi-geostrophic turbulence* called also simply *macroturbulence* (Panchev 1971; Vallis 2017). As known, small-scale motions are usually considered as turbulence in statistical sense according to the Kolmogorov's theory, while large-scale motions are described individually depending on structure and dynamics of a given phenomenon, where turbulence is also present. Since there is an enormous range of scales in atmospheric motion and turbulence, it is necessary to separate the scales of atmospheric turbulence from large-scale (non-turbulent) motions. In the theory, it can be done using deterministic equations of fluid dynamics and thermodynamics and a concept of *mean flow* (when fluid flow is divided into two components – mean and fluctuation parts); modern analysis of the problem is based on the CFD modeling.

Atmospheric turbulence at large scale is characterized by distinct (but not universal) properties, which can be briefly formulated as the following:

- The air motion in the PBL/ABL is essentially turbulent at high Reynolds up to Re = 10^7. The spectrum of large-scale turbulence closes to E(k) $\propto$ k^{-3} that corresponds to *enstrophy cascade range* of 2D turbulence (instead E(k) $\propto$ $k^{-5/3}$ for inverse energy inertial cascade range of 3D turbulence). Note: energy inertial range transfers energy to larger scales, and enstrophy inertial range transfers enstrophy to small scales.
- The large-scale turbulent flow is quasi-2D, inhomogeneous, nearly locally isotropic, statistically stably stratified, and assumed to be incompressible; it exhibits both balanced and unbalanced dynamics.

- Planetary waves or Rossby waves (inertial waves naturally occurring in rotating fluids) are the most important large-scale features in the Earth's atmosphere. Rossby waves appear from the conservation of potential vorticity; they are influenced by the Coriolis force and pressure gradient force at geostrophic balance. The simplest example of a Rossby wave occurs in a barotropic atmosphere. When Rossby waves break within horizontal critical layers, the resulting motion can be described as layerwise 2D highly inhomogeneous large-scale (or planetary-scale) *turbulent* regions known as *Rossby-wave surf zones* (region of Rossby-wave breaking), called sometimes *"wave-turbulence jigsaw puzzle."*
- Atmospheric tides are ubiquitous features of the Earth's atmosphere which cause turbulence. They are the persistent global oscillations whose periods are equal to or submultiples of the solar or lunar day. Tides are observed in all types of atmospheric fields, including wind, temperature and pressure, density, and geopotential height. Turbulence in the upper mesosphere arises from the unstable breakdown of atmospheric tides and gravity waves.
- Various types of atmospheric instabilities exist in PBL/ABL – convective instability, conditional instability of the second kind, barotropic instability, baroclinic instability, KH instability, Holmboe shear instability, dynamical instability, large-scale inertial instability (related to the *Taylor–Couette* problem), parcel instability (also called static instability), latent instability, and helical-vortex instability. Localized instabilities are associated mostly with the mean wind and/or thermal structure of the atmosphere (depending on altitude, season, and region); instabilities are rich energy sources of atmospheric (mesoscale) disturbances and produce vertical mixing triggering turbulent eddies of different scale.
- The large-scale mid-latitude circulation of the atmosphere is unstable and represents a large-scale turbulent flow, which is neither fully developed nor isotropic; the flow interacts with the Rossby waves that arise from the planetary rotation and might be regarded as a form of weak turbulence (Satoh 2014; Vallis 2017).
- Unstable zonally asymmetric flow produces a zonal jet which is a special organization of atmospheric turbulence. Jet streams are fast flowing, narrow, meandering air currents in the atmospheres of some planets, including Earth. It is assumed that atmospheric jets result from the interaction of large-scale atmospheric turbulence with arising β effect (which is well known in dynamical meteorology and defined as the meridional gradient of the Coriolis frequency $f_c = 2\Omega \sin \varphi$, where Ω is the planetary rotation rate and φ is latitude). The most important mechanism of the atmospheric turbulence-driven jet is Rossby-wave generation by baroclinic instabilities in the midlatitudes. Zonal jets play a critical role in the transfer of momentum and scalar substances (salt, heat, humidity, solids, gases, etc.). Zonal jets dominate the atmospheric circulation of all known rapidly rotating planets, including Earth, Mars, and the four giant planets (Galperin and Read 2019).
- Breaking gravity waves in the PBL are a preliminary source of large-scale turbulence in the Earth's atmosphere (Nappo 2013). The most common form of

atmospheric gravity waves is mountain waves. Under certain conditions, their amplitudes can grow as they move upward and break similar to surface waves. These breakdowns result in outbreaks of turbulence known as *clear air turbulence* (see Section 2.4.5). However, it is not always possible to distinguish atmospheric gravity waves and induced (or ambient) turbulence in real atmospheric conditions. Inertial gravity waves and turbulence are often observed to exist simultaneously in the stable PBL (Nappo 2013). It is assumed that the interaction of atmospheric wave and turbulence (known as *wave–turbulence coupling*) is associated with the following effects (Nappo 2013): 1) turbulence can be related to the local Richardson number, 2) turbulence is caused by mean and periodic components of wave flow, 3) turbulence extracts energy from the wave, thereby limiting its growth, and 4) turbulence modifies the mean fields. These processes, as a rule, may not be explained by linear theory; breakdown of atmospheric gravity waves is a purely nonlinear process.

Coherent flow structures (CFS), also known as self-organized turbulent structures, are complex chaotic turbulent quasi-periodic motions (large-scale and 3D) of concentrated vorticity (e.g., hairpin and/or horseshoe vortices, see also Section 1.3.3). CFS can exist in the field of velocity, pressure, density, or all of them; *"they are recognizable, despite randomness, by their common topological patterns, and that occur over and over again"* (Venditti et al. 2013). Flow visualization (including remote sensing imagery as well) is a common method, providing identification and analysis of CFS. It is commonly believed that CFS can result from the instability, e.g., KH instability, of initial laminar or turbulent or intermediate states. These instabilities are responsible for much of the production and dissipation of TKE. In particular, a most prominent type of environmental CFS is so-called *coherent structures in wall-bounded turbulence*. These structures appear in the lower atmosphere up to 20 km above ground level (i.e., near "wall"). There is a number of hypotheses (and conceptual models) concerning formation and evolution of the wall-bounded CFS in the atmosphere. The following causes and mechanisms are considered: 1) intermittency and interactions of eddies, 2) influence of the Reynolds stress field, 3) nonuniformity and/or instability of the mean flow velocity profile, 4) transient "shear-driven" bursts of vorticity in the viscous layer near the wall, and 5) mixing of different coherent structures. Origins of CFS near wall of a turbulent boundary layer have been studied extensively using CFD methods; however real-world observations, supported model data are limited. Typical manifestations of CFS in the Earth's environment include the atmospheric cyclonic and anti-cyclonic flow patterns, hurricane eye structures, as well as mesoscale cellular convection in the ABL. Comprehensive 3D modeling of these natural phenomena is tight because of large number of physical parameters involved and uncertainties in boundary conditions. Meanwhile, there is an idea to use CFS in the study of weather and climate and thereby to improve current forecasts. More information about CFS can be found in the book (Venditti et al. 2013) and in-text references.

Large-scale turbulence dynamics play a key role in atmospheric processes associated with planetary rotation and weak density stratification. The traditional methods of numerical modeling are commonly used for prediction of 2D or 3D turbulence;

however, it is impossible to resolve all the relevant length scales of motion at the high Reynolds numbers found in atmospheric flows (at very high Re $\geq 10^5$ turbulence "becomes" inherently 3D). To solve (or approach) the problem, different closure and filtering techniques are developed and incorporated into a numerical model in order to balance production and TKE dissipation rate. In the case of large-scale atmospheric turbulence, fluid dynamics equations operate with mean flow, which is the statistical average of velocity. The statistical average is defined as an average over a number of computer experiments under specific conditions. For example, popular Boussinesq approximation for mean flow is often used in order to investigate and model 2D atmospheric turbulence and effects of Rossby waves. Other technique is based on the *vorticity equation* which is applied in dynamical meteorology, particularly for studying zonal jets in geostrophic turbulence. Although large-scale and small-scale processes in fluid motions are coupled, characteristic features of atmospheric turbulence can be distinguished and studied (and modeled) separately.

2.4.3 SMALL-SCALE TURBULENCE

Small-scale turbulence is usually referred to as principally small spatial scales in the fluid motions (related to the size of eddy), which range from a few millimeters to a few hundred meters. The corresponding time scales are also short – from fractions of second to fractions of an hour. This type of turbulence is known also as microscale turbulence or dissipative *microturbulence* and represents a topic of micrometeorology (Arya 2001; Foken 2017). As follows from the geophysical literature, the study of microscale processes is connected with the study of the boundary layer. Thus, micrometeorology and boundary layer meteorology are very close disciplines and can be considered using the same theoretical approaches. The term "microturbulence" is often used in astrophysics to describe turbulence in the stellar atmosphere. Microturbulence is mechanism that causes broadening of the absorption lines in the stellar spectrum.

Microturbulence exists everywhere within the lowest region of the Earth's PBL/ABL; the boundary layer thickness is quite variable in time and space, ranging from hundreds of meters to a few kilometers. Structure of turbulence varies with height; therefore, the boundary layer is divided into a number of sublayers. The principal ones being the "outer layer" or "mixed layer" where large convective circulations form during the day, and the "inner layer" or "surface layer" whose characteristics are dominated by shear. Within the inner layer is a roughness sublayer where the turbulence is directly affected by the surface itself. The inner and outer layers are separated by a transition layer. The transition layer typically begins at one or two times the Monin–Obukhov length scale, $L_{MO} = -u_*^3 / (\kappa B_0)$, above ground level. The transition layer height varies from less than 10 m to more than 100 m depending on meteorological conditions (Lee et al. 2004).

The small scales include the dissipative range responsible for most of the energy dissipation and the inertial range; inertial-range scales are large compared to dissipative scales but small compared to the large scales. As known, the more intense the small-scale turbulence, the greater the rate of dissipation. Small-scale turbulence is, in turn, driven by the cascade of energy from the larger scales. Although there is no difficulty in defining these scale-ranges formally at high Reynolds numbers (e.g.

Batchelor 1953; Monin and Yaglom 1971, 1975), there is always some uncertainty in turbulence observations that prevents precise definitions, especially at moderate Reynolds numbers.

Unlike large-scale geophysical turbulence in the Earth's atmosphere, which is not homogeneous and not isotropic, microturbulence in the ABL is a classical case study, considered in terms of Kolmogorov (1941) and Obukhov–Monin (1954) theories as well as the Taylor's hypothesis on frozen turbulence (1938). All these theoretical concepts assume the homogeneity and isotropy of turbulence and, for this reason, they are widely applied in contemporary theoretical and particularly in experimental works on the microstructure of atmospheric turbulence. In the similarity theory, the small-scale turbulence exhibits universal properties and can be characterized by the energy spectrum in the inertial subrange.

Statistical description ignores geometrical properties and complexity of turbulent flow, and therefore, various numerical methods are invoked in order to investigate evolution of turbulent flow. LES is one of the most powerful computational tools today available to reproduce 3D turbulent fluid motions in the range of scales from 1 to 100 m and high Reynolds numbers up to Re~10^4. Atmospheric LES models use discrete models of the ABL (profiles of velocity, temperature, pressure) and Navier–Stokes solvers at various boundary conditions.

In the ABL, the dynamic behavior of air motion is strongly affected by turbulent friction, mainly generated by wind shear in the vicinity of the underlying surface. Following (Stull 1988)

> *We can define the boundary layer as that part of the troposphere that is directly influenced by the presence of the earth's surface, and responds to surface forcings with a timescale of about an hour or less. These forcings include frictional drag, evaporation and transpiration, heat transfer, pollutant emission, and terrain induced flow modification.*

Due to these surface forcings, ABL is generally turbulent and hence, turbulent transport processes (especially in vertical direction) determine exchanges of energy, mass, momentum, heat, and moisture between the surface and atmosphere i.e., earth-atmosphere exchange.

Briefly, geophysical factors affecting microscale (and mesoscale as well) air motions are the following 1) interior gravity waves, 2) wind shear, 3) thermals, 4) buoyancy, 5) boundary layer stability, 6) topography and complex terrain, and 7) solar radiation. These factors can lead to considerable modification of the ABL at all scales. Accordingly, several problems arise at computer modeling of microturbulence. The first one is adequate parameterization of the fluid mechanics equations by relevant variables, describing different environmental factors. The second one is the specification of the boundary conditions which should correspond to geometric characteristics of the surface – complex terrain and/or local surface inhomogeneities or profiles (such as land-water interfaces, plant canopies, vegetation, urban environment, etc.). This part may include the studies of surface roughness morphology, similarity and scaling of rough-wall boundary layers. The third one and may be the most important problem is the choice of an appropriate subgrid resolution to resolve fine scales of the flow and reveal turbulent features. The grid resolution can be at least 10 times smaller than the scale of the mean flow variations. In the case of microturbulence with the size of eddy 10–100 cm, the implementation of such a numerical study may be difficult due to limitations of current computational resources. Basically, as the mesh resolution is increased, smaller scale turbulent structures become visible.

2.4.4 WIND TURBULENCE

Surface wind is a fundamental meteorological variable, driven by pressure gradients, the Coriolis force, boundary layer mixing, and surface friction. The Reynolds number of the atmospheric wind speed is around $Re \approx 10^8$ that corresponds to a turbulent regime of air flow. Wind turbulence refers to as short-term fluctuations in wind speed in scales typically less than about ten minutes, especially for the horizontal velocity component. Turbulence arises as a result of dynamic and thermal instabilities. Dynamical instability occurs due to wind shear, whereas the thermal instability occurs due to the solar radiation (dependable on day/night time).

From meteorology (e.g., Ahrens and Henson 2019), we know that there are two main causes of turbulence occurrence in the ABL: 1) the friction of air flow due to interaction with the earth surface and 2) local variations of temperature (and density) in the atmosphere. The first one is associated with the frictional drag of the surface, which normally decreases with height and thereby wind speeds tend to increase with height above ground level. The ABL near the surface that is influenced by friction (i.e., *mechanical turbulence*) is called the *friction layer*. The average height of friction layer is of the order of 50–100 m. The second one is surface heating and instabilities that result to vertical motions of air flow and create *thermal turbulence*. Mechanical and thermal turbulence are usually interconnected – each magnifying the influence of the other.

Wind varies randomly in space and time and often is referred to as continuous gusts or stochastic gusts (wind gust is sudden and brief increase in speed of the wind that can create potential hazard). Site-specific wind characteristics include: 1) mean wind speed, 2) wind speed distribution, e.g., diurnal, seasonal, synoptic patterns, 3) turbulence, i.e., short-term fluctuations over the time period of few seconds up to an hour, 4) long-term fluctuations related to movements of large air masses across a given region, 5) distribution of wind direction, and 6) wind profile. Additionally, there is meteorological classification of winds (Figure 2.22), based on time-scale variability and configuration of wind patterns. Classification of wind speed is linked to the historical Beaufort scale.

Air flow in the lower ABL is generally turbulent and therefore exhibits both temporal and spatial variations. For statistical analysis of wind turbulence, the Reynolds decomposition is generally applied to divide the time-dependent instantaneous flow into mean and fluctuating components, i.e.,

$$u(t) = \bar{u} + u'(t), \tag{2.18}$$

$$\bar{u} \equiv \sqrt{\left(\bar{u}_x^2 + \bar{u}_y^2 + \bar{u}_z^2\right)}, \tag{2.19}$$

$$u' = \sqrt{\overline{u'^2}} \equiv \sqrt{\frac{1}{3}\left(u_x'^2 + u_y'^2 + u_z'^2\right)}, \tag{2.20}$$

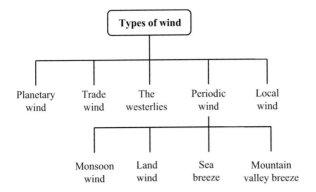

FIGURE 2.22   Types of wind. Meteorological classification.

where $\bar{u}$ and u' represent the mean and the fluctuating (turbulent) components of the wind speed, respectively. Time-averaging of the corresponding quantities over a period of time T is given by

$$\bar{u} = \frac{1}{T}\int_0^T u(t)\,dt,$$
(2.21)

$$\bar{u'} = \frac{1}{T}\int_0^T \left(u(t) - \bar{u}\right)dt,$$
(2.22)

$$\overline{\left(u'\right)^2} = \frac{1}{T}\int_0^T \left(u(t) - \bar{u}\right)^2 dt,$$
(2.23)

where the turbulent component is the difference between the instantaneous and the time-averaged speed, $u' = u - \bar{u}$.

The following important characteristics of atmospheric wind are known:

The two-parameter Weibull probability density function (pdf) of hourly average wind speeds (i.e., excluding turbulence) which is

$$f(u) = \frac{k}{c}\left(\frac{u}{c}\right)^{k-1}\exp\left[-\left(\frac{u}{c}\right)^k\right], \quad u > 0, \quad c > 0, \quad k > 0.$$
(2.24)

The corresponding cumulative distribution function is

$$F(u) = \int_0^u p(u)\,du = 1 - \exp\left[-\left(\frac{u}{c}\right)^k\right],$$
(2.25)

where u is the wind speed (m/s), c is a scale parameter (m/s), and k is a shape parameter. These two parameters are determined from experimental data. Their values may vary in the ranges $c \approx 1 \div 5$ m/s and $k = 1 \div 7$ depending on meteorological conditions. It interpolates between the exponential distribution (k = 1) and the Rayleigh

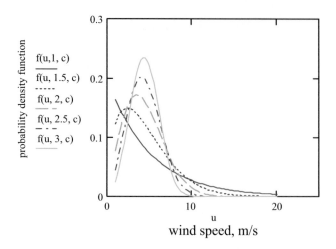

FIGURE 2.23 The Weibull distribution for various wind speeds. Parameters c = 5 m/s and k = 1, 1.5, 2, 2.5, and 3.

distribution (k = 2). The Weibull distribution for various wind speeds is shown in Figure 2.23. The Equation (2.25) is cumulative of relative frequency of each speed interval. There are other parametric pdf models (Raleigh, Gamma, Pearson, lognormal) to characterize wind regimes; however, Weibull pdf is commonly accepted as a good empirical representation of the surface wind over land and sea.

The vertical wind profile is a relationship between the wind speeds at one height, and those at another

$$\frac{u}{u_r} = \left(\frac{z}{z_r}\right)^\alpha, \tag{2.26}$$

where u is the wind speed (m/s) at height z (m) and u_r is the wind speed at a reference height z_r. The exponent α is empirically derived coefficient (the Hellmann exponent) that varies dependent on stability of the atmosphere. For neutral stability conditions, it sets as $\alpha \approx 1/7$ or 0.143.

The log wind profile is a semi-empirical relationship commonly used to describe the vertical distribution of horizontal mean wind speeds within the lowest portion of the PBL/ABL. The relationship is well described in the literature and given by

$$u(z) = \frac{u_*}{\kappa}\left[\ln\left(\frac{z-d}{z_0}\right) + \psi\left(z, z_0, L_{MO}\right)\right], \text{ for unstable conditions} \tag{2.27}$$

$$u(z) = \frac{u_*}{\kappa}\left[\ln\left(\frac{z-d}{z_0}\right)\right], \text{ for neutral stability} \tag{2.28}$$

where $u_* = C_D u$ is the friction velocity (m/s), κ is the von Kármán constant (~ 0.41), d is the zero-plane displacement above ground level (m), z_0 is the surface roughness or *aerodynamic roughness length* (m), and ψ is a stability term where L_{MO} is the Obukhov—Monin length scale. Under neutral stability, $z/L_{MO} = 0$ and ψ drops out.

Typical wind speed profiles in the ABL are shown in Figure 2.24. The values of parameters z_0 and C_D for different environmental objects are given in Table 2.5 (Wallace and Hobbs 2006).

The kinetic energy of turbulence, the mean and fluctuating (turbulent) parts

$$MKE / m \equiv \frac{1}{2}\left(\bar{u}_x^{\,2} + \bar{u}_y^{\,2} + \bar{u}_z^{\,2}\right), \quad \left[m^2 / s^2\right] \tag{2.29}$$

$$TKE / m \equiv e = \frac{1}{2}\left(u_x'^{\,2} + u_y'^{\,2} + u_z'^{\,2}\right), \quad \left[m^2 / s^2\right] \tag{2.30}$$

where u_x', u_y', and u_z' are the three fluctuating components of wind speed. In Equations (2.29)–(2.30), MKE is the kinetic energy of the mean flow per unit mass and TKE is the kinetic energy of the turbulent flow per unit mass (m is mass of flow). The instantaneous values of TKE can vary dramatically; therefore, it is often used a mean TKE value that is more representative of the overall air flow:

$$\bar{e} = \frac{1}{2}\left(\overline{\left(u_x'\right)^2} + \overline{\left(u_y'\right)^2} + \overline{\left(u_z'\right)^2}\right) = \frac{3}{2}\overline{u'^2}. \quad \left[m^2 / s^2\right] \tag{2.31}$$

Averaged term $\bar{e}$ can also be referred to as *turbulent kinetic energy* (TKE); it is one of the most important variables in micrometeorology.

The longitudinal turbulence intensity, I_u, (or turbulence level) is defined as the ratio of standard deviation, σ_u, of fluctuating wind speed to the mean wind speed, $\bar{u}$, in percentage

$$I_u = \frac{\sigma_u}{\bar{u}} = \frac{\sqrt{\overline{\left(u'\right)^2}}}{\bar{u}} = \frac{\sqrt{\frac{2}{3}\bar{e}}}{\bar{u}}, \quad [\%] \tag{2.32}$$

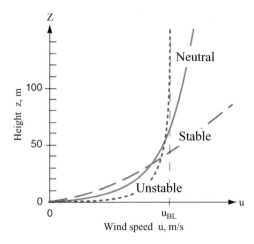

FIGURE 2.24 Typical wind speed profiles in atmospheric boundary layer ABL for different static stabilities. u_{BL} is average ABL wind speed. (Based on Stull 2017; Wallace and Hobbs 2006).

TABLE 2.5

The Davenport Classification: Aerodynamic Roughness Length z_0 and the Corresponding Drag Coefficient C_D

z_0 (m)	Classification	Landscape	C_D
0.0002	Sea	Calm sea, paved areas, snow-covered flat plain, tide flat, smooth desert.	0.0014
0.005	Smooth	Beaches, pack ice, morass, snow-covered field.	0.0028
0.03	Open	Grass prairie or farm field, tundra, airports, heather.	0.0047
0.1	Roughly open	Cultivated area with low crops and occasional obstacles (single bushes).	0.0075
0.25	Rough	High crops, crops of varied height, scattered obstacles such as tree or hedgerows, vineyard.	0.012
0.5	Very rough	Mixed farm field and forest clumps, orchards, scattered buildings.	0.018
1.0	Closed	Regular coverage with large size obstacles with open spaces roughly equal to obstacle heights, suburban houses, village, mature forest.	0.030
≥ 2	Chaotic	Center of large towns and cities, irregular forest with scattered clearing	0.062

(*Sources:* Wallace and Hobbs 2006).

where both u' and $\bar{u}$ are measured at the same point and averaged over the same period of time T. This definition can be applied to each component of the wind field u_x, u_y, u_z. The turbulence intensity clearly depends on the roughness of the surface and the height above ground level. It also depends on wind profile and stability of the atmosphere. With the height, the effect due to the increase of wind speed is predominant so that the turbulence intensity decreases with the height. There is some empirical scale of wind turbulence level, e.g., when $I_u < 2\%$ – light turbulence, $I_u \sim 5\%$ – moderate turbulence, and $I_u > 10\%$ – strong turbulence. The turbulence intensity is dependent on the height and the roughness of the terrain.

Wind turbulence spectral models describe the frequency content of wind speed variations. According to the Kolmogorov law, the turbulence spectrum approaches an asymptotic limit proportional to $S(f) \propto f^{-5/3}$ at high frequency (here f denotes the frequency in Hz). In the literature, we can find several more detailed expressions for the spectrum of the longitudinal component of turbulence. The most known are the *Kaimal* and *von Kármán* normalized wind turbulence spectra, which take the following forms (e.g., Burton et al. 2011):

$$\frac{f \cdot S_u(f)}{\sigma_u^2} = \frac{\left(4 \cdot f \cdot L_u / \bar{u}\right)}{\left(1 + 6 \cdot f \cdot L_u / \bar{u}\right)^{5/3}}, \quad \text{Kaimal spectrum} \quad (2.33)$$

$$\frac{f \cdot S_u(f)}{\sigma_u^2} = \frac{\left(4 \cdot f \cdot L_u / \overline{u}\right)}{\left(1 + 70.8 \cdot \left(f \cdot L_u / \overline{u}\right)^2\right)^{5/6}}, \qquad \text{von Kármán spectrum} \qquad (2.34)$$

where $S_u(f)$ is the power spectral density, $\overline{u}$ is the mean longitudinal component of wind speed, L_u is the length scale of turbulence, specific to the site, σ_u is the standard deviation, and f is the frequency in Hz. Several wind turbulence spectra are shown in Figure 2.25. The Kaimal spectrum has a lower, broader peak than the von Kármán spectrum. More recent work suggests that the von Kármán spectrum gives a good representation of atmospheric turbulence above about 150 m but has some deficiencies at lower altitudes (Burton et al. 2011).

There is also the general multiparameter expression of the wind spectrum in terms of the normalized frequency, $n = f \cdot z / \overline{u}$, given by

$$\frac{f \cdot S_u(f)}{u_*^2} = \frac{a \cdot n^{\gamma}}{\left(c + b \cdot n^{\alpha}\right)^{\beta}}, \qquad (2.35)$$

where $u_* = \kappa \overline{u} / \ln\left(z / z_0\right)$ is the friction velocity and a, b, c, α, β, and γ are six floating coefficients. Spectrum in the form (2.35) permits more flexibility and sensitivity to variations of the surface roughness and wind profile. Existing today experimental data (e.g., Burton et al. 2011) are able to validate this model anyway.

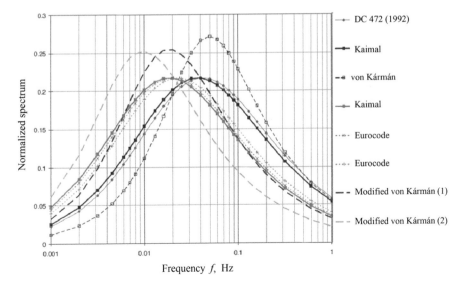

FIGURE 2.25 Normalized wind spectra (80 m height, 25 m/s, 50 degree latitude).

(*Source:* Burton et al. 2011).

FIGURE 2.26   Wind turbine.

Wind turbulence is an important problem not only in atmospheric science, mete-orology, and climate but also in the wind energy industry and wind power engineer-ing. Wind turbines, Figure 2.26 (https://www.raconteur.net/energy/renewable-energy/offshore-wind-energy/), usually are placed in regions with strong wind that automatically leads to situations where turbulence is a dominated factor. Wind tur-bulence has a significant effect on turbine power production and on the loads that cause wind turbine component fatigue. Today wind turbine turbulence is the subject of intensive experimental research and CFD modeling (Hölling et al. 2014; Burton et al. 2011).

2.4.5   Clear Air Turbulence

Clear air turbulence knows as CAT is generally referred to a *microscale* turbulence phenomenon. CAT encompasses a variety of irregular fluctuating motions of air mass in a clear atmosphere and/or cloudless regions at high altitude. The atmospheric region most susceptible to CAT is the high troposphere at altitudes of around 7,000–12,000 m (23,000–39,000 ft) above ground level. This type of turbulence tends to be localized in space and time with the vertical dimensions typically much smaller than the horizontal dimensions.

CAT is an extraordinarily challenging subject for aviation industry because of the significant impact on aircraft stability and flight safety. Often, pilots cannot forecast and avoid it, because CAT is invisible to the naked eye, unpredictable, and *undetectable* by conventional onboard sensors (radars). The physical impact of CAT on crew and passengers varies from discomfort for the lighter turbulence cat-egories to loss of flight control during the rare extreme turbulence event. In the most intense episodes, injuries, and in some very rare cases, fatalities have occurred (Ellrod et al. 2015).

Official document, issued by the Federal Aviation Administration, Advisory Circular AC 00–30C from March 22, 2016 gives us the following detailed description:

CAT is defined as "sudden severe turbulence occurring in cloudless regions that causes violent buffeting of aircraft." This term is commonly applied to higher altitude turbulence associated with wind shear. The most comprehensive definition is high-altitude turbulence encountered outside of convective clouds. This includes turbulence in cirrus clouds, within and in the vicinity of standing lenticular clouds and, in some cases, in clear air in the vicinity of thunderstorms. Generally, though, CAT definitions exclude turbulence caused by thunderstorms, low-altitude temperature inversions, thermals, strong surface winds, or local terrain features. CAT is a recognized problem that affects all aircraft operations. CAT is especially troublesome because it is often encountered unexpectedly and frequently without visual clues to warn pilots of the hazard.

(Internet http://www.faa.gov/regulations_policies/advisory_circulars and https://www. faa.gov/regulations_policies/advisory_circulars/index.cfm/go/document.information/ documentID/1029211)

Several causative mechanisms and scenarios of CAT have been suggested and discussed in scientific literature as well (see e.g., books Pao and Goldburg 2013; Panofsky and Dutton 1984; Sharman and Lane 2016). The most important causes are the following: 1) strong vertical wind shear, 2) jet stream (Figure 2.27), 3)

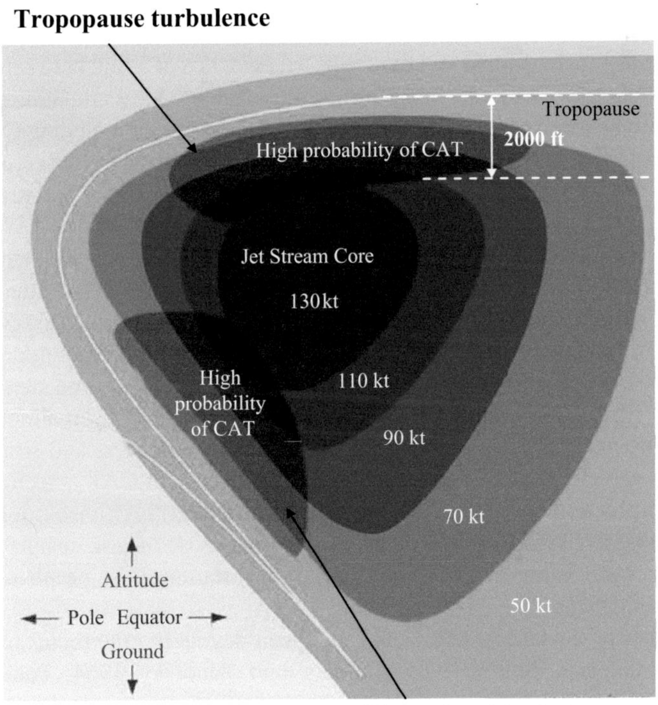

FIGURE 2.27   Conceptual model of CAT associated with jet stream in an upper-level front. (Based on Sharman and Lane 2016).

atmospheric gravity internal wave events, 4) propagation and breaking of mountain and/or lee waves, 5) strong mesoscale and synoptic-scale convection, and 6) strongly anticyclonic flows. These macroscale and microscale forcing processes create a favorable environment for CAT developments; however, dynamical properties of CAT are not yet well understood.

In view of fluid mechanics, it is commonly believed that large fraction of CAT is generated by the vertical shear KH instability over a flat surface or mountains through the amplification, steepening, overturning, and breaking of atmospheric (mountain) internal waves. This dynamical instability corresponds to typical temporal and spatial scales of CAT, which usually fit a few minutes and a few kilometers for high-altitude atmospheric conditions. Possible correlations of CAT with meteorological quantities have been investigated using the Richardson's turbulence criterion. Namely, KH instability requires $Ri < 0.25$, $Ri \equiv N^2 / u_z^2$ is the Richardson number, and at these conditions, the flow becomes unstable and turbulent. A value $0.25 \leq Ri \leq 1$ allows KH instability to grow in amplitude and result in turbulence as well. Although the Richardson number has been employed in CAT forecasting since the 1960s, it has not proven to be a "silver bullet" for reliable CAT diagnostics. The results of studying the accuracy of Ri as a CAT predictor are mixed (Sharman and Lane 2016).

A number of other criteria and algorithms have been considered for modeling and diagnostics of CAT. In particular, certain benefits can be achieved using coupling mesoscale-microscale "computer methodology" in turbulence prediction. Moreover, there are some observational and model-based evidence of a link between CAT indicators and *climate indices*. As many believed, global and local climatologic data can be used for a better diagnostics of turbulence of all categories including CAT as well. In particular, the season climatological patterns and changes, observed over the ocean, may be associated with atmospheric instabilities and/or jet streams – the major causes of CAT generation.

Many attempts have been made to observe CAT using various remote instruments such as radar-tracked balloons, ground-based radars, airborne (Doppler) lidars and infrared radiometers and others, mostly active sensors (see review Ellrod et al. 2015). Indeed, turbulence with space and time scales on the order of 10–100 m and 10–100 s, typical for CAT, requires microscale real-time measurements. Today this type of turbulence remains beyond conventional observational capabilities (and, perhaps, beyond high-resolution computer modeling and simulations as well). The achievement of the accuracy and timeliness of CAT detection is still a common problem. We believe that specialized (onboard) high-resolution *multisensor* active-passive system would be able to provide more reliable results in terms of operational forecasting and control of aviation turbulence. Some possibilities are discussed in Chapter 5.

A number of books and textbooks have been devoted to the topics of the atmospheric turbulence (Sutton 1955; Lumley and Panofsky 1964; Panchev 1971; Vinnichenko et al. 1980; Panofsky and Dutton 1984; Blackadar 1997; Lang and Lombargo 2010; Wyngaard 2010; Nappo 2013; Zilitinkevich 2013). The interested reader will find more insights beyond what this single section can expose.

2.5 TURBULENT JETS, PLUMES, AND WAKES

In fluid mechanics, jets, plumes, and wakes refer to a single free shear flow, i.e., flow which is not affected by bounded or submerged solid surfaces (also known as unbounded or wall-free turbulent flow). Free shear flow has been a subject of extensive investigations by many authors (see, e.g., books Tennekes and Lumley 1972; Townsend 1976; Pope 2000; Bernard and Wallace 2002; Kundu et al. 2016); see also historical review (Hunt and van den Bremer 2010).

Turbulent mixing, entrainment and transport processes, and vorticity are central to the behavior of jets, plumes, and wakes. These phenomena occur under the influence of natural and/or industrial internal or external sources (thermals, puffs, eruptions, explosions, engine exhausts, wind turbines, gas turbine combustors, etc.) and have been broadly studied over recent years in geophysics and fluid engineering technology. In particular, turbulent plumes induced by *submarine volcano eruption* are of great interest in geophysics (Woods 2010) and can be detected by satellites (see Section 5.4.11).

Analytic theory of submerged jets, plumes, and wakes has been initially considered in the mid 1930s on the basis of *similarity concept* (Abramovich 1963). Fundamentally, it is assumed that jets, plumes, or wakes exhibit self-similarity in the region beyond a certain downstream distance where a characteristic velocity of the flow can be scaled. The flow in this region, often turbulent, becomes independent on the initial conditions. Furthermore, the theory has been developed and applied for analysis of free and wall-bounded planar and round jets/wakes at both near- and far-flow turbulent regions. Thus, self-similar theory provides an assessment of jet/plume/wake statistical properties at many scales.

A self-similar solution for planar or round jets/wakes can be found for the time-averaged mean velocity and for the Reynolds shear stress from the Reynolds averaged Navier–Stokes equations (see Section 1.4.3). The self-similar region is those where the turbulent flow is fully developed. The flow is assumed incompressible and stationary. As a starting point, the Reynolds averaged Navier–Stokes equations are simplified through dimensional arguments. A self-similar solution is then sought by introducing a turbulent viscosity to close the governing equations.

The typical case study is free round jet/wake induced by axisymmetric source such as propeller, turbine, nozzle, solid disk, or screen. The Reynolds number, $Re = UD/\nu$, where U is the free stream velocity, D is the size of a flow-producing source, is the most important parameter, characterizing initial development of flow. It is commonly accepted that at Reynolds number $Re \leq 10^5$ the laminar–turbulent transition of the flow occurs, while at higher Reynolds number $Re \geq 10^5$ the flow becomes turbulent. Other critical parameters are streamwise velocity, steam width, momentum, temperature, turbulence intensity, mixing length, and turbulent length (the Taylor microscale).

Scaling of the normalized (dimensionless) mean axial, time-averaged velocity deficit of jet/plume/wake flow $\bar{U}_D$ can be expressed by a power-law relationship

$$\bar{U}_D = \frac{\Delta\bar{U}}{\bar{U}_\infty} = \frac{\bar{U}_w - \bar{U}_\infty}{\bar{U}_\infty} = A_0\left(\frac{x}{D}\right)^m, \tag{2.36}$$

where D is the diameter of the flow-producing source (e.g., rotor, propeller, nozzle), x is the streamwise distance from the source, $\bar{U}_\infty$ and $\bar{U}_w$ denotes the initial stream velocity and the wake velocity, respectively ($\Delta\bar{U} = \bar{U}_w - \bar{U}_\infty$ is the velocity deficit), A_0 and m are empirical coefficients.

Classical scaling asymptotics for far-field round jet/wake are given by (e.g., Tennekes and Lumley 1972; Bailly and Comte-Bellot 2015; Kundu et al. 2016):

$$\bar{U}_D = \frac{\bar{U}_w - \bar{U}_\infty}{\bar{U}_\infty} = A\left(\frac{x - x_0}{D}\right)^{-2/3} \propto x^{-2/3}, \qquad (2.37)$$

$$\frac{\delta_*}{D} = B\left(\frac{x - x_0}{D}\right)^{1/3} \propto x^{1/3}, \qquad (2.38)$$

$$\bar{U}_D(r) = C\exp\left(-\frac{r^2}{r_w^2}\right) \propto \exp\left(-\ln 2 \cdot r^2\right), \qquad (2.39)$$

where x_0 the virtual origin, A, B, and C are three constants linked to the source, δ_* is the width of jet/wake, and $\bar{U}_D(r)$ is the transverse (Gaussian) profile of the mean velocity deficit (x and r are axial and radial directions and r_w = const is the representative wake width). These results show that the velocity deficit $\bar{U}_D$ decays with the streamwise distance as $x^{-2/3}$ while the width of jet or wake δ_* increases with distance as $x^{1/3}$ and the Reynolds number degreases as $x^{-1/3}$. Figure 2.28 illustrates schematic diagram of plane turbulent jet.

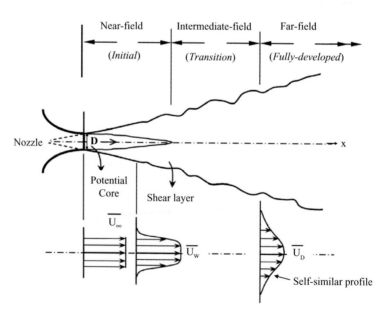

FIGURE 2.28 Schematic diagram of plane turbulent jet.

In practice, the most interesting wakes are those associated with axisymmetric bodies powered by jet or propeller engine. In these cases, the self-similar solution is often revisited using numerical CFD methods in order to explore 3D structure and dynamics of the induced turbulent flow. At present, it is a huge area of scientific research activity and we refer the interesting reader to the corresponding literature. Basic theory and applications of jets, plumes, and wakes can be found in books (Birkhoff and Zarantonello 1957; Abramovich 1963; Rodi 1982; Davies and Neves 1994; Yavorsky 1998; Lee and Chu 2003).

Turbulent jets, plumes, and wakes are of great importance in view of encompassing both civilian and military perspectives. In particular, wakes can be indicators of so-called unidentified fast-moving targets and/or localized *weakly emergent* events in the atmosphere and oceans; thereby these phenomena are relevant for remote sensing monitoring as well as for exploration purposes. Next, we will describe essential properties of jets, plumes, and wakes from this viewpoint.

2.5.1 Jets and Plumes

Turbulent jets and plumes play a fundamental role in a large variety of natural phenomena and industrial processes. Jets and plumes are turbulent shear flows driven by sources of momentum and buoyancy, respectively. The term "plume" has been used to describe flows, arising when buoyancy is supplied continuously, and the term "jet" has been used when momentum is supplied continuously. Flows which have the characteristics of both jets and plumes have been called "forced plumes" or "buoyant jets." Sometimes the term "buoyant plume" is used as well.

A turbulent buoyant jet (Figure 2.29a, b) is the turbulent flow generated by a source with both initial momentum and buoyancy fluxes. The basic mechanics of the buoyant jet is in principle the same as a momentum jet in the zone of flow establishment. Examples include smokestack emissions, fires and volcano eruptions, deep sea vents, thermals, puffs, sewage discharges, thermal effluents from power stations, ocean dumping of sludge, underwater oil blowout and many others. For example, volcanic plume is characterized by mixtures of volcanic particles, gases, and entrained air that are produced by a variety of explosive eruptions (Figure 2.29c). These plumes are generally able to disperse volcanic particles and gases over large areas by transport and fallout through the atmosphere.

The catastrophic event is underwater nuclear explosion (UNDEX) which creates vertical high-velocity water jet over the surface evolving then into a massive plume of bubbles and radioactive particles (Figure 2.29d). Collapsing plume generates water waves of considerable magnitude which can propagate over long distances. The most critical UNDEX is a very shallow-water explosion when explosively generated water waves and wave propagation process are affected by the bathometry (bottom slope).

In every case, a fluid with some momentum and/or buoyancy exits from a relatively narrow orifice and intrudes into a larger body of fluid with different characteristics, such as different speed, temperature, or contamination level. Knowledge of

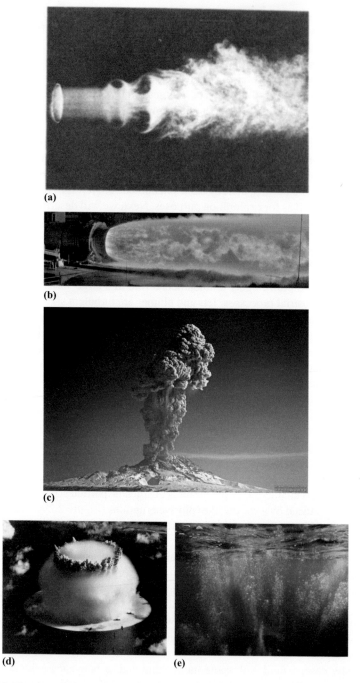

FIGURE 2.29 Examples of turbulent jet and plume. (a) Buoyant jets (After Rodi 1982), (b) the engine exhaust (Image from NASA), (c) volcanic ash plume, Shiveluch volcano (Kamchatka, Russia), (d) nuclear UNDEX plume above the surface, and (e) air bubble jets in sea water (natural scene).

turbulent mixing by jets and plumes is important for environmental control, impact, and risk assessment (Lee and Chu 2003). Today satellite imagery provides detection and monitoring of volcanic and fire plumes as well as thermal jets and plumes in coastal and estuary regions (see Chapter 5).

Several naturally occurring phenomena can be modeled as buoyant jets and plumes. In the atmosphere, these include flow above isolated heated regions of the earth's surface, above res and volcanoes, and in the ocean, flow below isolated regions of surface cooling and above fissures in mid-ocean ridges. In these cases, the flow tends to have a 2D character rather than an axisymmetric one (Rodi 1982). In particular, it has been pointed out that the properties of non-buoyant jets are much better understood than those of buoyant jets and plumes.

The rigorous theory for computing evolution of the turbulent buoyant plume is based on the Reynolds-averaged Navier–Stokes equations. The equations of mass continuity, momentum, and buoyancy (or reduced gravity) for unsteady turbulent flow are given by (e.g., Rodi 1982; Lee and Chu 2003)

$$\frac{1}{r}\frac{\partial(r\bar{u})}{\partial r}+\frac{\partial(\bar{w})}{\partial z}=0, \qquad\qquad \text{continuity} \qquad (2.40)$$

$$\frac{\partial\bar{w}}{\partial t}+\frac{1}{r}\frac{\partial(r\bar{w}\bar{u})}{\partial r}+\frac{\partial(\bar{w}^2)}{\partial z}=\bar{g}_r-\frac{1}{r}\frac{\partial\left(r\overline{u'w'}\right)}{\partial r}-\frac{\partial}{\partial z}\left(\overline{w'^2}\right), \qquad \text{momentum} \qquad (2.41)$$

$$\frac{\partial\bar{g}_r}{\partial t}+\frac{1}{r}\frac{\partial\left(r\overline{g_r u}\right)}{\partial r}+\frac{\partial\left(\overline{g_r w}\right)}{\partial z}$$

$$=\frac{g}{\rho_0}\left(\frac{\partial\rho_a}{\partial t}+\frac{\partial(\bar{w}\rho_a)}{\partial z}\right)-\frac{1}{r}\frac{\partial\left(r\overline{g_r' u'}\right)}{\partial r}-\frac{\partial\left(\overline{g_r' w'}\right)}{\partial z}, \qquad \text{buoyancy} \qquad (2.42)$$

where $\mathbf{u}(r;z;t)=u(t)\,\hat{\mathbf{r}}+w(t)\hat{\mathbf{z}}$ is the axisymmetric velocity fields in the cylindrical coordinate system ($\hat{\mathbf{r}}$ and $\hat{\mathbf{z}}$ demote unit vectors in the radial and vertical directions, respectively); here is also taken $u=\bar{u}+u'$, $w=\bar{w}+w'$, $g_r=\bar{g}_r+g_r'$, where $\bar{u}$, $\bar{w}$, $\bar{g}_r$ and u',w',g_r' denote the ensemble-averaged and fluctuation parts, respectively; $g_r=g(\rho_a-\rho)/\rho_0$ is the reduced gravity, where g is the gravitational acceleration, ρ and ρ_a are the density of the plume and ambient fluids, respectively, and ρ_0 is a density reference. Turbulence in the plume is responsible for the entertainment and mixing of the plume and ambient fluids that defines the ensuing dynamics of the system. Basically, Equations (2.40)–(2.42) provide CFD framework for modeling and simulations of round vertical buoyant plumes and jets from various sources (e.g., Woodhouse et al. 2016).

An important and interesting type of turbulent jet/plume is two-phase flow known as bubble jet or bubble plume in water (Figure 2.29e). Two-phase jets and plumes have environmental and technological applications in the oceans and the atmosphere. There are three different types of bubble jets/plumes distinguished by their sources – atmospheric, benthic, and cavitation. Benthic sources include hydro-thermal vents and seeps which pour out gases, mostly methane and carbon dioxide, arising from the sea floor. Cavitation occurs due to formation of small bubbles within a liquid at low-pressure regions, created by centrifugal pumps, water tur-bines, and ship propellers. Both benthic and cavitation gas flows are encountered in the upper ocean and perturb the air-sea interface creating sometimes dispersed gas-liquid turbulent flow.

The major atmospheric sources of bubble jets/plumes are surface breaking waves, which generate dispersed system comprising foam, whitecap, spray, and aerosol at the air-sea interface. At high winds, the system represents complex two-phase mixed turbulent flows of bubbles, jet drops and film drops, formed from the base of bursting bubbles. Massive bubble/droplet jets and plumes in the low atmosphere occur also due to UNDEX and other types of explosions in the ocean. All these volume nonuni-formities are detectable using remote sensing techniques.

The properties and dynamics of bubbles, droplets, and particles can be considered in terms of fluid mechanics (Clift et al. 1978; Chhabra, 2007; Brennen 2014). Recent review of the theory and applications of multiphase underwater jets and plumes can be found in (Boufadel et al. 2020). Turbulence in two-phase flows can be investigated using a common theory of turbulence in porous media (deLemos 2012).

2.5.2 WAKES

The wake is region of disturbed flow (often turbulent) downstream of a solid body moving through a fluid, caused by the flow of the fluid around the body and accom-panied by the flow separation (or boundary layer separation) from the body. Wake, produced by a streamlined flow exhibits strong mixing, turbulent entrainment, and chaotic motion involving various dynamical stages and swirling elements (eddies). The pressure within the wake is significantly smaller than that upstream of the body. Wake behind a body, placed in a free stream, manifests themselves in the form of a velocity deficit profile. Wake is also characterized by high Reynolds numbers and complex vortex structure. In relation to bluff bodies, the Reynolds number is deci-sive, where the length parameter is the transverse dimension of the bluff body (see below). A typical situation is transverse surrounding of a cylinder, when a quasi-periodic von Kármán vortex street occurs in the wake.

Specifically, we consider two categories of wakes: 1) in hydrodynamics, where *turbulent wake* is a region of disturbed free flow separated from a body, moving through the water and 2) in aerodynamics, where *wake turbulence* is a disturbance in the atmosphere, occurring behind an aircraft or other vehicle as it passes through the air with enough speed. Correspondently, wakes generated in the atmosphere and in the oceans are different by their spatiotemporal scales, dynamics, structural and geo-metrical characteristics.

The wake behind a moving body is a fundamental problem in turbulence with a history of more than 100 years. Analysis of the wake behavior even for a body of simple shape such as sphere, cylinder, ellipsoid, or cone is still a difficult theoretical problem. The reason is strong nonlinear dynamics and structural variability of the wake flow. This circumstance leads to the conceptual challenges in the developing overall models of the turbulent wake at different conditions; therefore, the most valuable data for a variety of shape design (e.g., an airfoil-shaped body) are obtained using numerical methods and solutions. Nowadays, CFD capabilities enable modeling and simulations of turbulent wake at high Reynolds numbers up to $Re{\sim}10^5 - 10^6$ (e.g., Kundu et al. 2016).

Wake in the atmosphere (or clear air) is often associated with *aerodynamic wake*, created by a large and heavy flying machine such as an aircraft, helicopter, unmanned aerial vehicle, and rocket. In the case of an aircraft passed in the air, aerodynamic wake is caused by a pair of cylindrical counter-rotating vortices, trailing from the tips of the wings; these vortices are called *wingtip vortices* (also known as trailing or lift-induced vortices). The existence of these vortices was demonstrated in 1907, just four years after the first powered aircraft flights by the Wright Brothers, by British aerodynamicist Frederick Lanchester (1868–1946). Details of his theory can be found in book (Kundu et al. 2016).

Aerodynamic wake – wake turbulence, is known as the phenomena resulting from the passage of an aircraft through the atmosphere (Figure 2.30a). The term includes wingtip vortices, thrust stream turbulence, jet blast, jet wash, propeller wash and/or rotor wash. Jet wash (or jetwash) refers to the rapidly moving disturbed mass of air pushed behind by the propulsion system (jet or propeller engine) which generates thrust. It is extremely turbulent, but of short duration. Wingtip vortices, occurring

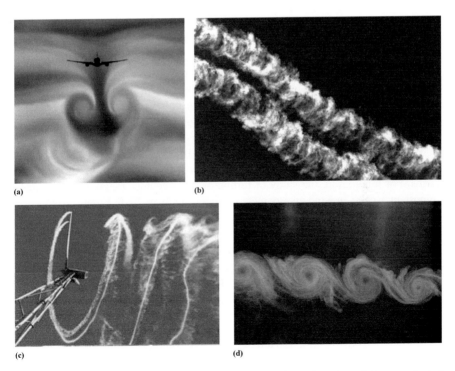

(a)   (b)

(c)   (d)

FIGURE 2.30   Examples of atmospheric wakes. (a) Aerodynamic wake turbulence and wind shear, (b) contrail vorticity, (c) wind turbine wake, and (d) wake swirling structure.

when a wing is generating lift are much more stable than jet wash and therefore, may not directly create a turbulent wake in the aerodynamic sense. The strength of wing-tip vortices depends on the weight, airspeed, shape, and wingspan of the aircraft. Wingtip vortices produce the primary and most dangerous turbulent event in aviation industry (Ginevsky and Zhelannikov 2009; Sharman and Lane 2016).

Atmospheric wake visible at clear air is well known as *condensation trail* or *contrail*. It is produced by aircraft, rocket, or missile engine exhaust (Figure 2.30b). Contrails or vapor trails are line-shaped clouds produced by aircraft engine exhaust or changes in air pressure at the altitude between 8,000 m (26,000 ft) and 12,000 m (40,000 ft) above ground level where low temperatures (between −50° and −60°C) ensure rapid condensation and freezing. Horizontal scales of contrails at these conditions may reach several tenths kilometers.

Contrails evaporate rapidly when the relative humidity of the surrounding air is low. If the relative humidity is high, however, contrails may persist for many hours. In this case, there must be sufficient mixing of the hot exhaust gases with the cold air to produce saturation. The release of particles in the exhaust can even provide nuclei on which ice crystals form. They may also form by a cooling process. The reduced pressure produced by air flowing over the wing causes the air to cool. This cooling can supersaturate the air, producing an *aerodynamic contrail*. This type of trail usually disappears quickly in the turbulent wake of the aircraft (Ahrens and Henson 2019).

Offshore wind turbines also produce atmospheric wake which is known as *turbine wake* (Figure 2.30c). Wind turbines extract energy from the wind, and create downstream wake turbulence decreasing the wind speed. With the distance from the rotor, the wake expands and returns to free laminar flow conditions. Turbine wakes create the cumulative effect on the energy production of a wind farm that results from changes in wind speed, caused by the impact of turbines on each other. It is important to evaluate the cumulative effect of turbine wakes on the energy production and to consider the possible impact of turbulence on wind farms to be built in the future.

Wind turbine farms contribute a considerable fraction to the production of renewable electrical energy. Turbine wakes can reduce the net power production of wind farms. For this reason, the study of turbine wakes becomes today one of the most important aspects in wind energy meteorology (Landberg 2016; Emeis 2018).

Atmospheric wakes can also be associated with environmental (meteorological) phenomena and/or disturbances occurring in the atmosphere. The most possible causes of the wake turbulence are: 1) hurricanes, cyclones, and typhoons; 2) air streamlines over complex terrain with isolated obstacles such as hills or urban objects, 3) interactions of airflow with mountain gravity lee waves, 4) air pollutions caused by exhaust gases and dust, and 5) atmospheric explosion and meteor trails.

Wake turbulence in the atmosphere can be measured using several remote sensing techniques. Among them, sound tomography and a high-resolution Doppler lidar are commonly applied methods. Data obtained are now commercially available. Optical observations are traditionally used for observations of atmospheric (wake) turbulence through measurements of variations in the refractive index (see Section 3.6.1).

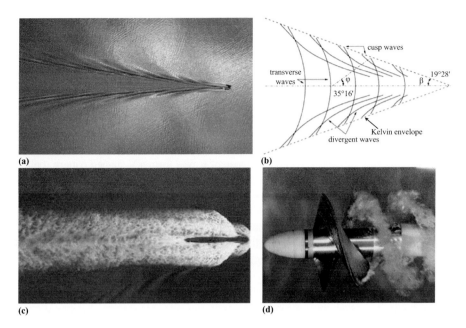

(a)   (b)

(c)   (d)

FIGURE 2.31   Turbulent wakes in the ocean. (a) Ship wake and (b) structure of the Kelvin wake, (c) surface wake behind moving submarine, (d) "super-cavitating" propeller wake.

Wake in the ocean can be generated by either moving (oscillating) source, or streamline flow. In particular, the wake induced by so-called *axisymmetric bluff body* is the most important case in the applied hydrodynamics. A bluff body can be defined as a body that, as a result of its shape, has separated flow over a substantial part of the body surface. An important feature of a bluff body flow is that there is a very strong interaction between the viscous and inviscid flow regions. Vortex dynamics play a key role in the developing turbulent wake flow of complex structure (Saffman 1992; Voropayev and Afanasyev 1994).

In hydrodynamics, the following types of turbulent wakes are known: the Kelvin wake, a von Kármán vortex street, wave breaking wake, the Bernoulli hump, internal wave wake, Rossby-wave wake, vortex wake, and cavitating wake. Last one represents two-phase turbulent flow which is typically formed behind a moving body (Fan and Tsuchiya 1990).

Wakes in the ocean may have more complicated structure and geometry than predicted by a fluid mechanics theory. Properties of the turbulent wake largely depend on parameters of a moving body and environmental conditions – wind, wave dynamics, boundary layer properties, upper ocean stratification, thermocline depth, ambient turbulence, bathometry, etc. Figure 2.31 illustrates different types of turbulent wakes in the ocean.

The Kelvin wake from the ship is the most remarkable surface wave phenomenon at sea, which has been described first by William Thomson (Lord Kelvin) in 1887. Using stationary phase arguments and the dispersion relation for surface gravity waves, he has proved that the wake behind a moving ship is delimited by an angle equal to 19.47°. Elementary theoretical explanation of the Kelvin wake pattern can be found in classical textbooks (Newman 1977; Lighthill 1978; Lamb 1997).

The ship wake and structure of the Kelvin ship wake are shown in Figure 2.31(a,b). Solid lines indicate wave crests for the divergent and transversal wave systems. The dashed line indicates the outer edges of the wake region. Note the phase shift between the divergent and transversal wave systems at the cusp line.

Transverse waves propagate approximately in the direction of motion of the ship in the angular range $0 \leq \phi \leq 35°16'$ and the divergent waves propagate outwards in the angular range $35°16' \leq \phi \leq 90°$, where ϕ is the angle between the ship track and the wave propagation direction. These two types of waves form a V-shaped pattern (called the Kelvin ship-wave wedge) of cusp waves with regular intervals along the wake at half angles (Kelvin angle) $\beta = \arcsin(1/3) \approx 19.47°$(or $19°28'$). The structure of the ship-wake pattern depends critically on the Froude number, $Fr = U / \sqrt{gL_s}$, where U is the speed and L_S is the ship length. For slow-moving vessels ($L_S > U^2/g$), the pattern tends waves have around the maximum possible wavelength of $2\pi U^2/g$ and propagate at small angles ϕ to the ship's pass; for speedboats ($L_S < U^2/g$), the waves are much shorter and propagate at large angles ϕ to the ship's pass, close to $\phi = 90°$ (Lighthill 1978).

Since Lord Kelvin until present, ship wakes have been getting great attention in hydrodynamics, naval applications, and remote sensing. Nowadays, numerical studies, CFD modeling, and observational data provide new insights to the wake dynamics, geometry, nonlinearity, and scaling. The interesting reader can find numerous references and data on this subject.

Ship wake can exhibit more complicated structure than predicted Kelvin wake pattern. Surface wakes of variable geometry are usually created by warships (e.g., carriers or fast-moving corvettes) and can be perfectly observed and recognized in aerospace optical images. Space radar (SAR) also enables the detection of ship wake signatures but their analysis requires enhanced data processing.

In general, the following physical principles in SAR ship detections are considered: 1) strong radar return from the ship body and its construction parts, 2) manifestation of the sea backscatter signatures of surface waves, 3) the contribution of breaking waves to the sea backscatter at high winds, and 4) radar observations of sea surface slicks generated by the ship engine and/or oil release. Detection performance is defined by characteristics of the SAR satellite system. At the same time, SAR measurements of a ship's turbulent wake (revealed as "dark-line" signatures) are limited because of the high speckle noise in the SAR ocean images and variable level of backscatter signals at the near-and far-field of the wake.

Automatic detection of ships at sea is the core application of SAR remote sensing at present. The problem is important for maritime surveillance, ship traffic monitoring, and management. Large datasets collected from the TerraSAR-X, RadarSat-2, and Sentinel-1 satellites are widely used for this goal. Various algorithms, based on computer vision, convolutional neural network (CNN), and machine learning are employed to improve ship detection performance. SAR wake signatures are usually not considered in automatic detection algorithms, although the wake itself is an important indicator of ship motion, maneuvering, and location.

As a matter of fact, space-based optical imagery of high spatial resolution (e.g., such as DigitalGlobe instruments with ~ 1 m resolution) provides better results in terms of geometrical and structural specification of the surface ship wakes than SAR observations; however, satellite high-resolution optical data are not always readily available for the public on a regular basis.

Wake in the ocean interior is a fascinating and rare event in deep ocean environment. Natural sources of deep-ocean wakes can be nonlinear internal waves and solitons (their breaking), bottom topography, underwater mountains, deep eddies, hydrothermal vents, and even swimming and/or decelerating fish. But the most challenging case is the wake, generated by a moving submerged nuclear submarine. The problem has received serious attention since the mid-1950s and has not been completely studied yet. Although this is still a delicate topic, today we have more information from open access publications.

Over the years, it was created more than 30 theoretical models and empirical approaches related to submarine-produced effects and disturbances. Among them the most known are hydrodynamic, thermodynamic, mechanical (propeller cavitation and aeration), electromagnetic, seawater electric conductivity, magnetohydrodynamic, chemical, biological, nuclear radiation, Cherenkov (Черенко́в) radiation, fluorescence, phosphorescence, ultrasound, and some other specific effects depending on submarine design.

Nowadays, hydrodynamic studies, utilizing CFD techniques attempt to predict wake pattern behind a submarine-type body in terms of the 3D compressible Navier–Stokes equations. This approach describes both transition regime and turbulent flow, near- and far-wakes with unsteady vortex structure. Shape, size, speed, and depth of a virtual moving body are the main tune parameters in such computer experiments. Eventually, the obtained numerical results are highly dependent on overall model specification and CFD capabilities.

In our notation, bulk properties of the turbulent wake behind a moving self-propeller body (known also as self-propeller wake) can be characterized by the following dimensionless numbers:

- The Reynolds numbers, $Re = UD/\nu$.
- The internal (body) Froude number, $Fr = U/ND$.
- The buoyancy (Brunt–Väisälä) frequency, $N = [g\Delta\rho/\rho_0 H]^{1/2}$ with mean density ρ_0 and density variation $\Delta\rho_0$ at mid-depth.
- The Strouhal number, $St = f \cdot (D/U)$ with f as the oscillation frequency.
- The Roshko number, $Ro = f \cdot (D^2)/\nu = St \cdot Re$.
- The characteristic time of downstream evolution of the wake, Nt.

Here D is the diameter, U is the speed, H is the depth of a moving body, and t is the time in seconds. A large number of model laboratory and numerical studies have been conducted during the years to explore turbulent wake from a self-propeller body moving through the water. As a whole, these data indicate that at typical values $Re \approx 10^4 - 10^5$, $Fr \gg 4$ (the range varies), and $Nt \sim 10^1 - 10^3$, the turbulent wake can collapse at a distance of about several hundred of body diameter D downstream depending on the speed of the body. This distance is also defined by the buoyancy frequency N and the dimensionless time Nt. Actually, laboratory experiments in the water tanks do not match real-world situations because of huge differences in the scales and geometry of the body itself, generated instabilities, motions velocities, and fluid stratification as well. Direct extrapolation of laboratory data to natural conditions is not valid anyway; however, it is commonly believed that physical mechanism of generation, evolution, and collapse of the turbulent wake could be similar to those produced by a submarine in real ocean environment; thereby wake phenomena make sense to investigate in laboratory especially at different regimes and various configurations of a moving body.

In nature, external ambient turbulence affects evolution and decay of the wake considerably. More detailed investigations show that the collapsing wake can generate weak internal waves and 2D turbulent spots (known as "pancake turbulence"), carrying the potential vorticity of the wake and propagating in vertical direction to the surface. At calm wind, the surface manifestations may appear in the form of a narrow slicking strip (or a few strips) or single or connected spots of variable roughness. This situation is potentially favorable for detection of the wake "signatures" by a sensitive remote sensor (under the term "signature" we understand any hydrodynamic disturbance or surface anomaly, occurring during the event).

As mentioned above, dimensionless analysis is commonly used in turbulence research; however, in the submarine's case study, we have to consider full-scale hydro-physical problem(s). Those include: vortexes, turbulence, internal waves, flow-body interactions, bluff body flows from submarine parts (hull, sail, control surfaces, etc.), hydrodynamic interactions, instabilities, and transport phenomena. Indeed, the submarine-produced pattern represents a 3D multi-structural flow region, consisting of the following principal components: 1) vortex wake, 2) hydrodynamic (drag) wake, 3) turbulent wake, 4) internal wave wake, 5) forerunner (known also as a *precursor*), and 6) cavitation (or bubbly) wake. After all, these components contribute conjunctively to the submarine-produced unsteady fluid flow and thereby we may qualify this entire wake event as a *complex wake system*.

There are some speculations or believes (by analogy with the ship's Kelvin wake) that a submarine, running close to the surface creates a V-shape wave pattern and the Bernoulli hump, which can be detected by conventional radar. At present, a number of model calculations and CFD simulations have been made in this regard. One (may be not the best) numerical example of a distinct pattern of "wake signature" is shown in Figure 2.32a.

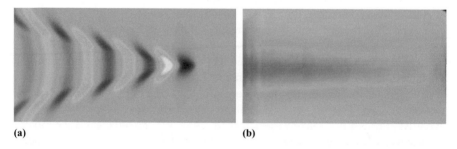

(a)   (b)

FIGURE 2.32 Hypothetic model realization of the surface patterns from moving submarine. (a) Typical deterministic V-shape wave pattern and (b) the same pattern with stochastic dominance.

The production of a Kelvin wake from a submerged moving body (sphere, cylinder, box, etc.) is a well-known phenomenon from the theory and laboratory experiments. However, the amplitude of the generated waves decreases exponentially with increasing depth and decreasing speed. For this reason, detection of moving submarines below about 30 m (~ 100 ft) by this means is highly unlikely. Moreover, ambient (inherent) turbulence, wind, surface waves and/or breaking waves affect significantly the submarine wake "signatures" in any case (e.g., according to research Benilov 1991). If we take these factors into account at least *phenomenologically*, distinct V-shape pattern will be completely destroyed and transformed into *extended spot*, losing all details (Figure 2.32b). The Bernoulli hump, supposedly appearing just above a moving submarine is also nonviable candidate for detection because of very narrow and small surface displacement (by the theory typically ~ 0.1–1.0 cm). Surface manifestations of internal lee waves could be more realistic scenario in this regard. Actually, sensitive radar may capture some elements of the predicted complex wake pattern (e.g., dark lines, spots, arks, or broken circles), but in reality, it will be difficult or even impossible to recognize and interpret these "signatures" in radar images without supported information or outstanding experiences. The impact of ocean environment on wake dynamics and the generation of "signatures" cannot be ignored anyway.

Moreover, as known (e.g., Renilson 2018), the next generation of silent submarine designs aims to reduce hydrodynamic drag and hull resistance using cloaking and/or stealth technology that, as supposed, will defeat active sonar and radar sensing. However, a *turbulent wake trail or centerline wake behind the body* (in Russian "кильватерный след" or "kil'vaternyĭ sled"), produced by a submarine propeller cannot be fully eliminated. The turbulent wake and its collapsing components (such as turbulent spots and/or density intrusions) can exist in the seawater long time, up to several hours and may reach the surface during natural convection. Thus, there is a potential possibility for non-acoustic detection of the wake "signatures" using sophisticated remote sensing technology. Today, however, this is more political than scientific problem...

2.6   CONCLUSIONS

This chapter has been a survey of atmospheric and oceanic turbulence as the most valuable and important part of GFD. This knowledge is absolutely necessary in order to provide detection and recognition of turbulence in the Earth's environments using remote sensing capabilities. It is also clear that there are still many tasks ahead to fully understand complexity of turbulence physics in nature.

As a matter of fact, geophysical turbulence is dynamical system where both forward cascade (from small to large scales) and inverse cascade (from large to small scales) can exist. Geophysical turbulence can also exhibit microscale motions (e.g., CAT). These cascades often lead to the occurrence of self-similar, self-organized (quasi-stochastic) flow patterns, which are close to coherent turbulent structures such as Lagrangian coherent structures, vortices, jets and wakes, wavy-vortex flows, etc. (see also Chapter 1). The generation, transport and dissipation of coherent flow structures are results of multiscale turbulent motions. In many geophysical processes and applications, all scales are of equal importance and objectivity, therefore they should be considered and investigated mutually.

Now, from the classical viewpoint, two principal factors distinguish geophysical turbulence from traditional fluid mechanics turbulence: the effects of stratification and rotation. Stratification exists in naturally occurring flows due to different densities of fluid. Here the gravitational force (or buoyancy force) is of great importance in the geophysical flows. An ambient rotation is associated with the Coriolis force and the centrifugal force. These forces have to be a crucial factor in geostrophic flow and turbulence. Geophysical flows are invariably turbulent that is characterized by high Reynolds number (Re > 10^5) with the dominant 3D instabilities; whereas geostrophic flow can be laminar or turbulent, barotropic or baroclinic, mostly 2D, depending on the balance between the pressure gradient, Coriolis, and buoyancy forces.

Both stratification and rotation play essential role in large-scale fluid motions in the atmosphere and the oceans resulting in spatiotemporal and structural variability of geophysical flows exhibited, as a rule, turbulent properties. It is commonly accepted, therefore, that the state of geophysical turbulence is typically steady (although not always), and most often it is studied as the statistically time-averaged flow, leaving aside all turbulent fluctuations – mathematical procedure, known as Reynolds averaging. From this viewpoint, geophysical turbulent flow can be considered as a stochastic dynamic system with a very large number of variables or "degrees of freedom" involving both large- and small-scale fluid motions. Correspondingly, mathematical descriptions and models of such a complex dynamic system are always limited by spatiotemporal predictability and data resolution.

Meanwhile, in geophysical turbulence research and applications specifically related to the atmosphere and the oceans, we are guided by well-established physical concepts which can be summarized as the following:

Geophysical turbulent flow is characterized by density stratification, stable or unstable, shear and rotation. Geophysical flows are governed by a set of coupled, nonlinear differential equations in four-dimensional space–time domain and exhibit a high sensitivity to boundary conditions and details. In mathematical sense, geophysical fluid system possesses chaotic properties, and the consequence its behavior is inherently unpredictable; such systems are also replete with instabilities of different type. Specifically, in the ocean-atmosphere system, geophysical turbulence phenomena are mostly associated with wave kinematics, wave breaking, wind and/or buoyancy forces (e.g., Vallis 2017; Zeitlin 2018).

Turbulent wall-bounded flow (i.e., boundary layer, pipe and channel flows, flow over topography, flow around a bluff body, etc.) is defined by geometry and structure of the surface. The presence of solid or liquid surfaces or interfaces (called "wall") leads to fundamental changes the nature of turbulent flow, first of all, in its inherent dynamics, scaling, streamwise structure, and shear stress effects.

Turbulent free flow is not confined by solid walls; usually it is free shear flow with mean velocity gradients that develop in the absence of boundaries. One of the most important features of free shear flows is entrainment when turbulence is continuously increased and spread throughout its lifetime. Other distinctive feature of turbulent free shear flows is a tendency to form coherent structures. They control the shape of the turbulent/non-turbulent interface and play an important role in the entrainment process and energy distribution. Finally, free shear flows become asymptotically independent of their initial conditions (or source conditions) at some distance that create free jets, plumes, wakes, and/or free shear layers often exhibited self-similar properties.

Turbulent mixing is the ability of turbulent flow to effectively mix entrained fluids to a molecular scale. Effective mixing of mass and momentum is one of the most important consequences of turbulence. Most of this mixing is accomplished by the stirring action of large, energetic eddies (Kundu et al. 2016). It is commonly assumed that turbulent mixing in geophysical flows is associated with the influence of hydrodynamic instabilities known as the Rayleigh–Taylor, Richtmyer–Meshkov, and Kelvin–Helmholtz instabilities as well as the Holmboe instability which can lead to a rapid irreversible turbulent mixing (see Section 1.2.4). These instabilities evolve through different stages before transitioning to turbulence, experiencing linear, weakly, and highly nonlinear states. Turbulent mixing is dominated by small-scale motions which are inherent in a turbulent boundary layer, shear flow, jet, or wake. Most theories consider the so-called passive scalar mixing (i.e., the mixing of scalars does not affect the flow itself). The classical equation for studying passive-scalar mixing is the advection–diffusion equation. This is the simplest paradigm of turbulent mixing.

Turbulent transport (turbulent diffusion) is fundamental phenomenon comprising dissipation, fluxes of conserved quantities, vorticity, intermittency, and mixing (Bird et al. 2002; Glasgow 2010; Tardu 2014). Turbulent transport of momentum and scalars include constituent gradients in the atmosphere and ocean such as water vapor, temperature, trace gases, particles, species, multiphase and multicomponent transport phenomena as well. In particular, turbulent transport is responsible for the exchange of heat, mass, and momentum between ocean and atmosphere; in this case, turbulent structure of the ABL is a key to assess heat/mass transfer effectiveness.

The CFD solvers of turbulence transport problems are based on the Reynolds stress transport equation and scalar flux transport equation describing fundamental transport mechanism of Reynolds stresses in both free and wall-bounded incompressible turbulent flow. The second-order closure models, also known as the second-moment or the Reynolds-stress closure models, are the most complex and physically the most realistic techniques for analysis of turbulent transport in fluid mechanics, geophysics, and engineering (Dewan 2011).

Turbulence modeling is still a challenging but vast area of activity in science and technology. At present, the following categories of mathematical models (but not software platforms) of geophysical turbulence are widely used: 1) empirical models based on statistical treatment of turbulence experimental data, 2) analytical models based on approximate solutions of simplified fluid mechanics equations, describing selected features of geophysical flow, 3) phenomenological statistical models, describing self-similarity, scaling, energy cascades, spectral and multifractal properties, and self-organized criticality of turbulence, 4) various numerical closure models based on CFD solvers of the governing equations of fluid mechanics, operated with RANS, LES, and DNS, and 5) combined models in which both statistical and numerical solutions are used in order to provide environmental engineering and technological geophysical applications involving turbulence.

Recent advances in CFD and computer science offer more realistic solutions of fluid mechanics problems where turbulence is one of the key factor affecting environmental changes. A number of commercial CFD solvers, developed in a friendly interface (e.g., CFX, X-FLOW, FLUENT, COMSOL), provide modeling and simulations of various turbulent phenomena in many environmental and industrial areas. However, it is not at all clear that increased computer hardware and software capabilities alone can ever make the complete solution of turbulence problems as a routine pastime. *Computer experiments may not always bring new physics.* We believe that both science and observations always will be equally demanded regarding this particular subject matter.

Over the past several decades, remote sensing observations have occasionally delivered incredible imaging datasets that give us an unprecedented global view into geophysical turbulence over broad ranges in spatial and temporal scales. They include oceanic and atmospheric vortices and jets, plumes, thermals, eddies, atmospheric convective and circulation cells, wave turbulence, ocean thermohaline structures, and some others. Selected materials will be considered and discussed in Chapter 5. Before all that, we refer to essential aspects of electromagnetic wave propagation and remote sensing technology that create a physical basis for turbulence detection (Chapter 3 and 4).

REFERENCES

Abramovich, G. N. 1963. *Theory of Turbulent Jets (translated from Russian, volume editor L. H. Schindel).* The MIT Press Classics, Cambridge, MA (In Russian: Абрамович Г. Н. Теория турбулентных струй. Государственное издательство физико-математической литературы. Москва, Наука, 1960).

Ahrens, C. D. and Henson, R. 2019. *Meteorology Today: An Introduction to Weather, Climate and the Environment,* 12th edition. Cengage Learning, Boston, MA.

Alford, M. H., MacKinnon, J. A., Simmons, H. L., and Nash, J. D. 2016. Near-inertial internal gravity waves in the ocean. *Annual Review of Marine Science,* 8(1):95–123. doi:10.1146/annurev-marine-010814-015746.

American Meteorological Society. 2018. *Turbulence. Glossary of Meteorology.* Available on the Internet http://glossary.ametsoc.org/wiki/turbulence.

Apel, J. R.1987. *Principles of Ocean Physics (International Geophysics Series, Vol. 38).* Academic Press, London, UK.

Apel, J. R., Ostrovsky, L. A., Stepanyants, Y. A., and Lynch, J. F. 2007. Internal solitons in the ocean and their effect on underwater sound. *The Journal of the Acoustical Society of America,* 121(2):695–722. doi:10.1121/1.2395914.

Arya, P. S. 2001. *Introduction to Micrometeorology*, 2nd edition. Academic Press, San Diego, CA.

Babanin, A. 2011. *Breaking and Dissipation of Ocean Surface Waves*. Cambridge University Press, Cambridge, UK.

Bailly, C. and Comte-Bellot, G. 2015. *Turbulence*. Springer International Publishing, Switzerland.

Banner, M. L. and Peregrine, D. H. 1993. Wave breaking in deep water. *Annual Review of Fluid Mechanics*, 25(1):373–397. doi:10.1146/annurev.fl.25.010193.002105.

Barenblatt, G. I. 1978a. Dynamics of turbulent spots and intrusions in a stably stratified fluid. *Izvestiya, Atmosphere and Oceanic Physics*, 14(2):139–145 (translated from Russian).

Barenblatt, G. I. 1978b. Self-similarity of temperature and salinity distributions in the upper thermocline. *Izvestiya, Atmosphere and Oceanic Physics*, 14(11):820–823 (translated from Russian).

Barenblatt, G. I. 1993. Scaling laws for fully developed shear flows. Part 1: Basic hypotheses and analysis. *Journal of Fluid Mechanics*, 248:513–520. doi:10.1017/s0022112093000874.

Barenblatt, G. I. and Prostokishin, V. M. 1993. Scaling laws for fully developed shear flows. Part 2. Processing of experimental data. *Journal of Fluid Mechanics*, 248:521–529. doi:10.1017/s0022112093000886.

Batchelor, G. K. 1953. *The Theory of Homogeneous Turbulence (Cambridge Science Classics)*. Cambridge University Press, New York.

Baumert, H. Z., Simpson, J., and Sundermann, J. (Eds.). 2005. *Marine Turbulence: Theories, Observations, and Models*. Cambridge UniversityPress, Cambridge, UK.

Benilov, A. 1991. *Soviet Research of Ocean Turbulence and Submarine Detection*. Delphic Associates Inc.

Bernard, P. S. and Wallace, J. M. 2002. *Turbulent Flow: Analysis, Measurement, and Prediction*. John Wiley & Sons, Hoboken, NJ.

Bird, R. B., Stewart, W. E., and Lightfoot, E. N. 2002. *Transport Phenomena*, 2nd edition. John Wiley & Sons, New York.

Birkhoff, G. and Zarantonello, E. H. 1957. *Jets, Wakes, and Cavities*. Academic Press, New York.

Biskamp, D. 2003. *Magnetohydrodynamic Turbulence*. Cambridge University Press, Cambridge, UK.

Blackadar, A. K. 1997. *Turbulence and Diffusion in the Atmosphere: Lectures in Environmental Sciences*. Springer, Berlin, Germany.

Bortkovskii, R. S. 1987. *Air-Sea Exchange of Heat and Moisture During Storms*. D. Reidel, Dordrecht, The Netherlands.

Boudreau, B. P. and Jorgensen, Bo B. (Eds.). 2001. *The Benthic Boundary Layer: Transport Processes and Biogeochemistry*. Oxford University Press, Oxford, New York.

Boufadel, M. C., Socolofsky, S., Katz, J., Yang, D., Daskiran, C., and Dewar, W. 2020. A review on multiphase underwater jets and plumes: Droplets, hydrodynamics, and chemistry. *Reviews of Geophysics*, 58(3) e2020RG000703. doi:10.1029/2020rg000703.

Brandt, A. and Fernando, H. J. S. 1995. *Double-Diffusive Convection (Geophysical Monograph Series, Vol. 94)*. American Geophysical Union, Washington DC.

Brekhovskikh, L. M. and Goncharov, V. 1994. *Mechanics of Continua and Wave Dynamics*, 2nd edition. Springer, Berlin, Germany.

Brennen, C. E. 2014. *Cavitation and Bubble Dynamics*. Cambridge University Press, Cambridge, UK.

Brown, G. L. and Roshko, A. 1974. On density effects and large structure in turbulent mixing layers. *Journal of Fluid Mechanics*, 64(04):775–816. doi:10.1017/s002211207400190x.

Brown, G. L. and Roshko, A. 2012. Turbulent shear layers and wakes. *Journal of Turbulence*, 13(51):1–32. doi:10.1080/14685248.2012.723805.

Bühler, O. 2014. *Waves and Mean Flows*, 2nd edition. Cambridge University Press, Cambridge, UK.

Burchard, H. 2002. *Applied Turbulence Modelling in Marine Waters*. Springer, Berlin, Germany.

Burton, T., Jenkins, N., Sharpe, D., and Bossanyi, E. (Eds.). 2011. *Wind Energy Handbook*, 2nd edition. John Wiley & Sons, Chichester, UK.

Clift, R., Grace, J. R., and Weber, M. E. 1978. *Bubbles, Drops, and Particles*. Academic Press, New York.

Chassignet, E. P., Cenedese, C., and Verron, J. (Eds.). 2012. *Buoyancy-Driven Flows*. Cambridge University Press, New York.

Chhabra, R. P. 2007. *Bubbles, Drops, and Particles in Non-Newtonian Fluids*, 2nd edition. CRC Press, Boca Raton, FL.

Craik, A. D. D. 1985. *Wave Interactions and Fluid Flows*. Cambridge University Press, Cambridge, UK.

Cushman-Roisin, B. and Beckers, J.-M. 2011. *Introduction to Geophysical Fluid Dynamics: Physical and Numerical Aspects*, 2nd edition. Elsevier – Academic Press, Amsterdam, The Netherlands.

Davies, P. A. and Neves, M. J. V. (Eds.). 1994. *Recent Research Advances in the Fluid Mechanics of Turbulent Jets and Plumes (Nato Science Series E: 255)*. Springer Science, Dordrecht, The Netherlands.

de Lemos, M. J. S. 2012. *Turbulence in Porous Media: Modeling and Applications*, 2nd edition. Elsevier, Waltham, MA.

Dewan, A. 2011. *Tackling Turbulent Flows in Engineering*. Springer-Verlag, Berlin, Heidelberg.

Ellrod, G. P., Knox, J. A., Lester, P. F., and Ehernberger, L. J. 2015. Aviation Meteorology| Clear Air Turbulence. In *Encyclopedia of Atmospheric Sciences*, 2nd edition, Vol. 1, 177–186. doi:10.1016/b978-0-12-382225-3.00104-3.

Emeis, S. 2018. *Wind Energy Meteorology: Atmospheric Physics for Wind Power Generation*, 2nd edition. Springer International Publishing, Switzerland.

Fan, L.-S. and Tsuchiya, K. 1990. *Bubble Wake Dynamics in Liquids and Liquid-Solid Suspensions*. Butterworth-Heinemann, Stoneham, MA.

Fedorov, K. N. 1978. *The Thermohaline Finestructure of the Ocean (translated from Russian by D. A. Brown)*. Pergamon Press, Oxford, UK.

Fedorov, K. N. and Ginsburg, A. I. 1992. *The Near-Surface Layer of the Ocean (translated from Russian by M. Rosenberg)*. CRC Press, Boca Raton, FL.

Foken, T. 2017. *Micrometeorology*, 2nd edition. Springer-Verlag, Berlin, Heidelberg.

Galperin, B. and Read, P. L. (Eds.). 2019. *Zonal Jets: Phenomenology, Genesis, and Physics*. Cambridge University Press, Cambridge, UK.

Gargett, A. E. 1989. Ocean turbulence. *Annual Review of Fluid Mechanics*, 21(1):419–451. doi:10.1146/annurev.fl.21.010189.002223.

Garratt, J. 1992. *The Atmospheric Boundary Layer*. Cambridge University Press, Cambridge. UK.

Garrett, C. and Munk, W. 1972. Space-time scales of internal waves. *Geophysical Fluid Dynamics*, 3(1):225–264. doi:10.1080/03091927208236082.

Gibson, C. H. 1988. Evidence and consequences of fossil turbulence in the ocean. *Elsevier Oceanography Series*, 46:319–334. doi:10.1016/s0422-9894(08)70555-5.

Gibson, C. H. 1999. Fossil turbulence revisited. *Journal of Marine Systems*, 21(1–4):147–167. doi:10.1016/s0924-7963(99)00024-x.

Ginevsky, A. S. and Zhelannikov, A. I. 2009. *Vortex Wakes of Aircrafts*. Springer, Dordrecht, The Netherlands.

Glasgow, L. A. 2010. *Transport Phenomena: An Introduction to Advanced Topics*. John Wiley & Sons, Hoboken, NJ.

Gregg, M. C. 1989. Scaling turbulent dissipation in the thermocline. *Journal of Geophysical Research*, 94(C7):9686–9698. doi:10.1029/jc094ic07p09686.

Gultepe, I. and Feltz, W. F. (Eds.). 2020. *Aviation Meteorology: Observations and Models*. Birkhäuser – Springer Nature, Switzerland.

Hölling, M., Peinke, J., and Ivanell, S. (Eds.). 2014. *Wind Energy – Impact of Turbulence (Research Topics in Wind Energy 2)*. Springer Science, Berlin, Heidelberg.

Huang, R. X. 2009. *Ocean Circulation: Wind-Driven and Thermohaline Processes*. Cambridge University Press, Cambridge, UK.

Hunt, G. R. and van den Bremer, T. S. 2010. Classical plume theory: 1937–2010 and beyond. *IMA Journal of Applied Mathematics*, 76(3):424–448. doi:10.1093/imamat/hxq056.

Huppert, H. E. and Turner, J. S. 1981. Double-diffusive convection. *Journal of Fluid Mechanics*, 106:299–329. doi:10.1017/s0022112081001614.

Kaimal, J. C. and Finnigan, J. J. 1994. *Atmospheric Boundary Layer Flows: Their Structure and Measurement*. Oxford University Press, New York.

Kantha, L. H. and Clayson, C. A. 2000. *Small Scale Processes in Geophysical Fluid Flows*. Academic Press, San Diego, CA.

Kitaigorodskii, S. A. 1973. *Physics of Air-Sea Interaction*. Israel Program of Scientific Translations, Jerusalem, Israel (translation from Russian).

Klymak, J. M. and Moum, J. N. 2007a. Oceanic isopycnal slope spectra. Part I: Internal waves. *Journal of Physical Oceanography*, 37(5):1215–1231. doi:10.1175/jpo3073.1.

Klymak, J. M. and Moum, J. N. 2007b. Oceanic isopycnal slope spectra. Part II: Turbulence. waves. *Journal of Physical Oceanography*, 37(5):1232–1245. doi:10.1175/jpo3074.1.

Koschmieder, E. L. 1993. *Bénard Cells and Taylor Vortices*. Cambridge University Press, Cambridge, UK.

Kraus, E. B. and Businger, J. A. 1994. *Atmosphere–Ocean Interaction*, 2nd edition. Oxford University Press, New York.

Kundu, P. K., Cohen, I. M., and Dowling, D. R. 2016. *Fluid Mechanics*, 6th edition. Elsevier – Academic Press, Amsterdam, The Netherlands.

Lamb, H. 1997. *Hydrodynamics*, 6th edition. Cambridge University Press, Cambridge, UK.

Landberg, L. 2016. *Meteorology for Wind Energy: An Introduction*. John Wiley & Sons, Chichester, UK.

Lang, P. R. and Lombargo, F. S. (Eds.). 2010. *Atmospheric Turbulence, Meteorological Modeling and Aerodynamics*. Nova Science Publishers, New York.

Langmuir, I. 1938. Surface motion of water induced by wind. *Science*, 87(2250):119–123. doi:10.1126/science.87.2250.119.

LeBlond, P. H. and Mysak, L. A. 1978. *Waves in the Ocean*. Elsevier, Amsterdam, The Netherlands.

Lee, J. H. W. and Chu, V. H. 2003. *Turbulent Jets and Plumes: A Lagrangian Approach*. Kluwer Academic Publishers, Dordrecht, The Netherlands.

Lee, X., Massman, W., and Law, B. (Eds.). 2004. *Handbook of Micrometeorology: A Guide for Surface Flux Measurement and Analysis*. Kluwer Academic Publishers, Dordrecht.

Leibovich, S. 1983. The form and dynamics of Langmuir circulations. *Annual Review of Fluid Mechanics*, 15(1):391–427. doi:10.1146/annurev.fl.15.010183.002135.

Levine, M. D. 2002. A modification of the Garrett-Munk internal wave spectrum. *Journal of Physical Oceanography*, 32(11):3166–3181. doi: 10.1175/1520-0485(2002)032<3166:AMOTGM>2.0.CO;2.

Lighthill, J. 1978. *Waves in Fluids*. Cambridge University Press, Cambridge, UK.

Lumley, J. L. and Panofsky, H. A. 1964. *The Structure of Atmospheric Turbulence. (Monographs and Texts in Physics and Astronomy Vol. XII)*. John Wiley & Sons, New York.

Marchuk, G. I. and Kagan, B. A. 1989. *Dynamics of Ocean Tides*. Kluwer Academic Publishers. Dordrecht, The Netherlands.

Massel, S. R. 2007. *Ocean Waves Breaking and Marine Aerosol Fluxes*. Springer, New York.

Massel, S. R. 2015. *Internal Gravity Waves in the Shallow Seas*. Springer, Switzerland.

McWilliams, J. C., Sullivan, P. P., and Moeng, C.-H. 1997. Langmuir turbulence in the ocean. *Journal of Fluid Mechanics*, 334:1–30. doi:10.1017/s0022112096004375.

McWilliams, J. C. 2006. *Fundamentals of Geophysical Fluid Dynamics*. Cambridge University Press, Cambridge, UK.

Miropol'sky, Y. Z. 2001. *Dynamics of Internal Gravity Waves in the Ocean (translated and edited from Russian by O. D. Shishkina)*. Springer, New York.

Monaldo, F. M., Li, X., Pichel, W. G., and Jackson, C. R. 2014. Ocean wind speed climatology from spaceborne SAR imagery. *Bulletin of the American Meteorological Society*, 95(4):565–569. doi:10.1175/bams-d-12-00165.1.

Monin, A. S. 1990. *Theoretical Geophysical Fluid Dynamics (translated from Russian by Ron Hardin)*. Kluwer Academic Publishers, Dordrecht, The Netherlands.

Monin, A. S. and Krasitskii, V. P. 1985. *Phenomena on the Ocean Surface*. USSR, Leningrad, Gidrometeoizdat (in Russian: Монин А. С., Красицкий, В. П. Явления на поверхности океана. Ленинград, Гидрометеоиздат, 1985).

Monin, A. S. and Ozmidov, R. V. 1985. *Turbulence in the Ocean*. D. Reidel Publishing Company, Dordrecht, Holland.

Monin, A. S. and Yaglom, A. M. 1971. *Statistical Fluid Mechanics, Volume 1: Mechanics of Turbulence (English translation, edited by J. L. Lumley)*. The MIT Press, Cambridge, MA.

Monin, A. S. and Yaglom, A. M. 1975. *Statistical Fluid Mechanics, Volume 2: Mechanics of Turbulence (English translation, edited by J. L. Lumley)*. The MIT Press, Cambridge, MA.

Morozov, E. G. 2018. *Oceanic Internal Tides: Observations, Analysis and Modeling: A Global View*. Springer, Cham, Switzerland.

Nappo, C. J. 2013. *An Introduction to Atmospheric Gravity Waves*, 2nd edition. Elsevier – Academic Press, Amsterdam, The Netherlands.

Nazarenko, S. 2011. *Wave Turbulence*. Springer, Berlin, Germany.

Newman, J. N. 1977. *Marine Hydrodynamics*. MIT Press, Cambridge, MA.

Nielsen, P. 1992. *Coastal Bottom Boundary Layers and Sediment Transport*. World Scientific, Singapore.

Nihoul, J. C. J.(Ed.). 1977. *Bottom Turbulence*. Elsevier, Amsterdam, The Netherlands.

Nihoul, J. C. J. and Jamart, B. M. (Eds.). 1989. *Mesoscale/Synoptic Coherent Structures in Geophysical Turbulence*. Elsevier Science Publishers, Amsterdam, The Netherlands.

Olbers, D., Willebrand, J., and Eden, C. 2012. *Ocean Dynamics*. Springer-Verlag, Berlin, Heidelberg.

Orlanski, I. 1975. A rational subdivision of scales for atmospheric processes. *Bulletin of the American Meteorological Society*, 56(5):527–530.

Osborn, T. R. 1980. Estimates of the local rate of vertical diffusion from dissipation measurements. *Journal of Physical Oceanography*, 10(1):83–89. doi:10.1175/1520-0485(1980)010<0083:eotlro>2.0.

Ozmidov, R. V. 1965. On the turbulent exchange in a stably stratified ocean. *Izvestiya, Atmospheric and Oceanic Physics*, 1(8):493–497 (translated from Russian).

Özsoy, E. 2020. *Geophysical Fluid Dynamics I: An Introduction to Atmosphere–Ocean Dynamics: Homogeneous Fluids*. Springer Nature, Switzerland.

Panchev, S. 1971. *Random Functions and Turbulence*. Pergamon Press, Oxford, New York.

Pandharinath, N. 2014. *Aviation Meteorology*. BS Publications, Hyderabad, India.

Panofsky, H. A. and Dutton, J. A. 1984. *Atmospheric Turbulence: Models and Methods for Engineering Applications*. John Wiley & Sons, New York.

Pao, Y.-H. and Goldburg, A. (Eds.). 2013. *Clear Air Turbulence and Its Detection*. Springer, New York.

Pedlosky, J. 1979. *Geophysical Fluid Dynamics*, 2nd edition. Springer, New York.

Perlin, A. A., Moum, J. N., Klymak, J. M., Levine, M. D., Boyd, T., and Kosro, P. M. 2005. A modified law-of-the-wall applied to oceanic bottom boundary layers. *Journal of Geophysical Research*, 110, C10S10. doi:10.1029/2004jc002310.

Phillips, O. M. 1980. *The Dynamics of the Upper Ocean*, 2nd edition. Cambridge University Press, Cambridge, UK.

Pope, S. B. 2000. *Turbulent Flows*. Cambridge University Press, Cambridge, UK.

Prants, S. V., Uleysky, M., and Budyansky, M. V. 2017. *Lagrangian Oceanography: Large-scale Transport and Mixing in the Ocean*. Springer International Publishing, Switzerland.

Radko, T. 2013. *Double-Diffusive Convection*. Cambridge University Press, Cambridge, UK.

Raizer, V. 2017. *Advances in Passive Microwave Remote Sensing of Oceans*. CRC Press, Boca Raton, FL.

Raizer, V. 2019. *Optical Remote Sensing of Ocean Hydrodynamics*. CRC Press, Boca Raton, FL.

Renilson, M. 2018. *Submarine Hydrodynamics*, 2nd edition. Springer, Cham, Switzerland.

Roberts, J. 1975. *Internal Gravity Waves in the Ocean*. Marcel Dekker, Inc., New York.

Rodi, W. (Ed.). 1982. *Turbulent Buoyant Jets and Plumes*. Pergamon Press, Oxford, UK.

Romero, L., Melville, W. K., and Kleiss, J. M. 2012. Spectral energy dissipation due to surface wave breaking. *Journal of Physical Oceanography*, 42(9):1421–1444. doi:10.1175/jpo-d-11-072.1.

Saffman, P. G. 1992. *Vortex Dynamics*. Cambridge University Press. Cambridge, UK.

Salmon, R. 1998. *Lectures on Geophysical Fluid Dynamics*. Oxford University Press, New York.

Satoh, M. 2014. *Atmospheric Circulation Dynamics and General Circulation Models (Springer Praxis Books)*, 2nd edition, Springer – Praxis, Chichester, UK.

Schmitt, R. W. 1994. Double diffusion in oceanography. *Annual Review of Fluid Mechanics*, 26(1):255–285. doi:10.1146/annurev.fl.26.010194.001351.

Schmitt, R. W. 2003. Observational and laboratory insights into salt finger convection. *Progress in Oceanography*, 56(3–4):419–433. doi:10.1016/s0079-6611(03)00033-8.

Sharman, R. and Lane, T. (Eds.). 2016. *Aviation Turbulence: Processes, Detection, Prediction*. Springer International Publishing, Switzerland.

Shibley, N. C. and Timmermans, M.-L. 2019. The formation of double-diffusive layers in a weakly turbulent environment. *Journal of Geophysical Research: Oceans*, 124(3):1445–1458. doi:10.1029/2018JC014625.

Shrira, V. and Nazarenko, S. (Eds.). 2013. *Advances in Wave Turbulence*. World Scientific Publishing, Singapore.

Siedler, G., Griffies, S. M., Gould, J., and Church, J. A. (Eds.). 2013. *Ocean Circulation and Climate: A 21st Century Perspective (Volume 103)*, 2nd edition. Elsevier – Academic Press, Amsterdam, The Netherlands.

Soloviev, A. and Klinger, B. 2001. Open ocean convection. In *Encyclopedia of Ocean Sciences*. (Eds. J. H. Steele, S. A. Thorpe, and K. K. Turekian). Elsevier – Academic Press, London, UK, pp. 2015–2022. doi:10.1006/rwos.2001.0118.

Soloviev, A. and Lukas, R. 2014. *The Near-Surface Layer of the Ocean: Structure, Dynamics and Applications*, 2nd edition. Springer, Dordrecht, The Netherlands.

Steele, J. H., Thorpe, S. A., and Turekian, K. K. (Eds.). 2009. *Elements of Physical Oceanography. A Derivative of Encyclopedia of Ocean Sciences*, 2nd edition. Elsevier – Academic Press, London, UK.

Stern, M. E. 1975. *Ocean Circulation Physics*. Academic Press, New York.

Stull, R. B. 1988. *An Introduction to Boundary Layer Meteorology*. Kluwer Academic Publishers, Dordrecht, The Netherlands.

Stull, R. 2017. *Practical Meteorology: An Algebra-based Survey of Atmospheric Science -version 1.02b*. University British Columbia, Canada. Free book available on Internet https://www.eoas.ubc.ca/books/Practical_Meteorology/.

Sutherland, B. R. 2010. *Internal Gravity Waves*. Cambridge University Press, Cambridge, UK.

Sutherland, P. and Melville, W. K. 2015. Field measurements of surface and near-surface turbulence in the presence of breaking waves. *Journal of Physical Oceanography*, 45(4):943–965. doi:10.1175/jpo-d-14-0133.1.

Sutton, O. G. 1955. *Atmospheric Turbulence*, 2nd edition. Methuen – John Wiley & Sons, New York.

Talley, L. D., Pickard, G. L., Emery, W. J., and Swift, J. H. 2011. *Descriptive Physical Oceanography: An Introduction*, 6th edition. Elsevier – Academic Press, London, UK.

Tampieri, F. 2017. *Turbulence and Dispersion in the Planetary Boundary Layer*. Springer International Publishing, Switzerland.

Tardu, S. 2014. *Transport and Coherent Structures in Wall Turbulence*. ISTE – John Wiley & Sons, Hoboken, NJ.

Tennekes, H. and Lumley, J. L. 1972. *A First Course in Turbulence*. The MIT Press, Cambridge, MA.

Thomson, J., Schwendeman, M. S., Zippel, S. F., Moghimi, S., Gemmrich, J., and Rogers, W. E. 2016. Wave-breaking turbulence in the ocean surface layer. *Journal of Physical Oceanography*, 46(6):1857–1870. doi:10.1175/jpo-d-15-0130.1.

Thorpe, S. A. 2004. Recent development in the study of ocean turbulence. *Annual Review of Earth and Planetary Sciences*, 32(1):91–109. doi:10.1146/annurev.earth.32.071603.152635.

Thorpe, S. A. 2005. *The Turbulent Ocean*. Cambridge University Press, Cambridge.

Thorpe, S. A. 2007. *An Introduction to Ocean Turbulence*. Cambridge University Press, Cambridge, UK.

Townsend, A. A. 1976. *The Structure of Turbulent Shear Flow*, 2nd edition. Cambridge University Press, New York.

Turner, J. S. 1973. *Buoyancy Effects in Fluids*. Cambridge University Press, Cambridge, UK.

Vallis, G. K. 2017. *Atmospheric and Oceanic Fluid Dynamics: Fundamentals and Large-scale Circulation*, 2nd edition. Cambridge University Press, Cambridge, UK.

Van Dyke, M. 1982. *An Album of Fluid Motion*, 14th edition. The Parabolic Press, Stanford, CA.

Velarde, M. G., Tarakanov, R., and Marchenko, A. V. (Eds.). 2018. *The Ocean in Motion: Circulation, Waves, Polar Oceanography*. Springer International Publishing, Switzerland.

Venditti, J. G., Best, J. L., Church, M., and Hardy, R. J. (Eds.). 2013. *Coherent Flow Structures at Earth's Surface*. Wiley – Blackwell, Chichester, UK.

Veron, F., Melville, W. K., and Lenain, L. 2009. Measurements of ocean surface turbulence and wave–turbulence interactions. *Journal of Physical Oceanography*, 39(9):2310–2323. doi:10.1175/2009jpo4019.1.

Vinnichenko, N. K., Pinus, N. Z., Shmeter, S. M., and Shur, G. N. 1980. *Turbulence in the Free Atmosphere (translated from Russian by F. L. Sinclair)*, 2nd edition. Springer Science, New York.

Vlasenko, V., Stashchuk, N., and Hutter, K. 2005. *Baroclinic Tides: Theoretical Modeling and Observational Evidence*. Cambridge University Press, Cambridge, UK.

Volyak, K. I., Lyakhov, G. A., and Shugan, I. V. 1987. Surface wave interaction. Theory and capability of oceanic remote sensing. In *Oceanic Remote Sensing* (Eds. F. V. Bunkin and K. I. Volyak), Nova Science Publishers, Commack, New York, pp. 107–145 (translated from Russian).

Voropayev, S. I. and Afanasyev, Y. D. 1994. *Vortex Structures in a Stratified Fluid: Order from Chaos*. Chapman & Hall/CRC Press, Boca Raton, FL.

Wallace, J. M. and Hobbs, P. V. 2006. *Atmospheric Science, Second Edition: An Introductory Survey (International Geophysics)*, 2nd edition. Elsevier – Academic Press, London, UK.

Woodhouse, M. J., Phillips, J. C., and Hogg, A. J. 2016. Unsteady turbulent buoyant plumes. *Journal of Fluid Mechanics*, 794:595–638. doi:10.1017/jfm.2016.101.

Woods, A. W. 2010. Turbulent plumes in nature. *Annual Review of Fluid Mechanics*, 42(1): 391–412. doi:10.1146/annurev-fluid-121108-145430.

Wyngaard, J. C. 2010. *Turbulence in the Atmosphere*. Cambridge University Press, Cambridge, UK.

Yavorsky, N. I. 1998. *Theory of Submerged Jets and Wakes*. Kutateladze Institute of Thermophysics, Novosibirsk (in Russian: Яворский, Н. И. Теория затопленных струй и следов. Новосибирск, Институт теплофизики СО РАН, 1998).

Zakharov, V. E. (Ed.). 1998. *Nonlinear Waves and Weak Turbulence*. American Mathematical Society, Providence, RI.

Zakharov, V. E., L'vov, V. S., and Falkovich, G. 1992. *Kolmogorov Spectra of Turbulence 1: Wave Turbulence*. Springer, Berlin, Germany.

Zeitlin, V. 2018. *Geophysical Fluid Dynamics: Understanding (almost) everything with rotating shallow water models*. Oxford University Press. Oxford, UK.

Zilitinkevich, S. S. 2013. *Atmospheric Turbulence and Planetary Boundary Layers*. Fizmatlit, Moscow (in Russian: Зилитинкевич С. С. Атмосферная турбулентность и планетарные пограничные слои. Москва, Физматлит, 2013).

3 Elements of Wave Propagation Theory

A method is more important than a discovery, since the right method will lead to new and even more important discoveries.

Lev Landau

3.1 INTRODUCTION

In the first two chapters, we have considered some important aspects of turbulence science. Turbulence is a complicated research subject involving many theories, observations, and a large scope of activities. In fact, there are only two major scientific milestones – fundamentals of fluid dynamics and statistical phenomenology. Although there is no rigorous mathematical proof of their relation, numerical modeling, along with spectral and statistical methods, has provided significant progress in the understanding of turbulence behavior.

In practice, the problem of turbulence has another challenge related to quantitative interpretation of observational data and extraction of relevant information, e.g., turbulent objects, structures, or features from dynamical background. The most common option available for all researchers and instrumentations is experimental measurements of the turbulence energy spectrum and/or its variations and further comparisons with famous Kolmogorov's *"five-thirds" law*. However, this traditional way (which is in statistical sense is definitely valid) does not provide needed information on structural properties and multiscale dynamics of turbulence being studied. Detection and prediction of turbulence require full picture, technically and literally. In this context, the theory of electromagnetic (EM) wave propagation provides the essential bridge between applied physics and the observation technology.

In the classical electrodynamics, the theory describes electromagnetic radiation as a flow of time-varying electric and magnetic fields traveling through free space or through a material medium with a universally fixed speed. These fields propagate in a form of harmonic (sinusoidal) EM waves, characterizing by its intensity, frequency, and polarization. Temporal evolution and propagation of monochromatic EM waves are completely described by Maxwell's equations. Based on these equations, the theory of electromagnetic radiation thereby provides a link between object's intrinsic properties and remote sensing measurements.

Wave propagation through random media is a fairly mature area of the research, having been studied extensively since the 1950s. Pioneering theoretical and some experimental works on light propagation have been conducted primarily in Soviet Union in the earlier 1960s. First monographs (Chernov 1960; Tatarski 1961), published in Russian, were immediately translated in English and triggered further studies of the problem known today as *wave propagation in turbulent media*. Consequently, a number of excellent monographs and textbooks have emerged over the years (Uscinski 1977; Strohbehn 1978; Rytov et al. 1989; Ishimaru 1991; Tatarskii et al. 1993; Andrews and Phillips 2005; Sasiela 2007) which cover various problems of wave propagation theory and practice. Although most studies and applications were considered in context with propagation of optical waves in atmosphere and laser beam phenomena, the theoretical basis created is valid for radiowaves and microwaves as well. In this context, the theory has practical relevance in many scientific areas dealing with scattering, absorption, emission, and fluctuations of EM waves, propagating in random media.

The wave propagation framework for turbulence research is traditionally based on a concept of random media and stochastic wave propagation approach. Random medium refers to a medium whose properties are random functions of time and space. Usually, it is associated with fluctuations of the physical parameters (temperature, density, or permittivity) of the medium and/or with the presence of random irregularities including the boundary as well. Such a medium is characterized by its statistical properties and the problem is to study the statistical characteristics of the propagated EM waves and fields. There are two different aspects in the wave propagation problem: the first is concerned with continuous macroscopic medium with variable (or random) refractive index or dielectric permittivity and the second is concerned with the discrete medium consisting of randomly distributed particles (scatterers). In both cases, wave propagation phenomena are described by stochastic equations.

More complicated case is when the medium represents complex flow with unpredictable properties and/or stochastic behavior. In this regard, the random media is characterized by structural variations and scale transformations (scaling) that is difficult or impossible to describe by classical mechanics or deterministic theories. For example, in geosciences and remote sensing, an efficient research tool is (multi) fractal and/or wavelet allowing the users to evaluate scaling, self-similarity, and irregularity of complex observational data.

This chapter is a brief description of the wave propagation theory. Classical stochastic wave equation and its approximate solutions will be discussed. Some aspects of wave propagation in complex moving media will be mentioned as well. We believe that this chapter can give the reader introductory knowledge on wave propagation in turbulent media. It is an essential step in the understanding of remote sensing capabilities in turbulence detection.

3.2   MAXWELL'S EQUATIONS

In the mid-19th century, the Scottish physicist James Clerk Maxwell (1831–1879) formulated a simple theory that had such far-reaching implications that is of fundamental importance to our world today. Maxwell's equations first appeared in the paper "A dynamical theory of the electromagnetic field" published in the *Philosophical Transactions of the Royal Society of London*, in 1865. These are the mathematical relationships describing the behavior of EM fields and waves in free space and material media at every point where the physical properties of the medium are continuous.

The set of partial differential equations for an EM field was established as a result of generalization of studies of electricity and magnetism. In modern notation, the inhomogeneous, macroscopic Maxwell's equations (in Gaussian units) are

$$\nabla \times \mathbf{E}(\mathbf{r},t) = -\frac{1}{c}\frac{\partial \mathbf{B}(\mathbf{r},t)}{\partial t}, \qquad \text{Faraday's Law} \qquad (3.1)$$

$$\nabla \times \mathbf{H}(\mathbf{r},t) = -\frac{1}{c}\frac{\partial \mathbf{D}(\mathbf{r},t)}{\partial t} + \frac{4\pi}{c}\mathbf{j}(\mathbf{r},t), \qquad \text{Ampère—Maxwell's Law} \qquad (3.2)$$

$$\nabla \cdot \mathbf{D} = 4\pi\rho(\mathbf{r},t), \qquad \text{Gauss's Law for electric field,} \quad (3.3)$$

$$\nabla \cdot \mathbf{B} = 0, \qquad \text{Gauss's Law for electric field,} \quad (3.4)$$

which are supplemented by material equations

$$\mathbf{j} = \sigma\mathbf{E}, \qquad \text{Ohm's law} \qquad (3.5)$$

$$\mathbf{D} = \mathbf{E} + 4\pi\mathbf{P}, \qquad (3.6)$$

$$\mathbf{B} = \mathbf{H} + 4\pi\mathbf{M} \qquad (3.7)$$

with the following quantities: $\mathbf{E}$ the electric field, $\mathbf{H}$ the magnetic field, $\mathbf{B}$ the magnetic induction, $\mathbf{D}$ the electric displacement, $\mathbf{P}$ the polarization field, $\mathbf{M}$ the magnetization field $\mathbf{j}$, the electric charge density, σ the specific conductivity, t denotes time, $\mathbf{r}$ coordinate vector, and c the speed of light (it can only be used in these equations when the EM wave is in a vacuum); the $\nabla \times$ symbol denotes curl operator and the $\nabla \cdot$ symbol denotes divergence operator. As explained by Born and Wolf (1999), $\mathbf{E}$, $\mathbf{D}$, $\mathbf{j}$, and ρ in this system are measured in electrostatic units, and $\mathbf{H}$ and $\mathbf{B}$ are measured in electromagnetic units. The use of Maxwell equations in this book is consistent with the tradition established by Born and Wolf.

In the linear approximation, the polarization field $\mathbf{P}$, the magnetization field $\mathbf{M}$, the electric susceptibility χ_e, and magnetic susceptibility χ_m are defined as

$$\mathbf{P} = \chi_e\mathbf{E}, \quad \mathbf{M} = \chi_m\mathbf{H}, \quad \varepsilon = 1 + 4\pi\chi_e, \quad \mu = 1 + 4\pi\chi_m, \qquad (3.8)$$

where ε is the permittivity and μ is the permeability and material equations take simple form

$$\mathbf{j} = \sigma\mathbf{E}, \quad \mathbf{D} = \varepsilon\mathbf{E}, \quad \mathbf{B} = \mu\mathbf{H}, \qquad (3.9)$$

Maxwell has also established purely theoretically that *light is an electromagnetic disturbance in the form of waves propagated through the luminiferous aether* with speed $c = 1/\sqrt{\varepsilon_0\mu_0} = 2.9979\times10^8$ m/s, where $\varepsilon_0 = (1/36\pi) \times 10^{-9}$ F/m (farad per meter) and $\mu_0 = 4\pi \times 10^{-7}$ H/m (henry per meter) are permittivity and permeability of free space, correspondently. In fact, Maxwell concluded that light is an electromagnetic wave having such wavelengths that it can be detected by the eye.

Maxwell's equations cover all aspects of electrodynamics and optics including wave propagation phenomena; the solutions are enormously varied depending on the problems being studied. A detailed historical development of electromagnetic theory and Maxwell's equations is given by (Elliott 1993; Born and Wolf 1999; Huray 2010).

3.3 ELECTROMAGNETIC WAVES

Definition: Electromagnetic waves or EM waves are waves that are created as a result of vibrations between an electric field and a magnetic field. In other words, EM waves are composed of oscillating magnetic and electric fields. The electric field and magnetic field of an EM wave are perpendicular (at right angles) to each other. They are also perpendicular to the direction of the electromagnetic wave propagation. EM waves are solutions of Maxwell's equations, which are the fundamental equations of electrodynamics.

Unlike mechanical waves, EM waves are "transverse" waves. This means that they are measured by their amplitude (height) and wavelength (distance between the highest/lowest points of two consecutive waves). EM waves travel with a constant velocity of $c = 2.9979 \times 10^8$ m/s in vacuum. They are deflected neither by the electric field, nor by the magnetic field.

In classical electrodynamics (Stratton 1941; Born and Wolf 1999; Jackson 1999), several mathematical descriptions (or models) of EM waves are commonly used in order to specify their geometrical and propagation characteristics in various media. These models are known as the following (Figure 3.1):

Plane wave is an unbounded frequency-constant wave whose wavefronts (surfaces of constant phase) are infinite parallel planes of constant peak-to-peak amplitude normal to the phase velocity vector. In complex notation, plane wave is expressed as:

$$\mathbf{E}(\mathbf{r},t) = \mathbf{E}_0 e^{\imath(\mathbf{k}\cdot\mathbf{r}-\omega t)}$$

Here $\mathbf{E}_0$ is the constant amplitude and $\phi_0 = \mathbf{k} \cdot \mathbf{r} - \omega t$ is the phase. Plane wave model is used as an approximation of a far wave field propagated at long distance from localized source of electromagnetic energy. For example, optical properties of natural sunlight can be described by plane waves.

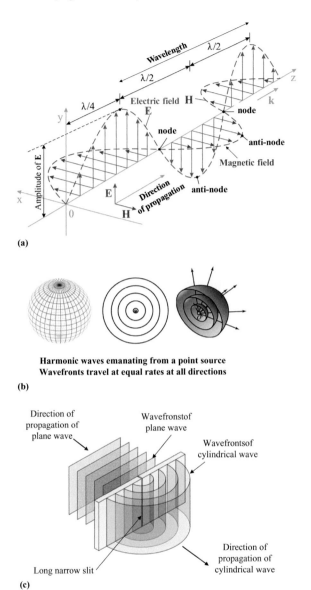

FIGURE 3.1 Types of electromagnetic waves. (a) Plane wave, (b) spherical wave. (c) cylindrical wave, and (d) circular polarization of electromagnetic.

Spherical wave is an unbounded wave emanating from a point source. A spherical wave is characterized by concentric spheres forming the equiphase surface. In complex notation, harmonic spherical wave is represented as:

$$\mathbf{E}(\mathbf{r},t) = \mathbf{E}_0 \frac{e^{i(\mathbf{k}\cdot\mathbf{r}-\omega t)}}{r}$$

Spherical wave propagates with spherical symmetry, i.e., and displacement $\mathbf{E}(\mathbf{r},t)$ of a harmonic spherical wave only depend on time t and the radial (scalar) distance r from the point source. In far field, spherical wave can be approximated by a plane wave; there are no radial components of spherical wave in the far field. A spherical wave model is widely used in the antenna theory, short-range propagation technology, near-field imaging techniques (e.g., based on the spherical-wave Mach–Zehnder interferometer), optical holography, and others specific areas of electrodynamics and acoustics.

Cylindrical wave is a wave in which the distribution of all quantities is homogeneous in some direction (which we take as the z-axis) and has complete axial symmetry about that direction distribution. In complex notation, cylindrical wave far away from the z-axis is expressed as

$$\mathbf{E}(\mathbf{r},t) = \mathbf{E}_0 \frac{e^{i(\mathbf{k}\cdot\mathbf{r}-\omega t)}}{\sqrt{r}}$$

Cylindrical waves are traveling outwards or inwards from the z-axis in the radial direction. On a given cylindrical surface centered on the z-axis, the electric field has constant phase across the surface. Conceptually, cylindrical wavefronts represent concentric cylinders created by an infinitely-long, straight, oscillating line charge. Cylindrical wave model is used in through-the-wall radar imaging techniques, ground-penetrating radar technology, and in study of propagation in fractal and complex media. Waves with cylindrical wavefronts are often encountered in underwater acoustics.

Circular wave is associated with special polarization state of EM wave; in this case, the net electric vector $\mathbf{E} = \{E_x, E_y, E_z\}$ rotates around the direction of propagation z as it moves along this direction and thereby traces out a circle pattern in the transverse plane. This type of polarization is known as circular polarization. The actual path of $\mathbf{E}$ is a circular (or elliptical) helix in space as the wave moves alone. The two components of electrical field $\mathbf{E} = \{E_x, E_y\}$ must have a time-phase difference of multiples of 90 degrees. Thus, circularly polarized wave can be mathematically represented as

$$\mathbf{E}(\mathbf{r},t) = E_0 \cos(\mathbf{k}\cdot\mathbf{r}-\omega t)\mathbf{e}_x + E_0 \sin(\mathbf{k}\cdot\mathbf{r}-\omega t)\mathbf{e}_y$$

$$\mathbf{E}_x(z,t) = E_{0x} \cos(kz-\omega t)\mathbf{e}_x$$

$$\mathbf{E}_y(z,t) = E_{0x} \cos\left(kz-\omega t+\frac{m\pi}{2}\right)\mathbf{e}_y$$

$$m = 0,\pm 1,\pm 2 \pm \ldots \text{ is odd number}$$

Circular polarization is not often encountered in nature; several instruments have been proposed for remote observations of circular polarization of gaseous and solid surfaces as well as for detection of organic compounds and biomass. Principles of circular polarization are widely used in microwave engineering including waveguide and antenna technologies (Nefyodov and Smolskiy 2019).

Beam wave is a wave of finite extent with focusing capabilities. In other words, a beam of light is a focused ray. The Gaussian-beam wave model is given by

$$E(x, y, z, t) = E_0 \exp\left(-\frac{x^2 + y^2}{w^2}\right) \exp\left(\iota(kz - \omega t)\right) \exp\left(\iota k\left(\frac{x^2 + y^2}{2R(z)}\right)\right) \exp\left(-\iota\varphi\right),$$

$$\underbrace{\qquad}_{\textit{Gaussian profile}} \quad \underbrace{\qquad}_{\textit{indirection wave term}} \quad \underbrace{\qquad}_{\textit{curvature of wavefront}}$$

where E_0 is the on-axis amplitude, w is the beam spot radius, R is its phase front radius of curvature, and $\varphi(z)$ is known as the *Gouy phase*. Beam wave propagation through inhomogeneous turbulent media is of growing practical interest in geosciences and remote sensing as well. First of all, we emphasize the importance of laser-based (or lidar) detection of turbulence in the ocean and atmosphere and measurements of wind shears and fluctuations above the ground or sea surface. In particular, Gaussian beam wave propagation models are widely used in atmospheric optics, laser imaging technology, and electronic engineering including telecommunication, beam waveguide antennas, microwave transmitters and receivers, and high-resolution radar.

3.4   THE WAVE EQUATION

In the case of an isotropic medium, Maxwell's equations are reduced to a single equation known as the wave equation which describes the propagation of light and EM waves through a nonuniform (turbulent) medium.

The wave equation is a linear second-order partial differential equation which describes the propagation of waves at a fixed speed in some substance. In geophysical electromagnetic theory, it is usually considered the case when there are no electric charge ($\rho = 0$) and currents ($|\mathbf{j}| = 0$). The vector wave equations for *inhomogeneous* medium are given by

$$\nabla^2 \mathbf{E} - \frac{\varepsilon\mu}{c^2}\frac{\partial^2 \mathbf{E}}{\partial^2 t} + \left(\text{grad}\ln\mu\right) \times \text{curl}\mathbf{E} + \text{grad}\left(\mathbf{E}\cdot\text{grad}\ln\varepsilon\right) = 0, \qquad (3.10)$$

$$\nabla^2 \mathbf{H} - \frac{\varepsilon\mu}{c^2}\frac{\partial^2 \mathbf{H}}{\partial^2 t} + \left(\text{grad}\ln\varepsilon\right) \times \text{curl}\mathbf{H} + \text{grad}\left(\mathbf{H}\cdot\text{grad}\ln\mu\right) = 0. \qquad (3.11)$$

For *homogeneous* medium Maxwell's equations reduce to

$$\nabla \times \mathbf{E}(\mathbf{r},t) = -\frac{\mu}{c}\frac{\partial \mathbf{H}(\mathbf{r},t)}{\partial t}, \tag{3.12}$$

$$\nabla \times \mathbf{H}(\mathbf{r},t) = -\frac{\varepsilon}{c}\frac{\partial \mathbf{E}(\mathbf{r},t)}{\partial t}, \tag{3.13}$$

$$\nabla \cdot \mathbf{E}(\mathbf{r},t) = 0, \tag{3.14}$$

$$\nabla \cdot \mathbf{H}(\mathbf{r},t) = 0, \tag{3.15}$$

and wave equations (gradln ε = gradln μ = 0) become

$$\nabla^2 \mathbf{E} - \frac{\varepsilon\mu}{c^2}\frac{\partial^2 \mathbf{E}}{\partial^2 t} = 0, \tag{3.16}$$

$$\nabla^2 \mathbf{H} - \frac{\varepsilon\mu}{c^2}\frac{\partial^2 \mathbf{H}}{\partial^2 t} = 0. \tag{3.17}$$

The basic solutions of the wave Equations (3.16) and (3.17) are written in the form of two uniform plane waves

$$\mathbf{E}(\mathbf{r},t) = \mathbf{E}_0 \exp\left[\imath(\mathbf{kr} - \omega t)\right], \tag{3.18}$$

$$\mathbf{H}(\mathbf{r},t) = \mathbf{H}_0 \exp\left[\imath(\mathbf{kr} - \omega t)\right], \tag{3.19}$$

where $\mathbf{E}_0$ and $\mathbf{H}_0$ are complex amplitudes of the electric and magnetic fields; k is the complex wave vector, k = k′ + $\imath$k″, ω is the angular frequency of the wave, and $\imath$ = $(-1)^{1/2}$ is the imaginary unit. Amplitudes $\mathbf{E}_0$ and $\mathbf{H}_0$ can be found using boundary conditions.

For a source-free, linear, isotropic, homogeneous, *lossy* medium (ρ = 0, |j| $\neq$ 0, σ > 0), the vector wave equation can be written in two forms

$$\nabla^2 \mathbf{E} + k^2 \mathbf{E} = 0, \quad \text{or} \quad \nabla^2 \mathbf{E} - \gamma^2 E = 0, \tag{3.20}$$

$$\nabla^2 \mathbf{H} + k^2 \mathbf{H} = 0, \quad \text{or} \quad \nabla^2 \mathbf{H} - \gamma^2 H = 0 \tag{3.21}$$

with complex wavenumber, k = k′ + $\imath$k″, or complex propagation constant, $\gamma = \alpha + \imath\beta$, which are given by

$$k^2 = \mu\varepsilon\omega^2 + \imath\mu\sigma\omega, \tag{3.22}$$

$$k' = \omega\sqrt{\frac{\mu\varepsilon}{2}\left[\sqrt{1+\left(\frac{\sigma}{\omega\varepsilon}\right)^2}+1\right]}, \quad k'' = \omega\sqrt{\frac{\mu\varepsilon}{2}\left[\sqrt{1+\left(\frac{\sigma}{\omega\varepsilon}\right)^2}-1\right]}, \tag{3.23}$$

$$\gamma = \alpha + \imath\beta = \sqrt{\imath\omega\mu(\sigma + \imath\omega\varepsilon)}, \tag{3.24}$$

$$\alpha = \omega \sqrt{\frac{\mu\varepsilon}{2}\left[\sqrt{1+\left(\frac{\sigma}{\omega\varepsilon}\right)^2}-1\right]}, \quad \beta = \omega \sqrt{\frac{\mu\varepsilon}{2}\left[\sqrt{1+\left(\frac{\sigma}{\omega\varepsilon}\right)^2}+1\right]}, \quad (3.25)$$

where α is the attenuation constant and β is the phase constant. The units are: γ (1/m), α (Np/m) (or dB/m in microwave engineering), and β (rad/m).

Equations (3.20) and (3.21) are also known as homogeneous vector Helmholtz equations or the *phasor* vector wave equation (symbol ∇^2 denotes the vector Laplacian operator). The propagation velocity of EM waves through a (non-vacuum) lossy medium $\left(\frac{\sigma}{\omega\varepsilon} \ll 1\right)$ is $\upsilon = \frac{\omega}{\beta} \approx \frac{1}{\sqrt{\mu\varepsilon}} = \frac{c}{\sqrt{\mu_r\varepsilon_r}} = \frac{1}{\sqrt{\mu_0\mu_r\varepsilon_0\varepsilon_r}}$ and the complex index of refraction (or the refractive index) is $n = \frac{c}{\upsilon} = \sqrt{\mu_r\varepsilon_r} = \frac{\sqrt{\mu\varepsilon}}{\sqrt{\mu_0\varepsilon_0}}$ (here $\mu_r = \mu/\mu_0$ and $\varepsilon_r = \varepsilon/\varepsilon_0$ are the *relative* permeability and the *relative* permittivity of the medium, respectively). The refractive index of a material medium is positive, $n > 1$, and a function of frequency $n = n(\omega)$. Most natural materials are non-magnetic at optical and microwave frequencies, that is $\mu_r \approx 1$, therefore $n \approx \sqrt{\varepsilon_r}$, where $\varepsilon_r = \varepsilon_r' + \iota\varepsilon_r''$ is the complex permittivity. In particular, variations (fluctuations) of the refractive index of air play a key role in atmospheric optical turbulence.

Practical solutions of the vector wave Equations (3.10) and (3.11) or reduced Equations (3.20) and (3.21) for random inhomogeneous media require rigorous calculations involving methods of computational electrodynamics (Cohen 2002; Lipatov 2002; Kampanis et al. 2019). Analytical methods for solving wave Equations (3.20)–(3.21) are very restricted and can be used only in very special cases, e.g., for planar layered media (Wait 1970; Brekhovskikh 1980). Therefore, various approximation techniques were developed and applied to solve the wave equation; the most known are the following: 1) method of Green's function, 2) Fourier transform, 3) Kirchhoff formula, 4) Wentzel–Kramers–Brillouin (WKB) approximation, 5) Born approximation, 6) Rytov approximation, 7) parabolic equation method, 8) extended Huygens–Fresnel principle, and a number of hybrid approaches.

The choice of an appropriate method for the solution of the wave equation depends on the problem being studied and also on the type of boundary and initial value problems. The generalized integral solution can be found through solving Cauchy problems for quasi-linear second-order partial differential equation related to the Helmholtz equation. This Cauchy problem is the pivotal problem in wave propagation phenomena. Detailed mathematical analysis of available methods and solutions is beyond the scope of this book and we refer the interested reader to the corresponding literature on electrodynamics and optics. The most important methods will be briefly overviewed in Section (3.7).

Meanwhile, direct numerical simulations of Maxwell's equations are of practical interest in geosciences and remote sensing, first of all, for computing scattering and emission signatures of complex natural media (soil, show, ice) and multiscale rough surfaces (sea waves, terrains, etc.). Wave propagation in natural media with strong fluctuations of physical parameters is also the case for direct electromagnetic modeling and simulations. An important geophysical application is detection of complex disturbances and surface features at highly turbulent OBL that may require the development of sophisticated techniques (Raizer 2017, 2019).

3.5 WAVE PROPAGATION PHENOMENA

Propagation of EM waves is encountered in many scientific and engineering disciplines. According to standard specification of frequency bands (Table 3.1), wave propagation includes three following modes:

1. Ground wave propagation (VLF, LF, MF).
2. Sky/ionosphere wave propagation (HF, VHF, UHF).
3. Space/troposphere wave propagation (2 MHz–30 MHz).

This common classification is used in communication and broadcasting technologies and does not clarify the area of remote sensing. Specifically in remote sensing, wave propagation is associated with principal mechanisms of interactions between EM waves and a material medium. The interactions result in various wave phenomena known as diffraction, reflection, refraction, diffusion, scattering, shadowing (self-shadowing), emission, transmission, absorption, attenuation, amplification, caustic, and foci. The occurrence of these phenomena strictly depends on configuration and properties of the medium (its structural nonuniformities) and characteristics of the interacting EM waves. For example, scattered waves are produced by rough surfaces, small objects, or by other irregularities of the medium. Electromagnetic response is defined by the properties and structure of turbulent medium in accordance with the specification of wave propagation problems (Figure 3.2).

TABLE 3.1
Classification of Radio Frequency Bands

Designation	Abbreviation Frequency Range	Wavelength	Characteristics	
Extremely low	ELF	3 Hz –29 Hz	10^5 km – 10^4 km	
Super low	SLF	30 Hz – 299 Hz	10^4 km – 10^3 km	
Ultra low	ULF	300 Hz – 2999 Hz	1000 km – 100 km	
Very low	VLF	3 kHz – 29 kHz	100 km – 10 km	
Low	LF	30 kHz – 299 kHz	10 km – 1 km	Ground waves
Medium	MF	300 kHz – 2999 kHz	1 km – 100 m	
High	HF	3 MHz – 29 MHz	100 m – 10 m	Sky waves
Very high	VHF	30 MHz – 299 MHz	10 m – 1 m	Space wave
Ultra high	UHF	300 MHz – 2999 MHz	1 m – 10 cm	
Super high	SHF	3 GHz – 29 GHz	10 cm – 1 cm	Satellite waves
Extremely high	EHF	30 GHz – 299 GHz	1 cm – 1 mm	

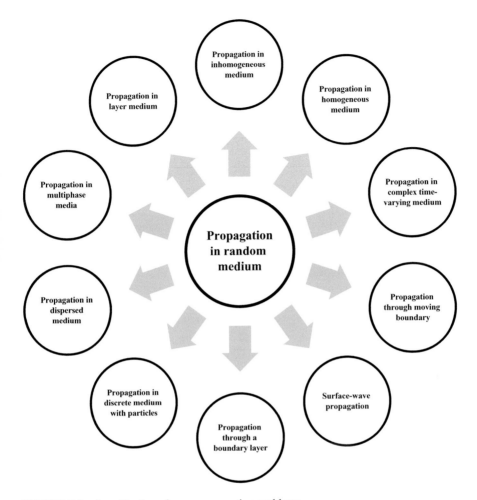

FIGURE 3.2   Specification of wave propagation problems.

In view of remote sensing capabilities, two basic categories of nonuniform macroscopic media – turbulent and non-turbulent media – can be distinguished. Turbulent medium is usually nonlinear; it is characterized by local spatiotemporal fluctuations of (effective) parameters; and non-turbulent medium is a linear medium with smoothly varying (effective) parameters. Such a classification allows us to use a unified theoretical approach, based on macroscopic wave Equations (3.10) and (3.11). Frequently, under the assumption of weak and non-local fluctuations, these equations are reduced to the Helmholtz Equation (3.20) or (3.21) which is formulated in terms of scalar quantities. Stochastic Helmholtz equation is a basic equation in wave propagation theory (Section 3.6.2).

This classic wave equation provides various solutions with many applications depending on the problem, boundary and initial conditions, and EM wavelength range being considered. It was realized for a long time that this problem is equivalent to an integral equation over the boundary. The possibility of finding analytic solutions is, however, restricted to very simple geometries; otherwise, a numerical method is necessary. For example, radio and optical propagation in sea water as well as the over-sea microwave propagation forecasting (that is important in submarine communications) are directly affected by ocean turbulence and surface dynamics that is difficult to predict by purely theory. In these cases, approximate analytical solutions may not always predict real-world effects. On the other hand, remote sensing observations can provide needed complementary information about environmental conditions specifically related to atmospheric and oceanic boundary layer turbulence.

3.6 PROPAGATION THROUGH TURBULENCE

Initially, the theory of wave propagation in turbulent media has been associated with acoustics. The history has begun in the late 1940s and the early 1950s when researchers developed general ray-optics formulas for propagation of sound in the ocean (Dashen et al. 1979; Flatté 1979, 1983; see also reviews Kallistratova 2002; Korotkova 2019). Later, the theory and applications have received much more attention in context with problems of light and laser beam propagation through a turbulent atmosphere (Tatarskii 1961, 1971; McCartney 1976; Strohbehn 1978; Zuev 1982; Weichel 1990; Andrews and Phillips 2005; Schmidt 2010). This subject is of renewed interest in aerospace industry including satellite communication, tracking, laser/lidar mapping, optical remote sensing, and engineering (Driggers 2003; Hemmati 2009).

The modern theory is based on the wave equation and its approximate solutions. Let's consider elements of the theory on the example of light propagation in atmosphere. Andrews and Phillips (2005) formulated this subject as the following:

> In the atmosphere, turbulent fluctuations in wind speed result in the mixing of atmospheric quantities such as temperature, water vapor, and the index of refraction. These quantities are called passive scalars because their dynamics do not affect the turbulence associated with velocity fluctuations. The most important of these quantities in optical wave propagation is the index-of-refraction fluctuations, commonly referred to as optical turbulence. Because it behaves like a passive additive, the theoretical framework of optical turbulence is based on the classical theory of turbulence concerning velocity fluctuations.

Basically, optical turbulence is an important microphysical effect which alone the propagation path can produce significant intensity fluctuations and/or variations in the direction the transmitting light beam propagates. These fluctuations of the intensity are called *scintillation* (Section 3.8).

3.6.1 OPTICAL TURBULENCE IN THE ATMOSPHERE

Atmospheric turbulence is the most problematic phenomenon in optical and microwave remote sensing including high-resolution multispectral imagery. Atmospheric turbulence is characterized by small-scale, irregular air motions caused by winds that vary in speed and direction. As mentioned above, atmospheric turbulence results in the random fluctuations of the refractive index that significantly impact on propagation of EM waves. These fluctuations are generally considered as random fields in both space and time. This causes the intensity of waves to be randomly distorted by propagation that is a challenging problem for theoretical analysis.

The Kolmogorov–Obukhov theory, proposed more than 60 years ago, remains the basic framework for the description of atmospheric turbulence. Standard atmospheric model of optical turbulence is defined by a set of the following semi-empirical formulas:

$$n_i(\mathbf{r},t) = n_0 + n_1(\mathbf{r},t) \approx 1 + 77.6 \times 10^{-6} \frac{P(\mathbf{r})}{T(\mathbf{r})}, \tag{3.26}$$

$$D_n(r_1,r_2) = \left[n(r_1) - n(r_2)\right]^2 = \begin{cases} D_n(r) = C_n^2 r^{2/3}, & \ell_0 \ll r \ll L_0, \\ \\ D_n(r) = C_n^2 \ell_0^{4/3} r^2 & r \ll \ell_0, \end{cases} \tag{3.27}$$

$$\Phi_n(k) = 0.033 C_n^2 k^{-11/3}, \quad 1/L_0 \ll k \ll 1/\ell_0, \tag{3.28}$$

where $n(\mathbf{r},t)$, n_0, and $n_1(\mathbf{r},t)$ represent a random index of refraction, its mean, and fluctuation parts, respectively; P is the air pressure in millibars, T is the temperature in Kelvin; $D_n(r_1,r_2)$ is the refractive-index structure function at two points r_1 and r_2 separated by distance $r = r_1 - r_2$, brackets $\langle \ldots \rangle$ denote an ensemble average. At sea level, typically $n - 1 \approx 3 \cdot 10^{-4} (n_0 = 1)$.

The refractive-index fluctuations mostly obey Kolmogorov statistics with the spectrum $\Phi_n(k)$ which is limited to the inertial subrange with outer scale (macroscale) L_0 and inner scale (microsacale) ℓ_0 of turbulence. Constant C_n^2 is the index-of-refraction structure constant (in units of $m^{-2/3}$), also called the *structure parameter*, and the inner scale is $\ell_0 = 7.4(\nu^3/\varepsilon)^{1/4}$, where ν is the kinematic viscosity and ε is the energy dissipation rate. Values of C_n^2 typically have a range from $10^{-17}\,m^{-2/3}$ or less for conditions of "weak turbulence" and up to $10^{-13}\,m^{-2/3}$ or more for "strong" turbulence. There are other models of the spectrum $\Phi_n(k)$ known as Tatarskii, von Kármán, Hill, "bump" models and Modified Atmospheric Spectrum model (Figure 3.3). As shown numerous investigations, the choice of the model depends on the type of experimental data and practical applications.

The classical Kolmogorov–Obukhov theory describes atmospheric turbulence as a statistical ensemble of eddies with various scales in the inertial interval $\ell_0 < r < L_0$ from the largest outer scale L_0 eddies down to the smallest, inner scale ℓ_0 eddies. In the Earth's atmosphere, the mean value is $\ell_0 \sim 0.1 \div 1$ cm that corresponds to typical fluctuations of the refractive index. Small-scale turbulence is usually isotropic, whereas large-scale turbulence mostly is anisotropic.

Thus, optical turbulence in the atmosphere is characterized by three parameters: the outer scale L_0, the inner scale ℓ_0, and the refractive-index fluctuations $n_1(\mathbf{r},t)$. However, there are some experimental optical measurements that the atmospheric turbulence may deviates from the Kolmogorov's 11/3 power law (3.28) especially at higher atmospheric layers – the troposphere and stratosphere. It is assumed that the

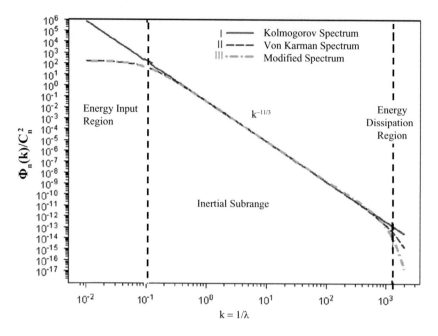

FIGURE 3.3 Spectral models of refractive index fluctuations in the atmosphere. (I) Kolmogorov spectrum, (II) von Kármán spectrum, and (III) Modified Spectrum.

power spectrum has a generalized form $\Phi_n\left(k\right) \approx \text{const} \cdot C_n^2 k^{-\alpha}$ with parameter, varying in the range $3 < \alpha < 5$ (Blaunstein et al. 2010; Korotkova 2014). Recently, oceanic non-Kolmogorov optical turbulence was investigated for spherical wave propagation (Yao et al. 2021).

Theoretically, the non-Kolmogorov turbulence and its effect on optical propagation can be investigated using the *Zernike polynomial expansion* (Boreman and Dainty 1996). The Zernike polynomials are common technique, used for describing classical aberrations of optical imaging systems. In particular, these polynomials provide significant correction of turbulent-corrupted wave front in optical imagery (Roggemann and Welsh 1996). Zernike polynomials and modes are also useful in adaptive optics and laser beam propagation theory (Lukin and Fortes 2002; Andrews and Phillips 2005).

3.6.2 STOCHASTIC HELMHOLTZ EQUATION

The stochastic wave equation is one of the fundamental stochastic partial differential equations of hyperbolic type. In the case of an isotropic random media, Maxwell's equations can be reduced to a single wave equation known as a scalar *inhomogeneous* stochastic Helmholtz equation given by

$$\nabla^2 U\left(\mathbf{r}\right) + k_0^2 n^2\left(\mathbf{r}\right) U\left(\mathbf{r}\right) = 0 \quad \text{or} \quad \nabla^2 U\left(\mathbf{r}\right) + k^2\left(\mathbf{r}\right) U\left(\mathbf{r}\right) = f\left(\mathbf{r}\right), \mathbf{r} \in D_{\pm}. \quad (3.29)$$

The stationary wave boundary value problem is formulated as

$$\lim_{|\mathbf{r}|\to\infty} |\mathbf{r}|^{\frac{n-1}{2}} \left(\frac{\partial}{\partial|\mathbf{r}|} - \imath k \right) U(\mathbf{r}) = 0, \tag{3.30}$$

where n = 2,3 is the dimension of the space. In Equations (3.29)–(3.30) U(**r**) is unknown stochastic wavefield, f(**r**) is the source function, and n(**r**) is the spatially varying refractive index; note that $k = k_0 n$ is the wavenumber in the medium with refractive index n and $k_0 = \dfrac{\omega}{c} = 2\pi/\lambda$ (λ is the EM wavelength) is the wavenumber in vacuum. The condition (3.30) is known as the *Sommerfeld radiation condition*; it is usually considered in relation with the scattering problem. It states that the scattered field consists of outgoing EM waves only. Solutions of the Helmholtz equation which satisfy the radiation condition are called radiating solutions or radiating functions.

Numerical solutions of the Helmholtz equation require a wide range of boundary conditions. Boundary conditions define the interface between the model geometry and its surroundings; correspondingly, two categories of boundary value problems are known: *interior* (i.e., solution in finite domain D_+) and *exterior* (i.e., solution in infinite domain D_-). An interior boundary is a dividing interface between two subdomains in the modeling domain, whereas an exterior boundary is an outer boundary of the modeling domain. There are several types of interior and exterior boundary conditions for computations of the Helmholtz equation; they are the following: 1) Dirichlet, 2) Neumann, 3) Robin, 4) Mixed, 5) Cauchy, and 6) Leontovich or impedance boundary conditions (for more detail see, e.g., books Towne 1967; Bleistein 1984). Wave phenomena, related to propagation in a turbulent medium are usually considered in connection with fluctuations of the refractive index n(**r**, ω), ω is the angular frequency, and specified exterior boundary conditions.

3.7 APPROXIMATE SOLUTIONS

As well known, many problems related to steady-state oscillations (mechanical, acoustical, thermal, electromagnetic, hydrodynamic, etc.) lead to the 2D Helmholtz equation. Thus, inhomogeneous Helmholtz Equation (3.29) is a time-independent linear partial differential equation. The interpretation of the unknown U(**r**, ω) and parameter n(**r**, ω) depends on the problem being modeled. In the case of wave propagation, U(**r**, ω) is the amplitude of a time-harmonic wave. Because there are difficulties in obtaining exact solutions of Equation (3.29) with random parameter n(r, ω) ≠ *const*, a number of mathematical methods and approaches have been suggested and applied for analysis of various wave propagation problems (Figure 3.4); most of them consider the case when the refractive-index variations are relatively small. Let's consider commonly used theoretical approaches.

3.7.1 Geometrical Optics

Geometrical optics (known also as ray optics) method uses a perturbation theory to solve the propagation problem (in vacuum). It ignores diffraction and polarization

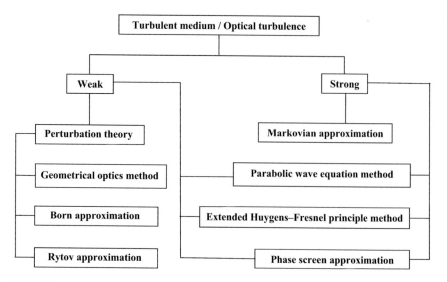

FIGURE 3.4 Analytic methods for solving wave propagation problem in turbulent medium.

effects and considers weak turbulent medium. Solution of Equation (3.29) is written in the form $U(\mathbf{r}) = A(\mathbf{r}) \exp[\imath k_0 \varphi(\mathbf{r})]$, where $A(\mathbf{r}) = A_0(\mathbf{r}) + \dfrac{A_1(\mathbf{r})}{\imath k_0} + \dfrac{A_2(\mathbf{r})}{(\imath k_0)^2} + \ldots$ is amplitude (assuming $\lambda \sim k_0^{-1} \to 0$) and $k_0 S(\mathbf{r})$ is the phase. Substituting $U(\mathbf{r})$ and $A(\mathbf{r})$ into (3.29) leads to a system of transport equations

$$\left(\nabla\varphi\right)^2 = n^2, \tag{3.31}$$

$$2\nabla\phi\cdot\nabla A_0 + A_0\nabla^2\phi = 0, \tag{3.32}$$

$$2\nabla\phi\cdot\nabla A_1 + A_1\nabla^2\phi = -\nabla^2 A_0, \tag{3.33}$$

$$2\nabla\phi\cdot\nabla A_m + A_m\nabla^2\phi = -\nabla^2 A_{m-1}. \tag{3.34}$$

The first Equation (3.31) is called *eikonal equation* that is the central result of geometrical optics. c and first-order perturbations A_0 and A_1 are defined from (3.32) and (3.33) once the eikonal is known. The surfaces $\varphi(\mathbf{r}) = \text{const}$ are called the geometrical wavefronts (Born and Wolf 1999). The geometrical optics approximation is restricted by the following conditions: $\lambda \ll \ell_0$, $\sqrt{\lambda L} \ll \ell_0$, and $\sigma_I^2 \ll 1$ (L is the wave propagation path and σ_I^2 is the normalized variance of intensity fluctuations).

The principles and methods of geometrical optics are used in many applications involving communication, remote sensing, imaging, holography, and optical instrument design as well. Geometrical optics is a basic theoretical approach which is considered at optical observations of the ocean and atmosphere. More about geometrical optics in the wave propagation and engineering problems can be found in books (Kravtsov and Orlov 1990; Wheelon 2001; Kravtsov 2005).

3.7.2   BORN APPROXIMATION

The Born approximation, also known as the "kinematical" or "single scattering" approximation, is a classical approach to solve wave equation using perturbation theory. The total wavefield is represented as a sum $U(\mathbf{r}, \omega) \approx U_0(\mathbf{r}, \omega) + U_1(\mathbf{r}, \omega)$, where $U_0(\mathbf{r}, \omega)$ is the unperturbed wavefield or "incident wavefield" and $U_1(\mathbf{r}, \omega)$ is the scattered wavefield.

The rigorous solution of Equation (3.29) for potential scattering is given by (Born and Wolf 1999, page 698)

$$U(\mathbf{r},\omega) = e^{ik_0 s_0 \cdot \mathbf{r}} + \int_V F(\mathbf{r}',\omega) U(\mathbf{r}',\omega) \frac{e^{ik_0|\mathbf{r}-\mathbf{r}'|}}{|\mathbf{r}-\mathbf{r}'|} d^3\mathbf{r}', \tag{3.35}$$

$$F(\mathbf{r},\omega) = \frac{1}{4\pi} k_0^2 \left[n^2(\mathbf{r},\omega) - 1 \right], \tag{3.36}$$

where the first term corresponds to incident field propagating in the direction specified by a real unit vector s_0 and the second term describes the behavior of the total field $U(\mathbf{r}, \omega)$ within the scattering volume V. Note that there are several approximate solutions of integral Equations (3.35)–(3.36). For example, the term $U(\mathbf{r}, \omega) \approx U_0(\mathbf{r}, \omega) + U_1(\mathbf{r}, \omega)$ under integral (3.35) can be replaced by $U_0(\mathbf{r}, \omega)$ and in this case, the approximate solution is given by (Born and Wolf 1999, page 700)

$$U(\mathbf{r},\omega) \approx U_1(\mathbf{r},\omega) \equiv e^{ik_0 s_0 \cdot \mathbf{r}} + \int_V F(\mathbf{r}',\omega) e^{ik_0 s_0 \cdot \mathbf{r}'} \frac{e^{ik_0|\mathbf{r}-\mathbf{r}'|}}{|\mathbf{r}-\mathbf{r}'|} d^3\mathbf{r}'. \tag{3.37}$$

The expression (3.37) is generally referred to as the first-order Born approximation; it describes *single scattering*. The first-order Born approximation is valid whenever the wave function $U(\mathbf{r}, \omega)$ is only slightly different from the incident plane wave $U_0(\mathbf{r}, \omega)$.

Higher order Born approximations also known as Born series are given by the recurrence relation

$$U_{n+1}(\mathbf{r},\omega) = U_0(\mathbf{r},\omega) + \int_V U_n(\mathbf{r}',\omega) F(\mathbf{r}',\omega) G(\mathbf{r}-\mathbf{r}') d^3\mathbf{r}', \tag{3.38}$$

$$G(\mathbf{r}-\mathbf{r}') = \frac{e^{ik_0|\mathbf{r}-\mathbf{r}'|}}{|\mathbf{r}-\mathbf{r}'|}, \tag{3.39}$$

where $U_0(\mathbf{r}, \omega) = e^{ik_0 s_0 \cdot \mathbf{r}}$. Equations (3.38)–(3.39) describe *multiple scattering*; the higher order extension process is also formulated by (Born and Wolf 1999). The Born approximation is the most useful method for calculating the scattering amplitude of a plane wave. The first-order Born approximation for scattering from a 3D isotropic homogeneous elastic medium of different properties is given by Equation (3.37).

3.7.3 RYTOV APPROXIMATION

The Rytov approximation (published first by Soviet radio-physicist S. M. Rytov in 1937 in Russian) or method of smooth perturbations provides a significant improvement both on geometrical optics approximation and Born approximation. This method takes into account weak scattering by assuming the heterogeneity perturbs the phase of the scattered wavefield. The Rytov approximation considers that the perturbation is multiplicative to the unperturbed field while the Born approximation considers it as additive.

In the Rytov approximation, the total wavefield in the Equation (3.29) is represented as $U = \exp(\psi)$, where the quantity ψ is the complex phase perturbation caused by a random medium with the complex refractive index $n(\mathbf{r}) = 1 + n_1(\mathbf{r})$, where $n_1(\mathbf{r})$ is its fluctuation part. The corresponding wavenumber is $k(\mathbf{r}) = k_0 n(\mathbf{r}) = k_0[1 + n_1(\mathbf{r})]$. Setting $U = \exp(\psi)$ in Equation (3.29) and expressing ψ as series result to

$$\nabla^2 \psi + \left(\nabla \psi\right)^2 + k_0^2 \left[1 + n_1\left(\mathbf{r}\right)\right]^2 = 0, \tag{3.40}$$

$$\psi = \sum_{m=0}^{\infty} \psi_m. \tag{3.41}$$

It is shown (Yura et al. 1983) that phases ψ_m satisfy the following transport equations:

$$\nabla^2 \psi_0 + \left(\nabla \psi_0\right)^2 + k_0^2 = 0, \tag{3.42}$$

$$\nabla^2 \psi_1 + 2\nabla \psi_0 \cdot \nabla \psi_1 + 2k_0^2 n_1\left(\mathbf{r}\right) = 0, \tag{3.43}$$

$$\nabla^2 \psi_2 + 2\nabla \psi_0 \cdot \nabla \psi_2 + k_0^2 n_1^{\,2}\left(\mathbf{r}\right) + \left(\nabla \psi_1\right)^2 = 0, \tag{3.44}$$

$$\text{---} \tag{3.45}$$

$$\nabla^2 \psi_m + 2\nabla \psi_0 \cdot \nabla \psi_m + \sum_{p=0}^{m-1} \nabla \psi_p \cdot \nabla \psi_{m-p} = 0, \qquad m = 3, 4, 5, \ldots. \tag{3.46}$$

Equations (3.42)–(3.46) are also known as the Rytov expansions or Rytov series. The solution for the first-order Rytov approximation is given by

$$\psi_1\left(\mathbf{r}, \omega\right) = -\frac{2k_0^2}{U_0\left(\mathbf{r}\right)} \int_V n_1\left(\mathbf{r}', \omega\right) G\left(\mathbf{r} - \mathbf{r}'\right) U_0\left(\mathbf{r}', \omega\right) d^3 \mathbf{r}', \tag{3.47}$$

$$G\left(\mathbf{r} - \mathbf{r}'\right) = -\frac{1}{4\pi} \left[\frac{\exp\left(ik_0 \left|\mathbf{r} - \mathbf{r}'\right|\right)}{\left|\mathbf{r} - \mathbf{r}'\right|}\right], \tag{3.48}$$

where $U_0 = \exp(\psi_0)$ and $G(\mathbf{r} - \mathbf{r}')$ is the Green function. The first-order Rytov approximation ψ_1 limits its scope of application by weak scattering assumption, and therefore, many authors have examined higher order Rytov series in order to improve the phase accuracy of the forward scattered waves (e.g., Wheelon 2003).

The Rytov approximation has been widely used for long-distance propagation of EM waves with forward scattering or small-angle scattering involved. Phase-change accumulation can be described by the Rytov transformation as well. At present, the Rytov approximation is one of the basic tools in wave propagation science and engineering including diffraction tomography, optical and microwave imaging, laser communication technology, seismology, and remote sensing.

There are many attempts to compare the Born approximation and the Rytov approximation for different cases and applications. As many assumed, the first-order Rytov approximation is best suited for modeling the transmitted or forward scattered field, whereas the first-order Born approximation is best suited for modeling the reflected or backscattered field. Numerical examples available from the literature demonstrate that the Rytov series ($m \geq 1$) give superior performance by comparison with the Born series for most practical cases especially related to long path propagation and inverse problems of scattering as well.

Note that both Born approximation (3.38) and Rytov approximation (3.47) consider only single scattering; they become identical in far field and correspond to weak turbulence theory. The range of their applicability is limited and both techniques break down at strong fluctuations. Strong fluctuation theory has evolved from several different approaches, such as the parabolic equation method and the extended Huygens–Fresnel principle (see below).

3.7.4 PARABOLIC WAVE EQUATION

Parabolic equation method provides a powerful modeling capability for propagation of EM waves over long distances in complex environments. The parabolic approximation was introduced first by Soviet radio-physicists Leontovich and Fock in the 1940s to treat the problem of diffraction of radiowaves around the Earth (Fock 1965). Since that time, a lot of literature has been devoted to investigation and application of the parabolic wave equation in various areas of applied physics.

The standard parabolic equation is derived from the 2D Helmholtz equation by separating the rapidly varying phase term to obtain an amplitude factor, which varies slowly in long range when the direction of propagation is predominantly along the z-axis (i.e., paraxial direction). In common case, the parabolic equation can be obtained with the substitution $U(\mathbf{r}, z) = u(\mathbf{r}, z)e^{iK_0 z}$ into the Equation (3.29), where $u(\mathbf{r}, z)$ is the slowly varying field envelope. Thus, the 2D scalar parabolic equation for the field envelope of the wave propagating in the positive direction of the z-axis is given by (e.g., Rytov et al. 1989; Ishimaru 1991)

$$2\iota k_0 \frac{\partial u(\mathbf{r},z)}{\partial z} + \nabla_T^2 u(\mathbf{r},z) + 2k_0^2 n_1(\mathbf{r},z) u(\mathbf{r},z) = 0, \qquad (3.49)$$

where $n^2(\mathbf{r}) \approx 1 + 2n_1(\mathbf{r})$ is the index of refraction and n_1 is its deviation from unit, $\nabla_T^2 = \dfrac{\partial^2}{\partial x^2} + \dfrac{\partial^2}{\partial y^2}$ is a transverse Laplacian operator, and $\mathbf{r} = \{x, y\}$ is the position vector. Equation (3.49) is the parabolic approximation to the wave equation (3.29). The small variations of the index of refraction, $n_1/n \ll 1$, and the absence of absorption, suggest that the effect from turbulence can be described by phase approximations. The complex envelope $u(\mathbf{r}, z)$ of the electric field $E(\mathbf{r}, z, t)$ in free space is introduced by $E(\mathbf{r}, z, t) = e^{\iota(k_0 z - \omega t)} u(\mathbf{r}, z)$.

The parabolic wave Equation (3.49) also known as forward scattering wave equation preserves the narrow-angle wave (e.g., laser beam) propagation for scattering. Parabolic equation for stochastic field was investigated by Tatarskii (1969, 1971) and can be solved analytically in the case of first-order and second-order field moments; last one is called the *mutual coherence function* (MCF) defined as $\Gamma_2(\mathbf{r}_1, \mathbf{r}_2, z) = \langle u(\mathbf{r}_1, z) u^*(\mathbf{r}_2, z) \rangle$, where $\mathbf{r}_1$ and $\mathbf{r}_2$ denote two points in the transverse plane at propagation distance z and brackets $\langle \cdots \rangle$ denote an ensemble average. The MCF is the statistically most significant quantity for applications and experiments that is considered in strong fluctuation theory for the cases of plane and spherical wave propagation and also for the more general beam wave case.

The parabolic Equation (3.49) for the MCF is given by

$$
\begin{aligned}
& 2\iota k_0 \frac{\partial \Gamma_2(\mathbf{r}_1, \mathbf{r}_2, z)}{\partial z} + \left(\nabla_{T1}^2 - \nabla_{T2}^2 \right) \Gamma_2(\mathbf{r}_1, \mathbf{r}_2, z) \\
& + 2k_0^2 \left[n_1(\mathbf{r}_1, z) - n_2(\mathbf{r}_2, z) \right] u(\mathbf{r}_1, z) u^*(\mathbf{r}_2, z) = 0,
\end{aligned}
\tag{3.50}
$$

where ∇_{T1}^2 and ∇_{T2}^2 are Laplacian operators in the variables $\mathbf{r}_1$ and $\mathbf{r}_2$, respectively. The Equation (3.50) can be solved exactly only for special cases of a plane wave or spherical wave (Tatarskii 1971; Ishimaru 1991). The resulting solutions for the MCF are available in the asymptotic regimes of weak turbulence and strong turbulence. The alternative theory combines the parabolic equation and the Markov approximation for the description of propagating wave field (Tatarskii 1969).

Nowadays, parabolic equation method is widely used in geosciences and remote sensing including radio, microwave, and optical wave propagation in atmosphere and troposphere, ground-based propagation over rough surfaces, terrains, obstacles, oversea propagation, and propagation through ducting environment. Books (Levy 2000; Apaydin and Sevgi 2017; Collins and Siegmann 2019) cover different aspects of parabolic equation methods, solutions, and applications. Parabolic equation method is a basic tool in acoustics for modeling of long-range sound propagation in inhomogeneous media (e.g., in the ocean and the atmosphere).

3.7.5 EXTENDED HUYGENS–FRESNEL PRINCIPLE

Extended Huygens–Fresnel principle is a popular technique used in analytical studying optical beam propagation in turbulence (see reviews: Gbur 2014; Charnotskii 2015). This method considers the turbulence as a phase distortion of the spherical wave in the standard Huygens–Fresnel principle, which is well known from the theory of diffraction (Born and Wolf 1999). This principle assumes that the field at any plane can be expressed by a superposition of the secondary spherical wave (called *Huygens wavelet*) with one another, originating from each point in the first plane.

According to the extended Huygens–Fresnel principle, the resulting monochromatic field $U(\mathbf{r}, L, \omega)$ in the Equation (3.29) after the propagation a distance $z = L$ is given byc

$$U(\mathbf{r},L,\omega) = -\frac{\imath k_0}{2\pi L}\exp(\imath k_0 L)\iint U_0(\boldsymbol{\rho},\omega)\exp\left[\frac{\imath k_0\,|\boldsymbol{\rho}-\mathbf{r}|^2}{2L} + \psi(\mathbf{r},\boldsymbol{\rho},L)\right]d^2\rho, \quad (3.51)$$

where $U_0(\boldsymbol{\rho},\omega)$ is the field in the source plane at $z = 0$, $\boldsymbol{\rho}$ and $\mathbf{r}$ are the transverse coordinates in the source and observation planes, respectively, and $\Psi(\mathbf{r},\boldsymbol{\rho},L)$ is the phase distortion of the Huygens wavelet due to turbulence, i.e., the complex phase perturbation caused by the random distribution of the index of refraction in the medium. Equation (3.51) is known also as extended Huygens–Fresnel integral.

It has been shown that the extended Huygens–Fresnel principle is applicable through first-order and second-order field moments in both regimes of weak or strong turbulence. In fact, it has been established that, up to second-order field moments, the parabolic equation method and the extended Huygens–Fresnel principle yield the same results (Andrews and Phillips 2005). There is the electromagnetic version of the extended Huygens–Fresnel principle and, hence, all the second-order field properties such as coherence and polarization can be predicted (Korotkova 2014). The extended Huygens–Fresnel principle method is widely used in optical coherence tomography for imaging in highly scattering (strongly turbulent) media; it is also a mathematical approach in the theory of scintillations.

3.8 SCINTILLATION

Scintillation – a flash or sparkle of light – refers to temporal or spatial modulations or fluctuations of the wave field, propagating thought a highly varied nonuniform medium. A familiar example is the twinkling of stars in the night sky which is caused by turbulence in the Earth's atmosphere and ionosphere. In the case of optical turbulence, the atmosphere exhibits random fluctuations in the index of refraction resulting to irradiance intensity fluctuations, or scintillation. In the case of propagating radiowaves or microwaves (e.g., in satellite communication technology), scintillation occurs mostly due to refractive scattering and is characterized by fluctuations of EM wave amplitude, phase, propagation direction, and polarization.

Scintillation is considered one from the most critical atmospheric effect that ultimately determines the performance and limitation of an optical system (Andrews et al. 2001). The scintillation effect in radar (SAR) imagery and target tracking is known as well (Skolnik 2008). Moreover, scintillation can affect the quality of the Global Navigation Satellite System (e.g., GPS/GNSS or GLONASS) signals, degrading the performance and position accuracy.

Electromagnetic scintillation has been studied since 1950s with discovery of pulsars and *interplanetary* medium using radio telescopes. Many areas of natural science – atmospheric physics, geophysics, ionospheric physics, ocean acoustics, astronomy and interplanetary physics, laser physics, and radiophysics – have benefitted during the scintillation research. Authors of the books (Tatarskii et al. 1993; Andrews et al. 2001; Wheelon 2001, 2003; Rino 2011) made significant contributions to this subject. In particular, these and many other publications demonstrate great attractiveness of scintillation phenomenology in turbulence detection using remote sensing methods, at least in statistical sense; therefore, in this section, we will outline some aspects of the scintillation theory and important results.

The theory of optical scintillation is based on the second-order and fourth-order field moments known, respectively, as the mutual coherence function (MCF) and fourth-order cross-coherence function. Fluctuations of the wave field intensity after passing through a turbulent medium are described by the *scintillation index* (normalized variance of irradiance fluctuations) given by

$$\text{SI}^2 = \sigma_I^2 = \frac{\langle I^2 \rangle - \langle I \rangle^2}{\langle I \rangle^2} = \frac{\langle I^2 \rangle}{\langle I \rangle^2} - 1, \quad \text{or} \quad \sigma_I^2 = \exp\left(4\sigma_\chi^2\right) - 1 \approx 4\sigma_\chi^2,$$
$$4\sigma_\chi^2 \ll 1, \quad \left[\text{unitless}\right] \tag{3.52}$$

where $I = |u(\mathbf{r})|^2$ is the irradiance intensity of the EM wave (e.g., the solution of the Helmholtz equation), the brackets $\langle \ldots \rangle$ denote an ensemble average (by time or space), and σ_χ^2 is the Rytov log-amplitude variance. According to (3.52), the scintillation index SI^2 is standard statistical measure of electromagnetic signal that is equivalent to the rms (root mean squared) intensity fluctuations (sometimes the scintillation index is denoted by the symbol $S_4 = \sqrt{\text{SI}^2}$ according to the fourth-order statistical moment of the wave field).

Based on the value of σ_I^2, fluctuations in the irradiance intensity can be identified as weak or strong that correspond to weak or strong turbulence. Relatively speaking, the weak fluctuation regime occurs at $\sigma_I^2 < 1$, the moderate-to-strong fluctuation regime is associated with $\sigma_I^2 \geq 1$, while there is the saturation regime when $\sigma_I^2 \to \infty$. For explanation of these criteria, we refer to the quotation by authors of the book (Andrews et al. 2001, page xi, Preface):

In most applications the irradiance fluctuations usually range from weak fluctuations (generally associated with shorter path lengths) up to the point where scintillation attains its peak values, the so-called focusing regime. Because of the potential for large values of scintillation, the focusing regime can be considered the most hostile to optical systems; unfortunately, the rigorously developed scintillation theory is not applicable in this regime.

Applications that involve very long propagation paths may cause the irradiance fluctuations to extend beyond the focusing regime into the saturation regime where scintillation begins to decrease toward a limiting value of unity. Asymptotic expressions (including inner scale effects) have been derived for the saturation regime, but recent findings show that these results do not compare well with either experimental or simulation data.

The corresponding theories have been developed in order to define the scintillation index for different fluctuation regimes and turbulent media (see, e.g., books Tatarskii et al. 1993; Rino 2011). Here is a brief description of the existing approaches.

3.8.1 THE RYTOV APPROXIMATION – WEAK TURBULENCE

The Rytov approximation describes log-amplitude and phase fluctuations of a wave, propagating through weak uniform turbulence. This approximation is based on the Rytov solution of the wave equation keeping the first two terms (3.42) and (3.43) in the expansion (3.41). This approach is also known as the weak fluctuation (or turbulence) theory. As a result, it was found theoretical formula for the irradiance variance, given by (Andrews et al. 2001)

$$\sigma_R^2(L) = 8\pi^2 k_0^2 L \int_0^1 \int_0^\infty k\Phi_n(k)\left[1 - \cos\left(\frac{Lk^2}{k_0}\xi\right)\right]dkd\xi \tag{3.53}$$

for an unbounded plane wave,

$$\beta_R^2(L) = 8\pi^2 k_0^2 L \int_0^1 \int_0^\infty k\Phi_n(k)\left\{1 - \cos\left[\frac{Lk^2}{k_0}\xi(1-\xi)\right]\right\}dkd\xi \tag{3.54}$$

for a spherical plane wave.

In the case of the Kolmogorov spectrum of refractive index fluctuations (3.28), relationships (3.53) and (3.54) lead to fundamental *Rytov variance*

$$\sigma_R^2 = 1.23C_n^2 k^{7/6}L^{11/6}, \quad \sigma_R^2 < 1 \quad \text{for an unbounded plane wave,} \tag{3.55}$$

$$\beta_R^2 = 0.4\sigma_R^2 = 0.5C_n^2 k^{7/6}L^{11/6}, \quad \beta_R^2 < 1, \quad \text{for a spherical plane wave,} \tag{3.56}$$

where $C_n^2\left(m^{-2/3}\right)$ is the index of refraction structure parameter, $k_0 = \frac{2\pi}{\lambda}$ and $k\,(m^{-1})$ are the electromagnetic and spatial wavenumbers, respectively, and $L\,(m)$ is the propagation path length between transmitter and receiver. The Rytov variance (3.55)–(3.56) represents the scintillation index for a weak fluctuation regime ($\sigma_R^2 < 1$ and $\beta_R^2 < 1$). However, experimental studies published in USSR journal in 1965 by (Gracheva and Gurvich 1966) and further works of a group of scientists from the A. M. Obukhov Institute of Atmospheric Physics, Moscow, have demonstrated the limitations of the weak fluctuation theory due to discovering so-called *scintillation saturation* phenomenon. It was shown later that neither the Born approximation nor

Rytov approximation can explain the scintillation saturation effect (see, e.g., Strohbehn 1978).

3.8.2 STRONG TURBULENCE

As mentioned above, the weak turbulence theory is valid only at $\sigma_R^2 < 1$ and/or $\beta_R^2 < 1$. When this limit is exceeded, the fluctuations are considered "strong" ($\sigma_R^2 > 1$ and/or $\beta_R^2 > 1$). In this case, the Rytov and Markov-based approaches are unable to explain the saturation of scintillation. Since the mid 1970s, several methods have been proposed to deal with strong turbulence problem (see, e.g., Strohbehn 1978). A number of scintillation-index models have been developed as well (Andrews et al. 2001; Rino 2011). Theory for strong fluctuations has also produced applicable results for certain second-order field statistics.

Today, strong turbulence theory of wave propagation is an efficient analytical tool in satellite communications, navigation, acoustics, and remote sensing. The most known techniques are based on 1) parabolic equation, 2) the extended Huygens–Fresnel principle, 3) Gaussian phase screen, 4) multiple-scattering theory, 5) Feynman path integral, and 6) heuristic model developed by (Hill and Clifford 1981). Although these methods and models provide significant improvements and advantages in the existing phenomenology, the theory of EM wave propagation under strong fluctuation conditions is not complete yet. For example, observations of interstellar and interplanetary scintillation (e.g., Strom et al. 2001) supposedly demonstrate that the division of environmental turbulence and/or scintillation into "weak," "moderate-to-strong," and "strong" categories may not be reliably balanced and/or relevant. The same situation may happen in remote sensing of the Earth environment as well.

The behavior of the scintillation index in the saturation regime can be predicted using so-called *asymptotic theory*. In this theory, the scintillation index for an unbounded plane wave or spherical wave can be expressed as (Andrews et al. 2001)

$$SI^2 = \sigma_I^2(L) = 1 + 32\pi^2 k_0^2 L \int_0^1 \int_0^\infty k\Phi_n(k) \sin^2\left[\frac{Lk^2}{2k_0} w(\xi,\xi)\right]$$

$$\times \exp\left\{-\int_0^1 D_S\left[\frac{Lk}{k_0} w(\tau,\xi)\right] d\tau\right\} dk d\xi, \tag{3.57}$$

$$w(\tau,\xi) = \begin{cases} \tau(1-\varepsilon\xi), & \tau < \xi, \\ \xi(1-\varepsilon\xi), & \tau > \xi, \end{cases} \tag{3.58}$$

where τ is a normalized distance variable, $D_S(\rho)$ is the structure function of the wave phase, the exponential function represents a low-pass spatial filter, and the parameter in (3.58) $\varepsilon = 0$ for a plane wave and $\varepsilon = 1$ for a spherical wave. Computations of (3.57)–(3.58) with the Kolmogorov spectrum $\Phi_n(k) = 0.033C_n^2 k^{-11/3}$ and $D_S(\rho) = 2.91C_n^2 k^2 L\rho^{5/3}$ lead to asymptotical formulas which account for saturation in the weak-to-strong fluctuation regime (Andrews et al. 2000; Andrews and Phillips 2005)

$$\sigma_{I,pl}^2 = \exp\left[\frac{0.49\sigma_R^2}{\left(1+1.11\sigma_R^{12/5}\right)^{7/6}} + \frac{0.51\sigma_R^2}{\left(1+0.69\sigma_R^{12/5}\right)^{5/6}}\right] - 1, \quad 0 \leq \sigma_R^2 < \infty \qquad (3.59)$$

for plane wave,

$$\sigma_{I,sph}^2 = \exp\left[\frac{0.49\beta_R^2}{\left(1+0.56\beta_R^{12/5}\right)^{7/6}} + \frac{0.51\beta_R^2}{\left(1+0.69\beta_R^{12/5}\right)^{5/6}}\right] - 1, \quad 0 \leq \beta_R^2 < \infty \qquad (3.60)$$

for spherical wave.

Formulas for the on-axis scintillation index (3.55)–(3.56) and (3.59)–(3.60) are obtained under the assumption that C_n^2 is constant. Propagation along a vertical path or slant path requires parameter $C_n^2(h)$ as a function of altitude h. Among known empirical models of $C_n^2(h)$, the most popular is the Hufnagel –Valley model given by

$$C_n^2(h) = 0.00594(v/27)^2 (10^{-5}h)^{10} \exp(-h/1000) + 2.7 \qquad (3.61)$$
$$\times 10^{-16} \exp(-h/1500) + A_0 \exp(-h/100),$$

where h is the altitude in (m), v is the rms wind speed (pseudowind) in (m/s), and $A_0 = C_n^2(0)$ is a ground-level value of C_n^2 in (m$^{-2/3}$). At wind speed v = 21 m/s it sets as $A_0 = 1.7 \times 10^{-14}$ m$^{-2/3}$. The value of $C_n^2(0)$ depends on environmental parameters – air temperature, pressure, and density along the altitude and usually is determined from experiments (that is not a simple task). With increased altitude h, the value of $C_n^2(0)$ decreases by a few orders of magnitude, up to an altitude ~ 10 km. The Hufnagel–Valley model (3.61) is also known as HV5/7 because it corresponds to an atmospheric coherence length of 5 cm and an isoplanatic angle of 7 μrad at the electromagnetic wavelength of 500 nm. The HV5/7 model is generally used to describe $C_n^2(h)$ during daytime.

Under weak turbulence, the normalized irradiance variance (the Rytov variance), integrated along a slant path z, is expressed as

$$\sigma_{RI}^2(L) = 2.25 \times k^{7/6} L^{5/6} \sec^{11/6}(\zeta) \int_0^L C_n^2(z)(L-z)^{5/6} dz, \qquad \text{for plane wave (3.62)}$$

$$\beta_{RI}^2(L) = 2.25 \times k^{7/6} L^{5/6} \sec^{11/6}(\zeta) \int_0^L C_n^2(z)(z/L)^{5/6}(L-z)^{5/6} dz, \qquad (3.63)$$

for spherical wave

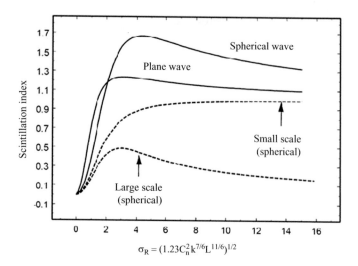

$$\sigma_R = (1.23 C_n^2 k^{7/6} L^{11/6})^{1/2}$$

FIGURE 3.5 The scintillation index of optical beam propagating in turbulent atmosphere versus the Rytov variance. The solid curves depict the scintillation index for the plane wave and spherical wave models in the absence of inner-scale effects. The dashed curves are the large-scale and small-scale scintillation for the spherical wave model. (Adapted from Andrews et al. 1999, 2001).

where ζ is the zenith angle. Note that under weak fluctuations, the scintillation index of an uplink spherical wave is the same as a downlink plane wave. Typical scintillation index is shown in Figure 3.5.

A scintillation model, developed by (Andrews et al. 2000) for uplink-downlink on-axes scintillation index, is applicable in moderate to strong fluctuation conditions that may arise under large zenith angles between transmitter and receiver. This model is used for optical communications but it can be adapted for remote sensing purposes as well. More information about scintillation theory and computations can be found in book (Andrews et al. 2001) and in the paper (Charnotskii 1994).

3.8.3 Aperture Averaging Factor

Scintillation Equations (3.55)–(3.60) correspond to a point receiver. If the receiver has a non-zero aperture diameter, on-axis intensity fluctuations will be the spatially averaged over the entire beam path. This effect, known as *aperture averaging*, has been investigated theoretically and experimentally by many researchers (see, e.g., book Andrews et al. 2001).

The parameter that is used to quantify the fading reduction by aperture averaging is called the *aperture averaging factor*, defined as the ratio of the normalized variance of the signal fluctuations (irradiance scintillation index) from a receiver with aperture diameter D to that from a receiver with a point receiver. The expression for the aperture averaging factor A is given by (Fried 1967)

$$A = \frac{\sigma_I^2(D)}{\sigma_I^2(0)}, \quad \text{and} \quad A = \frac{16}{\pi D^2} \int_0^D b_I(\rho) \left[\cos^{-1}\left(\frac{\rho}{D}\right) - \frac{\rho}{D}\left(1 - \frac{\rho^2}{D^2}\right)^{1/2} \right] \cdot \rho d\rho, \quad (3.64)$$

where $\sigma_I^2(D)$ and $\sigma_I^2(0)$ denote the scintillation index for a receiver with aperture diameter D and a "point receiver" with D = 0, respectively. In (3.64), ρ is the separation distance between two points of a wavefront transverse to the axis of propagation, and $b_I(\rho)$ is the normalized covariance function of irradiance fluctuations for a plane wave, defined through the second moment of the irradiance intensity Γ_4 and the mutual coherence function Γ_2. Under weak fluctuation regime with the Kolmogorov spectrum, the factor A can be approximated as (Andrews et al. 2001):

$$A = \left[1 + 1.06\left(\frac{kD^2}{4L}\right) \right]^{7/6} \quad \text{for plane wave,} \quad (3.65)$$

$$A = \left[1 + 1.33\left(\frac{kD^2}{4L}\right)^{5/6} \right]^{7/6} \quad \text{for spherical wave.} \quad (3.66)$$

The averaging effect pronounces when the diameter of the aperture is larger than the Fresnel zone, i.e., $D > \sqrt{\lambda L}$. The use of aperture averaging for reducing the effect of scintillation has been widely considered in the literature (see, e.g., Andrews and Phillips 2005). Substantial scintillation reduction can be obtained especially in the cases of strong turbulence. For example, accurate calculations (Churnside 1991) with the "bump" spectral model of turbulence have shown that the aperture averaging can reduce the value of the scintillation index considerably – within a factor of 2 of the measurements.

3.9 IMAGING THROUGH TURBULENCE

The distortion effect of atmospheric turbulence on optical imaging systems and image quality has been recognized since the 1950s. This phenomenon is well known from astronomy and observations of celestial objects through optical telescopes. Atmospheric turbulence results in reduction of spatial resolution of optical images. Turbulence degrades the images by inducing geometric distortion, defocus, warping, and blurring. This degradation process arises under the influence of many atmospheric factors but the most important impact on the imagery produces fluctuations of the index of refraction that leads to the distortion of the propagating wavefront. Figure 3.6 illustrates this effect.

An image formation is described by a linear convolution

$$I(x,y,t) = \iint O(x',y',t) P(x'-x, y'-y, t) dx'dy' + \eta$$
$$= O(x,y,t) \otimes P(x,y,t) + \eta, \quad (3.67)$$

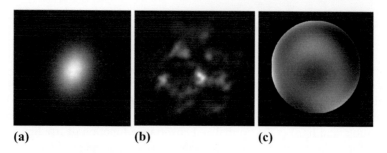

(a)   **(b)**   **(c)**

FIGURE 3.6 Example of laser beam distortion by atmospheric turbulence. (a) Original image, (b) degraded image, and (c) wavefront distortion.

where $I(x, y, t)$ is the observed image, $O(x, y, t)$ is the true or ideal image, $P(x, y, t)$ is the Point Spread Function (PSF), η denotes additive noise, and symbol "$\otimes$" denotes two-dimensional convolution. The problem of recovering the true image $O(x, y, t)$ from the given observed degraded image $I(x, y, t)$ is called *image restoration* in signal/image processing literature. In general, restoration methods require complete knowledge of the spatiotemporal PSF, which involves both instrument and medium parameters called also statistical or turbulent PSF. Analytic description of turbulent PSF can be found in paper (Charnotskii 2012). Eventually, the narrower the PSF, the less degradation occurs in the image-forming process; however, strong variations in turbulent PSF make the image restoration process difficult.

The performance of an optical system is characterized by the *modulation transfer function* (MTF) known from statistical optics (Goodman 2015). MTF is one of the best tools available to quantify the overall image formation and properties in terms of resolution and contrast. A total MTF of an optical system (including remote sensing electro-optical systems as well) is a multiplication of subsystem transfer functions associated with different acting factors, i.e., $\text{MTF}_{\text{total}} = \prod_{i=1}^{N} \text{MTF}_i$, where MTF_i is the MTF of the i-th factor and N is their number (Boreman 2001). Usually, MTF is a function of EM frequency.

The fundamental quantities for characterizing the performance of an optical imaging system in the presence of atmospheric-turbulence effects are 1) the structure constant, C_n^2, 2) the turbulence point spread function (PSF), 3) the optical transfer function (OTF) which is the Fourier transform of the normalized PSF, and 4) the turbulence MTF which is a measure of resolution in frequency domain limited by the Nyquist frequency. All these quantities are defined using both the system parameters and turbulence models (Boreman 2001; Blaunstein and Kopeika 2018).

In the spatial frequency domain, Equation (3.67) can be presented in terms of MTF using Fourier transform (denoted by tilde, noise term is omitted)

$$I = O(x, y, t) \otimes P(x, y, t) \rightarrow \tilde{F}(I) = \tilde{F}(O) \otimes \tilde{F}(PSF), \qquad (3.68)$$

$$OTF = \tilde{F}(PSF),$$
(3.69)

$$\tilde{F}(I) = \tilde{F}(O) \otimes OTF \quad \rightarrow \quad MTF = |OTF| \quad \text{and} \quad \tilde{F}(I) = \tilde{F}(O) \otimes MTF. \quad (3.70)$$

Under certain conditions, the total "optical system + atmospheric turbulence + aerosol" MTF_{total} can be written as the product of three components:

$$MTF_{total}(k) = MTF_{opt}(k) \times MTF_{trb}(k) \times MTF_{ars}(k),$$
(3.71)

$$MTF_{opt}(k) = \frac{2}{\pi}\left[\cos^{-1}(k/k_c) - (k/k_c)\sqrt{1-(k/k_c)^2}\right],$$
(3.72)
$$k/k_c < 1 \quad (\text{otherwise it is } 0),$$

$$MTF_{trb}(k) = \exp\left\{-3.44(\lambda f k / r_0)^{5/3}\left[1 - a(\lambda f k / D)^{1/3}\right]\right\},$$
(3.73)

$$MTF_{asl}(k) \approx \exp(-\tau_{sct}) + \left[1 - \exp(-\tau_{sct})\right]\exp\left[-(k/k_c)^2\right].$$
(3.74)

Terms $MTF_{opt}(k)$, $MTF_{trb}(k)$, and $MTF_{asl}(k)$ refer to a perfect diffraction limited optical imaging system (index "opt"), atmospheric turbulence (index "trb"), and atmospheric aerosol (index "asl"). In more detail, they are the following:

- $MTF_{opt}(k)$, Equation (3.72) is a function of the spatial frequency k (cycles/mm or line/mm), optical cut-off frequency $k_c = 1/(\lambda f\#)$, and the number (f#) describing the ratio between the focal length f of the objective and the diameter D of the entrance pupil, $f\# = f/D$.
- $MTF_{trb}(k)$, Equation (3.73), describes long exposure (a = 0) and short exposure (a = 1) cases. This term is defined by the *Fried parameter* (Fried 1966) which is

$$r_0 = \left[0.423 k_0^2 \sec(\zeta) \int_0^{h_{max}} C_n^2(z)\,dz\right]^{-3/5},$$
(3.75)

where ζ is the angle between the point direction and normal to the ground (i.e., zenith angle), $C_n^2(z)$ is vertical profile of the structure constant (called also atmospheric turbulence strength), h_{max} is maximal altitude, and $k_0 = 2\pi/\lambda$. The Fried parameter or atmospheric coherent length is defined as the diameter of the circular pupil for which the diffraction limited image and the seeing limited image have the same angular resolution. At visible range $r_0 \propto \lambda^{6/5}$, its typical value is 10–20 cm at good seeing conditions.

- $MTF_{asl}(k)$, Equation (3.74), includes the opacity τ_{sct} related to absorption and scattering by aerosol and/or dust particles in the atmosphere. The atmospheric opacity or mass absorption coefficient depends on EM frequency, gaseous concentrations (*greenhouse gas*), and particle properties (e.g., Kokhanovsky 2008).

Optical turbulence causes a variety of deleterious effects relevant to laser (lidar) beam propagation and remote sensing measurements. In particular, the most known optical distortions due to atmospheric turbulence are the following:

- Beam spreading – beam divergence of a plane wave due to diffraction.
- Beam wander and tilt– angular deviation of the centroid of a beam.
- Image dancing – fluctuation in the beam at the focal plane of the receiver.
- Beam scintillation – irradiance fluctuations within the beam cross section.
- Loss of spatial coherence – limits the effective aperture diameter in an imaging system.
- Reduction of spatial resolution and degradation of the image quality.
- Depolarization of light and temporal stretching of optical pulses.
- Astronomical seeing – blurring and twinkling of astronomical objects (stars).
- Distortion of incoming optical wavefront.

The impact of listed effects and phenomena on the optical and radio wave propagation and remote sensing observations can be significant. Thus, the angular resolution of an optical system is severely limited by atmospheric turbulence (aerosol impact is not considered). Instead diffraction-limited resolution $\alpha_{opt} \sim \lambda/D$, it will be $\alpha_{atm} \sim \lambda/r_0$. The angular resolution is limited by the optical system if its aperture $D < r_0$ and by atmospheric turbulence if $D > r_0$. The reduction is by factor D/r_0 that is a criterion of perfect or poor resolution of optical imagery. For example, the IKONOS satellite optical telescope has aperture $D = 0.7$ m that yields factor $D/r_0 \sim 0.7/(0.1 \div 0.2) \sim 3 \div 7$ and thereby atmospheric turbulence causes resolution loss similar to image blur. This becomes an issue in optical detection and recognition of small-scale ocean hydrodynamic features, associated, e.g., with (sub)surface wake signatures (Raizer 2019).

There are several techniques developed over the years in order to eliminate possible atmospheric effects and obtain high-quality image. The most known are 1) short-exposure imagery referred to a speckle imaging, 2) adaptive optics, and 3) hybrid imaging techniques that use elements of adaptive optics and image reconstruction (Roggemann and Welsh 1996). In particular, adaptive optics provide considerable enhancement of the capability of optical systems by reducing the effect of incoming wavefront distortions (Lukin 1995; Lukin and Fortes 2002; Tyson 2016). Adaptive optics at the transmitter and/or the receiver is also used to compensate turbulence effects, reduce scintillation loss, and improve overall quality of turbulence-corrupted imagery that is important for space-based remote sensing observations.

3.10 PROPAGATION IN TIME-VARYING MEDIA

The media whose parameters such as permittivity, conductivity, permeability vary with time are called the *time-varying media*. Wave propagation in time-varying media is the relevant research topic in plasma physics, astrophysics, acoustics, and electromagnetics (Davis 1990; Kalluri 2010; Nerukh et al. 2012). It is based on classical electromagnetic theory (Stratton 1941, Born and Wolf 1999; Jackson 1999), governed by Maxwell's equations for a time-varying field. Time-varying electromagnetic phenomena including reflection and transmission have been investigated theoretically by Ginzburg and Tsytovich (1990).

Real turbulent medium is a moving medium. All properties are changed in time and space randomly. It means that we have to deal with inhomogeneous macroscopic media with time-dependent properties and nonsteady state of EM fields. We may refer this particular topic to *electrodynamics of moving media* (but not a body!) or *nonstationary electrodynamics*. In any case, Maxwell's equations require general solutions in space and time domains.

Interaction of EM waves with moving uniform medium can be investigated using *acoustic-electromagnetic analogy* (e.g., Brekhovskikh and Godin 1999; Carcione 2015; Ostashev and Wilson 2016). This analogy involves many similar phenomena such as Fresnel reflection and transmission, refraction, waves in elastic media, scattering, pulse propagation, frequency conversation, etc.

Time-varying approach is applicable to remote sensing problems as well. Realistic examples include EM wave propagation through a temporal boundary between two dielectric media (when permittivity of one medium changes with time instantly). Well-known examples are propagation in time-modulated medium or a time-varying dielectric slab. In these cases, the Fresnel-type formulae for reflection and transmission in time domain become complicated for analytic analysis and we refer the interested reader to special literature.

3.11 CONCLUSIONS

The goal of this chapter is to give the reader the initial theoretical knowledge on wave propagation in random (or turbulent) media. The material presented provides a physical basis for developments of remote sensing techniques and capabilities for detection of dynamical environment involving turbulence. The overall mathematical approach relies only on fundamental properties of the wave equation, derived from Maxwell's equations but not on a scientific intuitions or phenomenological assumptions.

The applicable theory of wave propagation describes variations of the wave field in a quasi-steady-state manner only. In many applications, fluctuations in the refractive index (or permittivity) are assumed to be small and represented by some stationary random process, such as a Gaussian process, which contribute directly to optical turbulence. The scattering by turbulence is also considered as a result of spatial fluctuations of the medium's refractive index. This standard approach does not count time-varying physical fields which are important sources of geophysical turbulence and nonstationary processes.

The theoretical understanding of the wave propagation effects grew out from the Kolmogorov theory of turbulence that has initiated further development of models and techniques for the description of wave propagation phenomena. The scheme is simple. First, a semi-empirical model (mostly related to atmospheric and oceanic turbulence) of the refractive index is selected in accordance with medium's properties. Second, the wave equation is utilized in some way. Third, the model of the refractive index is incorporated into approximate (or numerical) solution of the wave equation. As a result, both mean and phase characteristics of the propagating through turbulence wave can be investigated for different (may be hypothetic) turbulence scenarios, e.g., for so-called "weak" or "strong" turbulence.

Ultimately, turbulence can deform the wavefront resulting to irradiance (intensity) fluctuations, referred to as scintillation. In the case of laser beam propagation through atmosphere or ocean, other effects such beam spreading, wander, distortion, pulsations, depolarization, etc. occur. All these phenomena are caused by the impact of stochastic spatiotemporal inhomogeneities, which actually is difficult (or impossible) to predict or measure in real-world environment.

From this viewpoint, we may conclude that wave propagation in turbulent media is a tough and very flexible research problem, involving two practically independent concepts: on the one hand, phenomenological statistical theory of turbulence and, on the other hand, rigorous (numerical) or approximate solvers of Maxwell's equations or derived wave equation.

Because wave propagation effects are largely dependent on local properties of turbulent medium; for this reason, classical wave propagation theory, operated mostly with *mean quantities* and correlation functions, may not be able to resolve a variety of complex multiscale turbulent phenomena, occurred (and observed) in real world. Approximate solutions usually provide the treatment of low-dimensional dynamic structures that may not fit properties of natural turbulent flows. Nevertheless, this speculative assumption should be addressed in further study. There is value in pointing-out the obvious here.

3.12 NOTES ON THE LITERATURE

Electromagnetic theory and wave propagation in various media are discussed in books (Papas 1988; Davis 1990, 2000; Ghosh 2002; Jin 2004; Someda 2006; Zhdanov 2018).

Stochastic theory of wave propagation and applications are considered in books (Sobczyk 1985; Klyatskin 2005, 2015).

Essential information on radiowave propagation in geophysical media can be found in books (Kukushkin 2004; Armand and Polyakov 2005; Levis et al. 2010; Saakian 2011).

Problems of EM wave propagation in complex and time-varying media are discussed in books (Davis 1990; Kalluri 2010; Nerukh et al. 2012).

Principles of image formation and image transfer theory are considered in books (Zege et al. 1991; Blahut 2004).

REFERENCES

Andrews, L. C. and Phillips, R. L. 2005. *Laser Beam Propagation through Random Media*, 2nd edition. SPIE Press, Bellingham, WA.
Andrews, L. C., Phillips, R. L., and Hopen, C. Y. 2001. *Laser Beam Scintillations with Applications*, SPIE Press, Bellingham, WA.

Andrews, L., Phillips, R., and Young, C. 2000. Scintillation model for a satellite communication link at large zenith angles. *Optical Engineering*, 39(12):3272–3280. doi:10.1117/1.1327839.

Andrews, L. C., Phillips, R. L., Hopen, C. Y., and Al-Habash, M. A. 1999. Theory of optical scintillation. *Journal of the Optical Society of America A*, 16(6):1417–1429. doi:10.1364/josaa.16.001417.

Apaydin, G. A. and Sevgi, L. 2017. *Radio Wave Propagation and Parabolic Equation Modeling*. IEEE Press – John Wiley & Sons, Hoboken, NJ.

Armand, N. A. and Polyakov, V. M. 2005. *Radio Propagation and Remote Sensing of the Environment*. CRC Press, Boca Raton, FL.

Blahut, R. E. 2004. *Theory of Remote Image Formation*. Cambridge University Press, Cambridge, UK.

Blaunstein, N. and Kopeika, N. (Eds.). 2018. *Optical Waves and Laser Beams in the Irregular Atmosphere*. CRC Press, Boca Raton, FL.

Blaunstein, N., Arnon, S., Zilberman, A., and Kopeika, N. 2010. *Applied Aspects of Optical Communication and LIDAR*. CRC Press, Boka Raton, FL.

Bleistein, N. 1984. *Mathematical Methods for Wave Phenomena (Computer Science & Applied Mathematics)*. Academic Press, Orlando, FL.

Boreman, G. D. 2001. *Modulation Transfer Function in Optical and Electro-optical Systems*. SPIE Press, Bellingham, WA.

Boreman, G. D. and Dainty, C. 1996. Zernike expansions for non-Kolmogorov turbulence. *Journal of the Optical Society of America A*, 13(3):517–522. doi:10.1364/josaa.13.000517.

Born, M. and Wolf, E. 1999. *Principles of Optics*, 7th (expanded) edition. Cambridge University Press, Cambridge, UK.

Brekhovskikh, L. 1980. *Waves in Layered Media*, 2nd edition *(translated from Russian by R. T. Beyer)*. Academic Press, New York.

Brekhovskikh, L. M. and Godin, O. A. 1999. *Acoustics of Layered Media II: Point Sources and Bounded Beams*, 2nd edition. Springer-Verlag, Heidelberg, Berlin.

Carcione, J. M. 2015. *Wave Fields in Real Media: Wave Propagation in Anisotropic, Anelastic, Porous and Electromagnetic Media*, 3rd edition. Elsevier, Amsterdam, The Netherlands.

Charnotskii, M. I. 1994. Asymptotic analysis of finite-beam scintillations in a turbulent medium. *Waves in Random Media*, 4(3):243–273.

Charnotskii, M. 2012. Statistical modeling of the point spread function for imaging through turbulence. *Optical Engineering*, 51(10):101706-1–101706-11. doi:10.1117/1.oe.51.10.101706.

Charnotskii, M. 2015. Extended Huygens–Fresnel principle and optical waves propagation in turbulence: Discussion. *Journal of the Optical Society of America A*, 32(7):1357–1365. doi:10.1364/josaa.32.001357.

Chernov, L. A. 1960. *Wave Propagation in a Random Medium (translated from Russian by R. A. Silverman)*. McGraw-Hill, New York (reprinted by Dover Publications, Mineola, New York,1988).

Churnside, J. H. 1991. Aperture averaging of optical scintillations in the turbulent atmosphere. *Applied Optics*, 30(15):1982–1994. doi:10.1364/ao.30.001982.

Cohen, G. 2002. *Higher-Order Numerical Methods for Transient Wave Equations*. Springer-Verlag, Heidelberg, Berlin.

Collins, M. D. and Siegmann, W. L. 2019. *Parabolic Wave Equations with Applications*. Springer-Science, New York.

Dashen, R., Munk, W. H., Watson, K. M., Zachariasen, F., and Flatté, S. M. (Eds.). 1979. *Sound Transmission through a Fluctuating Ocean (Cambridge Monographs on Mechanics)*. Cambridge University Press, Cambridge, UK.

Davis, J. L. 1990. *Wave Propagation in Electromagnetic Media*. Springer-Verlag, Berlin, Heidelberg.

Davis, J. L. 2000. *Mathematics of Wave Propagation*. Princeton University Press, Princeton, NJ.

Driggers, R. G.(Ed.). 2003. *Encyclopedia of Optical Engineering*. Marcel Dekker, New York.

Elliott, R. S. 1993. *Electromagnetics: History, Theory, and Applications*. IEEE Press Series on Electromagnetic Waves, New York.

Flatté, S. M. 1983. Wave propagation through random media: Contributions from ocean acoustics. *Proceedings of the IEEE*, 71(11):1267–1294. doi:10.1109/proc.1983.12764.

Flatté, S. M. 1979. *Sound Transmission through a Fluctuating Ocean*. Cambridge University Press, Cambridge, UK.

Fock, V. A. 1965. *Electromagnetic Diffraction and Propagation Problems* (translation from Russian). Pergamon Press, Oxford.

Fried, D. L. 1966. Optical resolution through a randomly inhomogeneous medium for very long and very short exposures. *Journal of the Optical Society of America*, 56(10):1372–1379. doi:10.1364/JOSA.56.001372.

Fried, D. L. 1967. Aperture averaging of scintillation. *Journal of the Optical Society of America*, 57(2):169–175. doi:10.1364/josa.57.000169

Gbur, G. 2014. Partially coherent beam propagation in atmospheric turbulence [Invited]. *Journal of the Optical Society of America A*, 31(9):2038–2045. doi:10.1364/josaa.31.002038.

Ghosh, S. N. 2002. *Electromagnetic Theory and Wave Propagation*, 2nd edition. CRC Press, Boca Raton, FL.

Ginzburg, V. L. and Tsytovich, V. N. 1990. *Transition Radiation and Transition Scattering*. Adam Hilger, Bristol, New York (translation from Russian).

Goodman, J. W. 2015. *Statistical Optics*, 2nd edition. John Wiley & Sons, Hoboken, NJ.

Gracheva, M. E. and Gurvich, A. S. 1966. Strong fluctuations in the intensity of light propagated through the atmosphere close to the earth. *Soviet Radiophysics*, 8(4):511–515 (translated from Russian: *Izvestiya Vysshikh Uchebnykh Zavedenii Radiofizika*, 8(4):717–724, 1965). doi:10.1007/bf01038327.

Hemmati, H. (Ed.). 2009. *Near-Earth Laser Communications*. CRC Press, Boca Raton, FL.

Hill, R. J. and Clifford, S. F. 1981. Theory of saturation of optical scintillation by strong turbulence for arbitrary refractive-index spectra. *Journal of the Optical Society of America*, 71(6):675–686. doi:10.1364/josa.71.000675.

Huray, P. G. 2010. *Maxwell's Equations*. IEEE Press – John Wiley & Sons, Hoboken, NJ.

Ishimaru, A. 1991. *Electromagnetic Wave Propagation, Radiation, and Scattering*. Englewood Cliffs, Prentice Hall, NJ.

Jackson, J. D. 1999. *Classical Electrodynamics*, 3rd edition. John Wiley & Sons, New York.

Jin, Ya-Q. (Ed.). 2004. *Wave Propagation: Scattering and Emission in Complex Media*. Science Press – World Scientific, Beijing, Singapore.

Kallistratova, M. A. 2002. Acoustic waves in the turbulent atmosphere: A review. *Journal of Atmospheric and Oceanic Technology*, 19(8):1139–1150. doi: 10.1175/1520-0426(2002) 019<1139:awitta>2.

Kalluri, D. K. 2010. *Electromagnetics of Time Varying Complex Media: Frequency and Polarization Transformer*. CRC Press, Boca Raton, FL.

Kampanis, N. A., Dougalis, V., and Ekaterinaris, J. A. 2019. *Effective Computational Methods for Wave Propagation*. CRC Press, Boca Raton, FL.

Klyatskin, V. I. 2005. *Stochastic Equations through the Eye of the Physicist: Basic Concepts, Exact Results and Asymptotic Approximations*. Elsevier Science, Amsterdam, The Netherlands.

Klyatskin, V. I. 2015. *Stochastic Equations: Theory and Applications in Acoustics, Hydrodynamics, Magnetohydrodynamics, and Radiophysics. Vol. 1: Basic Concepts, Exact Results, and Asymptotic Approximations and Vol. 2: Coherent Phenomena in Stochastic Dynamic Systems (translated from Russian by A. Vinogradov)*. Springer International Publishing, Switzerland.

Kokhanovsky, A. A. 2008. *Aerosol Optics: Light Absorption and Scattering by Particles in the Atmosphere*. Springer, Praxis Publishing, Chichester, UK.

Korotkova, O. 2014. *Random Light Beams: Theory and Applications*. CRC Press, Boca Raton, FL.

Korotkova, O. 2019. Light propagation in a turbulent ocean. *Progress in Optics*, 64:1–43. doi:10.1016/bs.po.2018.09.001.

Kravtsov, Y. A. 2005. *Geometrical Optics in Engineering Physics*. Alpha Science International Ltd., Harrow, UK.

Kravtsov, Yu. A. and Orlov, Yu. I. 1990. *Geometrical Optics of Inhomogeneous Media. (Springer Series on Wave Phenomena vol 6)*. Springer-Verlag, Berlin, Heidelberg.

Kukushkin, A. 2004. *Radio Wave Propagation in the Marine Boundary Layer*. Wiley-VCH, Weinheim, Germany.

Levis, C., Johnson, J. T., and Teixeira, F. L. 2010. *Radiowave Propagation: Physics and Applications*. John Wiley & Sons, Hoboken, NJ.

Levy, M. 2000. *Parabolic Equation Methods for Electromagnetic Wave Propagation*. The Institution of Electrical Engineers, London, UK.

Lipatov, A. S. 2002. *The Hybrid Multiscale Simulation Technology: An Introduction with Application to Astrophysical and Laboratory Plasmas*. Springer-Verlag, Heidelberg, Berlin.

Lukin, V. P. 1995. *Atmospheric Adaptive Optics (translated from Russian by P. F. Schippnick)*. SPIE Press, Bellingham, WA.

Lukin, V. P. and Fortes, B. V. 2002. *Adaptive Beaming and Imaging in the Turbulent Atmosphere (translated from Russian by A. B. Malikova)*. SPIE Press, Bellingham, WA.

McCartney, E. J. 1976. *Optics of the Atmosphere: Scattering by Molecules and Particles*. Wiley, New York.

Nefyodov, E. I. and Smolskiy, S. M. 2019. *Electromagnetic Fields and Waves: Microwave and mmWave Engineering with Generalized Macroscopic Electrodynamics (Textbooks in Telecommunication Engineering)*. Springer International Publishing, Switzerland.

Nerukh, A., Benson, T., Sakhnenko, N., and Sewell, P. 2012. *Non-stationary Electromagnetics*. Pan Stanford Publishing, Singapore – CRC Press, Boca Raton, FL.

Ostashev, V. E. and Wilson, D. K. 2016. *Acoustics in Moving Inhomogeneous Media*, 2nd edition. CRC Press, Boca Raton, FL.

Papas, C. H. 1988. *Theory of Electromagnetic Wave Propagation*. Dover Publications, New York.

Raizer, V. 2017. *Advances in Passive Microwave Remote Sensing of Oceans*. CRC Press, Boca Raton, FL.

Raizer, V. 2019. *Optical Remote Sensing of Ocean Hydrodynamics*. CRC Press, Boca Raton, FL.

Rino, C. L. 2011. *The Theory of Scintillation with Applications in Remote Sensing*. IEEE Press – John Wiley & Sons, Hoboken, NJ.

Roggemann, M. C. and Welsh, B. M. 1996. *Imaging through Turbulence*. CRC Press, Boca Raton, FL.

Rytov, S. M., Kravtsov, Yu. A., and Tatarskii, V. I. 1989. *Principles of Statistical Radiophysics 4: Wave Propagation Through Random Media 4 (translated from Russian by A. P. Repyev)*. Springer-Verlag, Berlin, Heidelberg.

Saakian, A. 2011. *Radio Wave Propagation Fundamentals*. Artech House, Norwood, MA.

Sasiela, R. J. 2007. *Electromagnetic Wave Propagation in Turbulence: Evaluation and Application of Mellin Transforms*, 2nd edition. SPIE Press, Bellingham, WA.

Schmidt, J. D. 2010. *Numerical Simulation of Optical Wave Propagation with Examples in MATLAB*. SPIE Press, Bellingham, WA.

Skolnik, M. I. (Ed.). 2008. *Radar Handbook*, 3nd edition. McGraw-Hill, New York.

Sobczyk, K. 1985. *Stochastic Wave Propagation*. Elsevier, Amsterdam, The Netherlands

Someda, C. G. 2006. *Electromagnetic Waves*, 2nd edition. CRC Press, Boca Raton, FL.

Stratton, J. A. 1941. *Electromagnetic Theory*. McGraw-Hill (reprinted by John Wiley & Sons–IEEE Press, 2007).

Strohbehn, J. W. (Ed.). 1978. *Laser Beam Propagation in the Atmosphere*. Springer-Varlag, Berlin, Heidelberg.

Strom, R., Bo, P., Walker, M., and Rendong, N. (Eds.). 2001. *Sources and Scintillations: Refraction and Scattering in Radio Astronomy IAU Colloquium 182*. Springer Science, Dordrecht.

Tatarskii, V. I. 1961. *Wave Propagation in a Turbulent Medium (translated from Russian by A. Silverman)*. McGraw-Hill, New York.

Tatarskii, V. I. 1969. Light propagation in a medium with random refractive index inhomogeneities in the Markov random process approximation. *Soviet Physics JETP*, 29(6): 1133–1138 (translated from Russian by L. M. Matarrese). Available on the Internet http://www.jetp.ac.ru/cgi-bin/dn/e_029_06_1133.pdf.

Tatarskii, V. I. 1971. *The Effects of the Turbulent Atmosphere on Wave Propagation*. Israel Program for Scientific Translations, Jerusalem (reproduced by National Technical Information Service, Springfield, VA, USA, TT68–50464).

Tatarskii, V. I., Ishimaru, A., and Zavorotny, V. U. (Eds.). 1993.*Wave propagation in random media (Scintillation)*. *Proceedings of the conference held 3—7 August, 1992 at the University of Washington*, Seattle, USA. SPIE, Bellingham, WA and Institute of Physics Publishing, Techno House, Bristol, England.

Towne, D. H. 1967. *Wave Phenomena*. Dover Publication, New York.

Tyson, R. K. 2016. *Principles of Adaptive Optics*, 4th edition. CRC Press, Boca Raton, FL.

Uscinski, B. J. 1977. *The Elements of Wave Propagation in Turbulent Media*. McGraw-Hill, New York.

Wait, J. R. 1970. *Electromagnetic Waves in Stratified Media: Revised Edition Including Supplemented Material*. Pergamon Press, Oxford, NY.

Weichel, H. 1990. *Laser Beam Propagation in the Atmosphere*. SPIE Optical Engineering Press, Bellingham, WA.

Wheelon, A. D. 2001. *Electromagnetic Scintillation: Volume 1, Geometrical Optics*. Cambridge University Press, Cambridge, UK.

Wheelon, A. D. 2003. *Electromagnetic Scintillation: Volume 2, Weak Scattering*. Cambridge University Press, Cambridge, UK.

Yura, H. T., Sung, C. C., Clifford, S. F., and Hill, R. J. 1983. Second-order Rytov approximation. *Journal of the Optical Society of America*, 73(4):500–502. doi:10.1364/josa.73.000500.

Zege, E. P., Ivanov, A. P., and Katsev, I. L. 1991. *Image Transfer through a Scattering Medium*. Springer, New York.

Zhdanov, M. S. 2018. *Foundations of Geophysical Electromagnetic Theory and Methods*, 2nd edition. Elsevier, Amsterdam, The Netherlands.

Zuev, V. E. 1982. *Laser Beams in the Atmosphere (translated from Russian by S. Wood)*. Consultants Bureau, New York.

Yao, J.-R., Wang, H.-T., Zhang, H.-J., Cai, J.-D., Ming-Yuan Ren, M.-Y., Zhang, Yu, and Korotkova, O. 2021. Oceanic non-Kolmogorov optical turbulence and spherical wave propagation. *Optics Express*, 29(2):1340–1359. doi:10.1364/OE.409498.

4 Remote Sensing Tool

It is difficult to say what is impossible, for the dream of yesterday is the hope of today and the reality of tomorrow.

Robert H. Goddard

4.1 INTRODUCTION

This chapter is an introductory overview of remote sensing principles and capabilities. Not all, but most of them which, we believe are important to know. We emphasize on satellite remote sensing that is more informative than most of other methods in view of global observations of turbulence. Miscellaneous aspects related to remote sensing will be discussed as well.

Numerous popular textbooks and monographs have been devoted to remote sensing problems (see our note on the literature, Section 4.12); nevertheless, the subject matter deserves more attention especially in context with the title of our book.

Remote sensing is a vast area of natural science and technology which is incredibly attractive, progressive, highly producible, demandable, and well organized across the world. The reason is a great opportunity to explore our environments from the distance – air or space, providing global geophysical monitoring and ecological forecasting that cannot be achieved using *in situ* observations.

Today, remote sensing is the principal tool for studying and predicting environmental changes – in the atmosphere, the oceans, the land surfaces, and the near space of the planet as well. More specifically, remote sensing offers unique opportunity for scientists and researchers to explore the origin, formation, and evolution of catastrophic events and natural disasters such as hurricanes, tsunami, tropical cyclones, earthquakes, storms, and weather and atmospheric anomalies. In such extreme climate events, air and fluid masses exhibit global and local chaotic motions and geophysical flows become violently unstable and practically unpredictable. All these perturbations can be characterized by the simple word "turbulence" (in a broad sense). Many remote sensing instruments and airspace missions have been dedicated to environmental studies; however, the history shows that most of them were not directly implemented to detection of Earth's turbulence.

4.2 A BRIEF MODERN HISTORY

A modern history of remote sensing has begun with photography. Aerial photography became a valuable observational tool during the First World War and came fully into its own during the Second World War. With the advent of Soviet Sputnik in 1957, the possibility of putting film cameras on orbiting spacecraft was realized. Cameras were mounted on meteorological and military satellites to obtaining black and white images of the Earth's surface and selected ground objects/targets as well. Later, the

DOI: 10.1201/9781003217565-4

first cosmonauts and astronauts carried out hand photo and video cameras to document interesting atmospheric phenomena and/or geographical features with opportunity as they circumnavigated the globe. In the early stages (in the 1960s), visual analysis, "false color" filtering, and analog photogrammetric methods were the main options for providing an assessment and interpretation of aerial and space photographic films. Basic civil applications included agriculture, geology, hydrology, and cartography.

Other types of remote sensing instruments became available during the Cold War, with the development of a number of well-resolution *reconnaissance aerial photo cameras*, radars, sonars, and thermal infrared detection systems. A new era of spy satellites and space reconnaissance has begun (Kupperberg 2003). These special satellites have become an integral component of the military planning efforts especially in the USA and Russia providing a variety of services such as military communication services, electronic reconnaissance, missile early warning, nuclear explosion detection, and gathering intelligence optical and radar imagery data.

The most known military missions at that time were US Corona satellites operated in the time period 1960–1972 and Soviet Zenit 2–8 series satellites (Kosmos series) operated from 1961 to 1994. The Corona satellites used special 70 mm film with a 24-inch (610 mm) focal length camera, manufactured by Eastman Kodak. Zenits used the SA-20 camera with a focal length of 1 m and the SA-10 camera with a focal length of 0.2 m (http://www.astronautix.com/z/zenit-2satellite.html). Unlike Corona program, Zenit program has provided the return capsule carried both the film and the cameras; it was an advantage at that time. Among other historic military missions, we can mention series of photographic reconnaissance satellites KH-9 Hexagon carried several film cameras with resolution about 0.6 m (2 ft) and Soviet "Yantar" (the Russian word for "amber") intelligence satellite carried film photo camera and later electro-optical camera with resolution about 2 m.

Since the late 1960s, space satellite technology made a revolution in remote sensing offering data collection across the globe using almost the entire range of electromagnetic spectrum. The first satellite multispectral imagery has been conducted during TIROS, Landsat, and Skylab missions in the 1970s. Significant contribution to remote sensing was made by two USSR satellites "Kosmos-243" (launched in 1968) and "Kosmos-384" (launched in 1970), which carried four passive microwave radiometers at wavelengths 0.8, 1.35, 3.4, and 8.5 cm. Since that time, the era of passive microwave radiometry and imagery has started.

Over several decades, since the late 1970s, a number of radar systems known as synthetic aperture radar (SAR) and airborne side-looking real aperture radar (SLAR) of Ku-, X-, C-, S-, L-, and P-bands have been operated from several (non-military) well-equipped aircraft-laboratories (NOAA and NASA P-3 and DC-8, C-130, Antonov An-24 with SLAR "Toros" and "Nit," Tupolev Tu-134Skh with SLAR) and satellite platforms (Seasat, ERS-1 and 2, Almaz-1 known as Kosmos1870, Radarsat-1). These early radar systems allowed the researchers first-ever obtain sophisticated information about interactions between ocean surface dynamics and deep-ocean wave processes including internal wave manifestations. The most valuable multisensor observations at that time were conducted during the *Joint U.S./*

Russia Internal Wave Remote Sensing Experiment, JUSREX'92 in 1992 (in more detail see book Raizer 2017).

Satellite multiband passive microwave radiometers (SMMR, SSM/I, DMSP, TRMM, WindSat, AMSU, AMSR-2, AQUA series, Aquarius, SMOS, Russian Meteor-M No. 2-2 with MTVZA-GY and the others), optical imagers of high resolution (Spot series, EarlyBird, IKONOS, QuickBird, WorldView, GeoEye, Russian Resurs series and *Kanopus* series), and hyperspectral optical and infrared systems (AVHRR, Terra/MODIS, Sentinel-2) provided systematic observations of the Earth's surfaces, ocean, and atmosphere.

All these and other remote sensing missions have demonstrated great capabilities not only for accurate (digital) imagery and mapping of the Earth's surfaces, but also for the obtaining hydro-meteorological information associated with the ocean-atmosphere interaction and heat and mass transfer. In particular, stochastic features and motions of complex geometry – vortexes, eddies, fronts, jets, wakes, and convective cells, occasionally registered (and visible) in the space optical and radar images can be perceived as direct manifestations of geophysical turbulence (Chapter 5).

How to observe and investigate turbulence from space? This problem was not in serious demand so far; however, an idea of remote sensing of turbulence in the *broadest sense* (that we call "turbulence detection") comes true with the appearance of high-resolution optical imagery. Optical observations have unique advantages over the other (radar, radiometer, lidar) remote sensing methods due to documental character of video information, offering detailed investigation of spatial structure and dynamics of environmental processes and events. For example, our experiences show that detection and recognition of *localized hydrodynamic features* – turbulent flows in the ocean — is quite possible if use sophisticated multispectral (or better *multisensor*) imaging technology (Raizer 2017, 2019).

4.3 DEFINITIONS, PRINCIPLES, AND OBJECTIVES

In natural science, the term "remote sensing" is most commonly associated with the gauging of the Earth's electromagnetic radiation and obtaining geophysical information. The word "remote" means at distance or faraway, and the word "sensing" means intuitive or acquired perception or ability to estimate something. Thus, remote sensing can be defined by several ways. Here are some of them:

> *The term remote sensing means the sensing of the Earth's surface from space by making use of the properties of the electromagnetic wave emitted, reflected or diffracted by the sensed objects, for the purpose of improving natural resource management, land use and the protection of the environment.*

(United Nations 1986, Internet https://undocs.org/pdf?symbol=en/A/RES/41/65)

> *The term remote sensing refers to methods that employ electromagnetic energy, such as light, heat, and radio waves, as the means of detecting and measuring target characteristics.*

(Sabins 2007)

Remote sensing is the science and art of obtaining information about an object, area, or phenomenon through the analysis of data acquired by a device that is not in contact with the object, area, or phenomenon under investigation.

(Lillesand et al. 2015)

Remote sensing is defined as the acquisition of information about an object without being in physical contact with it.

(Elachi and van Zyl 2006)

Remote sensing is a scientific technology of data acquisition, based on fundamental principles of electromagnetic radiation.

(Raizer 2019)

According to these and others definitions, remote sensing involves several key components – collection, assessment, utilization, and application of data of the interest (more detailed specification is shown in Figure 4.1). The observational components include: 1) *Target*: the object or material that is being studied, 2) *Energy source*: natural or artificial (e.g., sun, terrestrial radiation, pulse from radar, laser/lidar), 3) *Sensor*: is the device that measures the reflected and emitted signal from the target, and 4) *Transmission path*: the passage of radiation from a source of energy from the target to the sensor. Thus, we have to employ all these (and many others) components in order to obtain desired results and achieve the goals being sought.

Remote sensing is based on the registration and measurement of the electromagnetic radiance, scattered, reflected, or emitted from the object (target) being studied. Remote sensing is practiced across the whole electromagnetic spectrum (Figure 4.2) from low-frequency radio waves through the microwave, submillimeter, far infrared, near infrared, visible, ultraviolet, x-ray, and gamma-ray regions. However, only two major portions of the electromagnetic spectrum, referred to optical/infrared and microwave ranges, are commonly used at regular monitoring of the Earth's environment (optics – for ocean colour, bioactivity, pollutions; microwave – for land, vegetation, soil moisture, sea surface temperature and salinity, the near-surface wind, atmospheric parameters, ice coverage). The performance of these observations is defined by the sensor's capability (information capacity), operational electromagnetic wavelengths, and purpose of the mission as well.

Correspondingly, remote sensors are divided into two major categories: active and passive.

The active sensor transmits electromagnetic waves at selected frequency and polarization and then measures the scattered or reflected signal from the object. Active remote sensors are radar, scatterometer, interferometer, altimeter, Global Position System (GPS) tracker, sonar, and lidar. The measurements contain information about geometrical characteristics of the object, its physical (and sometimes biochemical) properties; at special measurements, the distance from the object to the sensor is defined as well. The passive sensor does not transmit electromagnetic waves, but it measures natural energy that is reflected or emitted from the object itself or by its surrounding areas. Reflected sunlight is the most common external source

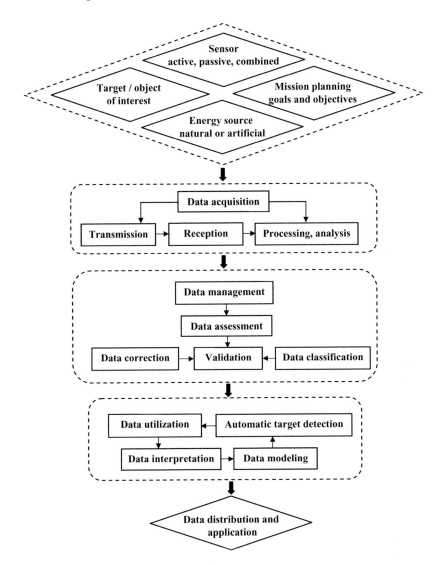

FIGURE 4.1 Components of remote sensing.

of radiation sensed by passive instruments. Passive remote sensors are microwave and infrared radiometers, spectroradiometers, optical spectrometers, analog or digital or TV optical cameras, and ultraviolet spectrophotometer. The measurements contain integral information about thermodynamic, structural, and physical properties of the object. There is a combined active and passive instrument operated with a single antenna and known as *radiometer-scatterometer*. Such a system provides the complementary information contained in the emission and backscattering signatures of the object (e.g., land and/or ocean targets).

The difference between active and passive, including both optical and microwave sensors, is also in day/night observation capabilities. Some microwaves can pass

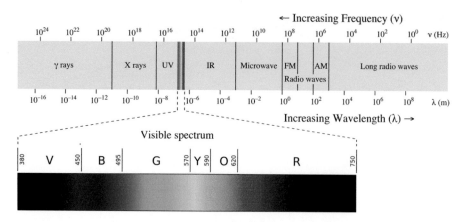

FIGURE 4.2 The electromagnetic spectrum and spectral bands used in remote sensing. V – Violet, B – Blue, G – Green, Y – Yellow, R – Red.

through the atmosphere and clouds and have almost all-weather capability, which is their principal advantage. Optical sensors (especially operated in the *atmospheric transmission windows*) provide much better spatial resolution and quality of data/images than microwave sensors and that is an important advantage for Earth's observations and surveillance.

Remote sensing is based on science and technology, involving different disciplines – electromagnetic theory, applied physics, computer science, electronics and communications. All of them are integrated into a *remote sensing process* which consists of many operations and developments; the most important stages are the following (see also Figure 4.1):

1. Specification of the source of electromagnetic energy (microwave, sun, light).
2. Characterization of electromagnetic radiation (type, frequency, polarization).
3. Interaction of radiation with an object (land, ice, ocean, atmosphere, etc.).
4. Propagation radiation through the atmosphere.
5. Receiving and recording of radiation by remote sensor.
6. Transmission, reception, and processing of the recorded by the sensor data.
7. Assessment, statistical analysis, and classification of the obtained remotely sensed data; thematic (geophysical) interpretation and application – the final and most important step of the remote sensing process.

As a result of these operations, the user (researcher) should be able to obtain relevant information and generate scientific or commercial product. In practice, however, such a multistage scheme may not always be achieved because of the influence of a large number of *interfering factors*. Possible causes are: unknown nature of the observed object, variable background, environmental and/or instrument noise, data acquisition errors, etc. Sometimes collected remotely sensed data turn out to be incomplete, even confused, and/or unable to interpret at all. In such situation, more and more additional resources, efforts, and scientific studies are required to obtain needed product and many may face here technological challenges. Actually, remote sensing of dynamical stochastic systems like natural turbulence is always a challenge.

To provide overall interpretation of remotely sensed data, a preliminary theoretical analysis of the electromagnetic radiation (including an observation process as well) is needed. As a rule, this complimentary and very important part of remote sensing consists of independent and motivated scientific research that can be done using various electromagnetic models, radiative transfer theory, semi-empirical approximations, and/or other appropriate (hybrid) approaches. Indeed, modeling and simulations of radiation characteristics not only provide a better understanding and interpretation of experimental data but also guide further trends and developments in remote sensing research.

Briefly, science objectives of remote sensing can be formulated as the following:

1. Measure electromagnetic radiance, reflected, scattered, or emitted by the object at selected wavelengths; evaluate radiance signatures of the object being observed.
2. Define parameters, properties, and behavior of the object by its radiance signatures.
3. Make significant contribution in study, monitoring, and prediction of environmental processes and event.

To accomplish these (and some others, more specific and *ambitious*) objectives in remote sensing, both efficient observation methodology and innovative information technology (IT) are necessary to apply in a most creative and constructive manner. This option creates possibilities for *comprehensive remote sensing* which is a quintessence of scientific, engineering, and technological problems involving also so-called *big data* analysis and management.

Overall performance of remote sensing products is estimated in terms of various quantities which will be considered in the next section.

4.4   QUANTITIES IN REMOTE SENSING

Different sensors collect different types of data. Correspondently, there are quantities which are used in remote sensing to evaluate the collected information. These quantities are referred to instrumental measurements. Common nomenclature and definition of different quantities are shown in Table 4.1. There are also so-called "environmental quantities of interest" related to remote sensing applications and geographic information system (GIS) and known as geophysical, meteorological, biological, hydrological, geographical quantities, etc. Usually they represent statistical datasets of parameters derived from remote sensing observations. Environmental quantities are widely used in global monitoring of ecosystems and forecasting.

According to common classification of remote sensors and their operational frequency bands within the electromagnetic spectrum, we may distinguish the following types of *instrument-related quantities*:

TABLE 4.1
Nomenclature of Quantities Commonly Used in Remote Sensing

Concept	Symbol	Equation	Measures Unit
Radiant energy	Q_e		joules (J)
Radiant flux	Φ_e	$\Phi_e = \dfrac{\partial Q_e}{\partial t}$	watts (W)
Radiant exitance	M_e	$M_e = \dfrac{\partial \Phi_e^{emitted}}{\partial A}$	$W \cdot m^{-2}$
Irradiance	E_e	$E_e = \dfrac{\partial \Phi_e^{received}}{\partial A}$	$W \cdot m^{-2}$
Radiant intensity	I_e, Ω	$I_{e,\Omega} = \dfrac{\partial \Phi_e}{\partial \Omega}$	$W \cdot sr^{-1}$
Radiance	$L_{e,\Omega}$	$L_{e,\Omega} = \dfrac{\partial^2 \Phi_e}{\partial \Omega \partial A \cos\theta}$	$W \cdot m^{-2} \cdot sr^{-1}$
Spectral radiance in wavelength	$L_{e,\Omega,\lambda}$	$L_{e,\Omega,\lambda} = \dfrac{\partial L_{e,\Omega}}{\partial \lambda}$	$W \cdot m^{-3} \cdot sr^{-1}$
Emissivity, hemispherical	κ	$\kappa = \dfrac{M_e}{M_e^o}$	
Reflectance, hemispherical	R	$R = \dfrac{\Phi_e^r}{\Phi_e^i}$	
Absorptance, hemispherical	α	$\alpha = \dfrac{\Phi_e^a}{\Phi_e^i}$	
Transmittance, hemispherical	T	$T = \dfrac{\Phi_e^t}{\Phi_e^i}$	
Reflectance, directional spectral	$R_{\Omega,\lambda}$	$R_{\Omega,\lambda} = \dfrac{L_{e,\Omega,\lambda}^r}{L_{e,\Omega,\lambda}^i}$	
Absorptance, directional spectral	$\alpha_{\Omega,\lambda}$	$\alpha_{\Omega,\lambda} = \dfrac{L_{e,\Omega,\lambda}^a}{L_{e,\Omega,\lambda}^i}$	
Transmittance, directional spectral	$T_{\Omega,\lambda}$	$T_{\Omega,\lambda} = \dfrac{L_{e,\Omega,\lambda}^t}{L_{e,\Omega,\lambda}^i}$	
Emissivity, directional spectral	$\kappa_{\Omega,\lambda}$	$\kappa_\Omega = \dfrac{L_{e,\Omega,\lambda}}{L_{e,\Omega,\lambda}^o}$	
Attenuation coefficient, hemispherical	μ	$\mu = -\dfrac{1}{\Phi_e} \dfrac{d\Phi_e}{dz}$	m^{-1}
Attenuation coefficient, directional spectral	$\mu_{\Omega,\lambda}$	$\mu_{\Omega,\lambda} = -\dfrac{1}{L_{e,\Omega,\lambda}} \dfrac{dL_{e,\Omega,\lambda}}{dz}$	m^{-1}
Brightness temperature, spectral (low Frequency, the Rayleigh–Jeans law)	$T_{B\lambda} = \kappa_{\Omega,\lambda} T_0$		Kelvin

(Continued)

TABLE 4.1 (Continued)
Nomenclature of Quantities Commonly Used in Remote Sensing

Concept	Symbol	Equation	Measures Unit
Radar cross section (RCS)		$\sigma = \lim\limits_{r \to \infty} 4\pi r^2 \dfrac{|E_s|^2}{|E_i|^2}$	m²
		$\sigma_{dB} = 10\log_{10}(\sigma/1m^2)$	dBsm
Polarization extinction ratio (PER)		$PER = 10\log_{10}\dfrac{P_{principal}}{P_{orthogonal}}$	dB

M_e^o exitance of a black body
Φ_e^i incident flux
Φ_e^r reflected flux
Φ_e^a absorbed flux
Φ_e^t transmitted flux
A an area on the surface
Ω the solid angle
θ angle between flux direction and the normal
$P_{principal}$ principal linear polarization mode
$P_{orthogonal}$ orthogonal polarization mode
E_s intensity of incident electric field
E_i intensity of scattered electric field
T_0 thermodynamic temperature of the surface
z vertical coordinate

Photometric optical quantities by definition refer to the measurement of light, weighted by spectral response of the human eye. Therefore, photometric quantities are restricted to the wavelength range from about 360 to 830 nanometers (nm; 1000 nm = 1 μm). Typical photometric units include luminance, luminous flux, and luminous intensity.

Radiometric optical quantities by definition refer to the measurement of electromagnetic radiation within the frequency range between 3×10^{11} and 3×10^{16} Hz. This range corresponds to wavelengths between 0.01 and 1000 micrometers (μm) and includes spectral regions commonly called the ultraviolet, the visible, and the infrared. Typical radiometric units include radiance, radiant flux, and radiant intensity.

Radiometric microwave quantity refers to the measurement of thermal microwave radiation in the frequency range from 1 GHz to 200 GHz (0.15–30 cm wavelength range). Thus, frequency-dependent thermal microwave radiation of black-body target is defined by the Planck's law either the Rayleigh-Jeans approximation, or the Wien's displacement law depending on frequency range. Black-body radiation is an ideal theoretical concept. Thermal microwave radiance unit is the brightness temperature expressed in Kelvin.

Infrared (IR) thermal quantity refers to the measurement of emitted, reflected, or transmitted electromagnetic radiation in the wavelength range from 3 to 14 μm. The application mostly is associated with thermal IR imagery; however, detectors are available for all thermal bands (SWIR, MWIR, and LWIR). Thermal imaging is simply the process of converting IR radiation (heat) data into a visible image (or map) of physical temperatures of the scene being observed.

Radar quantities refer to the measurements of scattered or reflected radiance; it is associated with radar and/or *scatterometer* observations. Typical radar frequencies (in remote sensing) are from 0.3 GHz to 300 GHz (from 1 m to 1 mm wavelength range). Absolute measure of the target size as observed by radar is called the *radar cross section* (RCS) and is given in units of area (square meters). RCS unit can also be expressed in decibels relative to a square meter (dBsm), i.e., $\sigma_{dBsm} = 10\log_{10}(\sigma/1\,m^2)$, where σ is the RCS in m^2.

Polarimetric quantities refer to the measurement of polarization (or *polarisation)* characteristics of the object being observed. The control of the coherent polarization properties is known as *ellipsometry* (in optical observations) and *polarimetry* (in microwave radar and/or radiometer observations). For example, the degree of linear polarization is quantified with the *polarization extinction ratio* (PER), defined as the ratio of optical powers of two perpendicular polarizations. In radar remote sensing, it is often used *cross polarization ratio* associated with cross-polarized returns.

Listed quantities are relevant for remote sensing measurements in optical (visual and IR) and microwave ranges of the electromagnetic spectrum depending on the instrument design and implementation.

Other type of quantities – *radiation quantities*, – are used to characterize various electromagnetic wave propagation processes. In the theory, radiation quantities can be roughly classified according to six major wave propagation phenomena: 1) reflection, 2) scattering, 3) transmission, 4) absorption, 5) attenuation, and 6) emission. Radiation quantities appear in wave propagation equations and solutions in terms of the dimensionless reflection, scattering, transmission, absorption, and emission coefficients. They can be measured by a certain remote sensor – lidar, radar, or radiometer.

In order to convert remotely sensed data and/or radiation quantities (listed above) into the principle *physical quantities* of the interest (that is supposed to be an ultimate goal of remote sensing missions), it is necessary to create and apply scientifically reasonable mathematical models and/or retrieval algorithms and then complete *information analysis* of remotely sensed data. As a rule, this process consists of many steps including data assessment, digital processing, theoretical or empirical analysis, and/or even full-scale numerical studies. Thus, interpretation of remotely sensed information can be qualified as an *integrating scientific activity* allowing the researcher to create relevant product and/or provide further thematic applications.

There are also *non-relevant* quantities related to instrument offset, noise, signal-degrading artifacts, or errors in data acquisition system. These issues may occur, e.g., in the case of very high-resolution optical imagery of the Earth's surfaces when fluctuations in wave propagation characteristics (e.g., due to optical turbulence) cause the degradation of the pixel values and thereby reduce overall quality of images. Complementary models and techniques can help to reveal possible uncertainties and instrument errors. For this, it is necessary to determine first which approach(s) and/or procedure(s) will be most appropriate for a given instrumentation and task.

4.5   INSTRUMENT CONCEPT

The successful implementation of remote sensing missions depends largely on both an instrument specification (type) and an observation concept. A basic classification of remote sensing instruments is shown in Figure 4.3. Instrument concept is an important part of any remote sensing mission. Briefly, the instrument concept includes the following technological components: 1) instrument design, 2) instrument characteristics, 3) payload platform, 4) observation capability, 5) acquisition system performance, and 6) implementation approach.

The instrument concept is usually considered in the context with goals and objectives of remote sensing missions; however, many instruments are designed for multipurpose civil and military applications (e.g., satellite optics of high resolution). In this case, the efficiency of remote sensing observations largely depends on the choice and number of spectral channels and their spectral sensitivity characteristics which are defined in terms of *spectral response* and *spectral sensitivity* of the remote sensing system (this terminology refers mostly to optical systems). Correspondingly, "spectral response" is a range of EM waves in which the sensor has a significant responsivity, and "spectral sensitivity" is the relative efficiency of detection as a function of the frequency or wavelength of the signal.

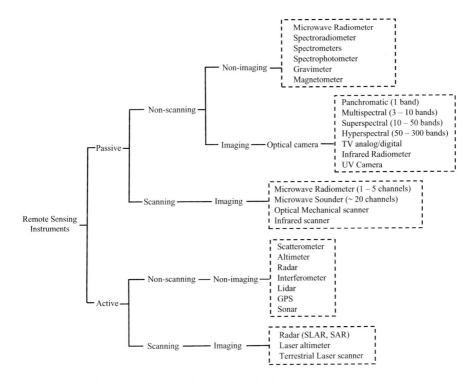

FIGURE 4.3   Basic classification of remote sensing instruments.

Today, there are different instrument concepts in use, ranging from a one-frequency concept to a multifrequency (also known as multispectral and hyperspectral) concept. These concepts differ by observational capabilities and characteristics; the most important characteristics are the following: a) spectral range and/or *bandwidth*, b) polarization, c) scanning, swath, and observing cycle; d) spatial resolution (IFOV, pixel, MTF), e) radiometric resolution, f) accuracy. Other differences are associated with technical parameters such as sensor's mass, dimensions or volume, power consumption, signal-to-noise ratio and stability (for microwave devices), end-to-end calibration cycle, and the instrument cost as well.

One-frequency system such as a weather radar, scatterometer, or passive microwave radiometer provides measurements at specified EM wavelength. The choice may not be always optimized by spectral sensitivity characteristics needed or recommended (e.g., by the theory) for reliable retrieval of the object's specific properties. For example, remote sensing observation of sea surface temperature (SST) requires the use of passive microwave radiometers, operated at C-band (4–8 GHz or 7.5–3.75 cm wavelength range), whereas observations of sea surface salinity (SSS) are optimal at L-band (1–2 GHz or 30–15 cm wavelength range). Information about the water-vapor and liquid-water content in the atmosphere can be obtained by measuring thermal radiation at the high-frequency range from 20 to 60 GHz (1.5 to 0.5 cm wavelengths).

Better performance can be reached using multifrequency systems such as the 7-channel Special Sensor Microwave Imager (SSM/I), the 24-channel Special Sensor Microwave Imager Sounder (SSMIS), which provide observations of ocean and atmosphere parameters simultaneously. Other remarkable examples are the 13-channel Global Precipitation Measurement Microwave Imager (GMI) and Russian Meteor-M No. 2-2 satellite launched in 2019 and carried 29-channel microwave imager/sounder MTVZA-GY (e.g., Mitnik et al. 2017).

More specialized multispectral (MS) optical imagers of high spatial resolution (~ 4 m for spectral and 1 m for panchromatic channels) such as DigitalGlobe's satellite series (IKONOS, QuickBird, GeoEye-1,WorldView-1, 2, 3, 4) are capable to detect small ground objects and dynamical features in the ocean such as ship wakes, roughness change, jets, etc.; however, such high-resolution information is available mostly for commercial and government use.

Hyperspectral (HS) optical imaging systems provide the unique color signature of individual objects and surfaces. For example, modern HS imagers of medium spatial resolution (~ 30 m) such as Earth Observing-1 (EO-1) NASA, PRISMA (Italy), and EnMAP (Germany) carried 200+ spectral bands are promising tools for systematic and multipurpose monitoring of the Earth's environments.

Satellite observations evolve the most important source of data for monitoring and understanding the environmental changes. In today's scenario, more than 150 remote sensing satellites are orbiting the Earth and supporting various geophysical studies. The most widely operated remote sensing satellites which are potentially capable of *turbulence detection* are listed in Table 4.2. To understand better the capabilities, we will consider basic satellite observation principles.

TABLE 4.2

A Short List of Currently Operated Satellites Capable for Turbulence Detection

Mission/Sensor	Agency/Country	Launch	Bands	Spectral Range	Resolution	Other Specifications
Optics:						
Landsat-8, -9 OLI, OLI-II	USGS/NASA	2013, 2020	11	VIS, NIR SWIR, TIRS, Cirrus	30 m	16-day revisit
Sentinel 2A, 2B	ESA	2015, 2017	13	VIS, NIR, SWIR, Cirrus	10/20/60 m	10-day revisit,
EnMAP	DLR Germany	2017	230	VNIR, SWIR	30 m	4-day revisit
Geo-Kompsat 2B GOCI-II	KIOST Korea	2019	12	UV, VIS, NIR	250/1000 m	Geostationary over NE Asia
Terra, Aqua MODIS	NASA	1999 2002	36	VIS, NIR, SWIR	250/500/1000 m	2-day revisit
Suomi NPP, JPSS-1 VIIRS	NOAA/NASA	2018	22	VIS, IR	750 m	2-day revisit
SPOT-6, -7	ESA	2012, 2014	5	VIS (PAN+MS), NIR	1.5/ 6 m	1–3 days revisit
WorldView-2	DigitalGlobe USA	2009	9	VIS (PAN+MS), NIR	0.5/2 m	~ 1 day revisit
WorldView-3	DigitalGlobe USA	2014	9	VIS (PAN+MS), NIR, SWIR	1.24/3.7/30 m	< 1 day revisit
Passive Microwave:						
ATMS JPSS-1	NOAA	2011	22	23 to 183 GHz	16/32/75 km	20 day repeat cycle
GPM/GMI	NASA/JAXA	2014	13	10.65 to 183.3 GHz	4 – 32 km	2.5 days revisit
AMSR-2	NASA/ JAXA	2012	7	6.93 to 89 GHz	5 – 60 km	2 days revisit
SSMIS	DMSP USA	2003	24	19 to 183 GHz	14 – 75 km	2 days revisit
Meteor-M2-2 MTVZA-GY	Roscosmos Russia	2019	29	10.6 to 183.3 GHz	16/32/48 km	
SAR:						
RADARSAT-2	Canada	2007	FP	C	3 – 100 m	average daily revisit
TerraSAR-X/TanDEN-X	DLR, Germany	2010	SP, DP	X	1/2/3/16 m	11-day repeat cycle
COSMO SkyMed 1 – 4	Italy	2007–2010	SP, DP	X	3/5/15 m	16 days revisit
Sentinel-1A	ESA, Copernicus	2014	SP, DP	C	5/20/40 m	12-day repeat cycle
ALOS PALSAR-2	JAXA, Japan	2014	SP, DP, FP	L	3/6/10/100 m	14 days revisit

(Continued)

TABLE 4.2 (Continued)
A Short List of Currently Operated Satellites Capable for Turbulence Detection

Mission/Sensor	Agency/Country	Launch	Bands	Spectral Range	Resolution	Other Specifications
Kompsat-5	KARI, Korea	2013	DP	X	1/3/20 m	28 days revisit
Capella X-SAR	Capella Space, USA	2018–2021	SP, DP	X	0.5 x 0.5 m	1–3 hours revisit

SP: Single polarization

DP: Dual polarization

FP: Full polarization (quad)

H –Horizontal, V –vertical polarization

HH or VV,

HH + HV or VV + VH

HH+HV+VH+VV

4.6   USING SATELLITES

Remote sensing technology operates with a wide variety of platforms, instruments, and techniques. Compared to aircraft or other low-altitude platforms, satellites have advantages and many unique characteristics which make them primary tool for global, systematic, and efficient observations of the Earth's environments. Some of these advantages include:

1. Continuous acquisition of data.
2. Frequent and regular re-visit capabilities resulting in up-to-date information.
3. Broad coverage area.
4. Good spectral resolution.
5. Semi-automated/computerized processing and analysis.
6. Ability to manipulate/enhance data for better image interpretation.
7. Accurate data mapping.

From this viewpoint, remote sensing satellites are sometimes referred to as "eyes in the sky."

Nowadays, remote sensing satellites provide a wide range of applications and services that require the development of flexible satellite networking technology. The network consists of four major components: 1) space segment, 2) management segment, 3) ground segment and 4) user segment. Satellite network differs from a *terrestrial network* and includes several additional components such as communication, telemetry, tracking and command systems as well as ground station (in more detail see, e.g., book Maini and Agrawal 2014). Eventually, future development and application of space technology should include the seamless integration of satellite and terrestrial networks, which can bring more efficient service.

Remote sensing missions and applications demand primary specifications of satellite observation systems: instrument payload, orbital parameters, data acquisition systems, data assessment and management. For example, specifications of MS optical imagers include camera focal length, field of view, aperture size, number of bands, band width, detector size, quantisation bits, altitude, revisit time, etc. These parameters are associated with four different resolutions known as spatial, spectral, temporal, and radiometric resolution. Resolutions depend on instrumental characteristics and can vary with the goals and applications of remote sensing missions. Capabilities of space-based observations are defined by satellite characteristics, which we will consider in the next section.

4.6.1   SATELLITE CLASSIFICATION

According to common classifications, satellites can be divided into five principal types: 1) research, 2) communications, 3) weather, 4) navigational, and 5) applications. Specifically:

1. *Research satellites* provide measurements of fundamental properties of outer space: magnetic fields, the flux of cosmic rays and micrometeorites, and properties of celestial objects that are difficult or impossible to observe from the earth.

2. **Communications satellites** provide worldwide linkup of radio, telephone, television, Internet, and commercial and military services.

3. **Weather satellites**, or meteorological satellites, are primarily used to monitor the weather and climate of the Earth. They also provide information about environmental events such as hurricanes, wildfires, dust storms, snow cover, sea ice, and atmospheric and oceanic conditions as well.

4. **Navigation satellites** are used for providing autonomous geo-spatial positioning, navigation, and tracking. Today, the Global Position System (GPS) and the Global Orbiting Navigation Satellite System (GLONASS) are fully operational navigation systems widely used in civil, commercial, and military services.

5. **Applications satellites** are designed initially for testing and improving satellite technology itself. These satellites also have been used for a number of military purposes, including satellites for tracking missile launches and detecting nuclear explosions, a large number of various spy satellites, electronic warfare satellites, and other satellites aiming surveillance and reconnaissance. Sometimes these satellites have subsequently proved to have civilian benefits, such as commercially available satellite photographs of very high resolution showing surface objects, urban structures, and maritime infrastructure in great detail.

Remote sensing satellites can be referred to all listed types of satellites because they already used to provide geophysical information, in one way or another. Large category of satellites is dedicated to certain remote sensing missions – global monitoring of natural resources and environmental changes, measurements of ocean and atmosphere parameters, land mapping, GIS applications, etc. Synergy of satellite- and ground-based observations is very beneficial for forecasting models. In any case, satellite observation concept (in which we focus on) is defined uniquely by a set of key parameters which are briefly considered below.

4.6.2 OBSERVATIONAL PARAMETERS

The performance of satellite missions largely depends on the choice of instrument payload, observation, and orbital parameters. Observation parameters are closely related to the instrument characteristics such as spectral range, resolution, scanning, and swath. It is the evolution of the satellite instrument components that is complicated and increasingly the focus of research. Let's consider major observational parameters in more detail.

1. **Spectral range and number of frequency channels** are the most important characteristics of any remote sensing observing system. Spectral range (or the *wavelength range*) is specific portion of the electromagnetic spectrum in which sensors can measure the radiance. Number of frequency channels defines the amount of independent spectral information collected by the sensor. For example, microwave radiometer/sounder may consist of up to 20+ channels, whereas hyperspectral (HS) optical imager may carry 200+ channels. The quality and volume of data depend on both spectral range and number of channels of the sensor.

2. **Resolution** is the smallest measure that can be discriminated or resolved by a sensor. There are four types of resolution:

- *Spatial* – refers to the space size; it is "*a measure of the smallest angular or linear separation between two objects that can be resolved by the sensor*" (Jensen 2014).
- *Spectral* – refers to the ability of a sensor to resolve and distinguish fine wavelength intervals. Spatial and spectral resolutions are linked with each other; a better spatial resolution leads to a coarser spectral resolution and vice versa.
- *Radiometric* – refers to the ability of a sensor to define smallest differences in intensity of radiance. *Coarse* radiometric resolution corresponds to a few brightness levels (bits), whereas *fine* radiometric resolution corresponds to large number of brightness levels. Regular sensors record 8-bit data; advanced sensors record 16-bit data (8 bits = 2^8 = 256 levels, i.e., from 0 to 255; 16 bits = 2^{16} = 65,536 levels, i.e., from 0 to 65,535). The intrinsic radiometric resolution of a sensor depends mainly on the signal-to-noise ratio of the detector.
- *Temporal* – refers to time and frequency of data collection; the higher the temporal resolution, the shorter time between data acquisitions. Temporal resolution of satellites can vary from about 16 days to 15 minutes and depends primarily on the platform, payload, swath, and orbital characteristics.

3. ***Scanning, swath, and revisit time*** define an observation process and data sampling.
   - *Scanning* refers to the data acquisition through a narrow-view sweeping over the area being observed. In remote sensing, there are several types of scanning (Figure 4.4) including two main modes known as *across-track scanning* and *along-track scanning*. Across-track scanners measure series of lines oriented perpendicular to the direction of motion of the sensor platform. Along-track scanners also use the forward motion of the platform to record successive scan lines using a linear array of detectors. Differences between *whiskbroom* and *pushbroom* sensors are given in Table 4.3.
   - *Swath* is a pattern of scans on the ground scene. Swath width D can be defined through the *instantaneous field of view* (IFOV), $\alpha = D/H$, where α is the solid angle given in units of micro radians (µrad), and H is the altitude, or the *ground-projected instantaneous field of view* (GIFOV), GIFOV = 2H tan (α/2), given in the units of meters. Spatial resolution – ground surface area represented by one pixel – is defined by GIFOV. Figure 4.5 illustrates standard geometric characteristics of sampling: FOV, IFOV, GIFOV, GSD, and swath width.
   - *Revisit time* (or revisit interval or revisit period, also associated with the *repeat cycle* of the satellite) refers to the time elapsed between successive observations of the same target area. Revisit time depends on orbit, target location, and swath.

4.6.3   ORBITAL PARAMETERS

Orbital parameters are used to describe satellite orbit. The performance of satellite missions largely depends on the choice of orbital parameters. Six Keplerian elements

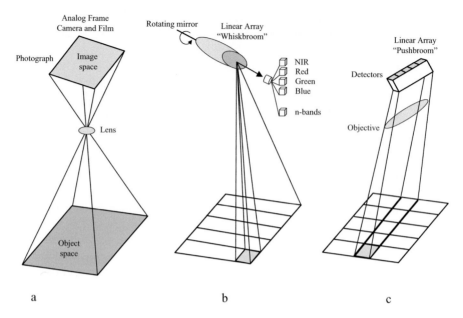

FIGURE 4.4 Types of scanning. (a) Frame, (b) whiskbroom, and (c) pushbroom.

TABLE 4.3
Difference between Whiskbroom and Pushbroom Scanners

Whiskbroom versus Pushbroom

Whiskbroom	Pushbroom
Visible / NIR / MWIR / TIR	Mainly visible / NIR
Wide swath width	Narrow swath width
Complex mechanical system	Simple mechanical system
Rotating mirror	Array of sensitive sensors
Simple optical system	Complex optical system
Shorter dwell time	Longer dwell time
Pixel distortion	No pixel distortion

(Figure 4.6) describe the size, shape, and orientation of an orbit as well as the location of a satellite in the orbit. The main orbital parameters are the following:

1. *Orbital altitude* refers to the height (vertical distance) of the satellite above the earth's surface. Remote sensing satellites operated at an altitude of 500–1,000 km have large swath width and low spatial resolution. Optical multispectral (MS) imagers of high resolution (e.g., IKONOS and QuickBird) operate at orbital altitudes of between 450 and 770 km.

2. *Orbital inclination* is defined as the angle (in degrees) between the orbital plane and the earth's equatorial plane. The inclination angle sets in accordance with satellite launch mission and trajectory design. For inclination 0°–90°, the satellite moves in the same direction as the rotation of the earth; for inclination 90°–180°, the satellite moves in the direction, opposite to the rotation of the earth.

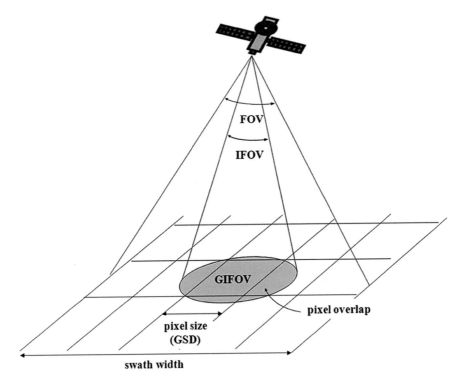

FIGURE 4.5 Observation concept: FOV, IFOV, GIFOV, GSD, and swath.

3. **Orbital period** is the time (in minutes) that takes a satellite to complete one full revolution around the earth. Orbital period T is calculated from the Kepler's Third Law, $T = 2\pi\sqrt{a^3 / GM}$, where a is the semi-major axis of the orbit (satellite altitude + earth radius, ~ 6380 km), G is the gravitational constant, and M is the mass of the earth (GM = 3, 986 × $10^{14} m^3 s^{-2}$). The orbital period of most remote sensing satellites ranges between 90 and 100 minutes.

4. **Repeat cycle** is defined as the time (in days) between two successive revolutions of the satellite on the same orbits. The difference between *revisit time* and the *repeat cycle* is that the first parameter depends on the orbit only, but the second parameter is relevant to payload of satellite as well. There is always a trade-off between repeat cycle and spatial resolution. Satellites with high repeat cycles mostly produce low spatial resolution data, and vice versa. For example, IKONOS has repeat cycle 14 days (max) and revisit time 1–3 days.

4.6.4 TYPES OF ORBITS

Different orbits provide various results related to continuous monitoring, global mapping, select coverage, or repeat coverage. Orbits are grouped into several categories according to main purposes of remotes sensing missions, which can be scientific,

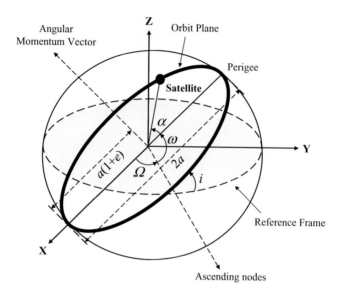

α	true anomaly (instantaneous angle from satellite to perigee)
ω	argument of perigee (twist)
Ω	longitude of the ascending node (pin)
a	semi-major axis of the elliptical orbit (size)
e	eccentricity of the orbital ellipse (shape)
i	inclination of the orbital plane (tilt)

FIGURE 4.6 Six Keplerian elements. (Based on Capderou 2005).

commercial, or military. Satellite orbits are categorized by shape, by inclinations, and by altitude. Altitude classification of satellite orbits includes several types of orbits known as

1. *Geostationary orbit* (GEO) is orbit with attitude 36,000 km above earth surface and inclination angle of zero. Most satellites in GEO are used for meteorological, telecommunication, and military purposes. A constellation of three equally spaced satellites can provide full coverage of the Earth, except Polar Regions. The NOAA Climate Prediction Center uses five GEO satellites: GOES-11, GOES-13, MSG-2, Meteosat-7, and MTSAT-2.
2. *Low earth orbit* (LEO) is circular orbit with altitude 500–1,500 km and can be as low as 160 km above the Earth. Satellites in LEO have an orbital period in the range of 90–120 minutes. LEO most commonly used for remote sensing, military purposes (high-resolution imagery) and for human spaceflight missions. It is the orbit used for the International Space Station (ISS).
3. *Medium low earth orbit* (MEO) or Intermediate Circular Orbit (ICO) is orbit with altitude from 2,000 km to of 35,786 km (below GEO) above sea level. The orbital periods of MEO satellites range from 2 to 12 hours. MEO is

particularly suited for constellations of satellites mainly used for navigation, communication, and space environment science.

4. *Highly elliptical orbit* (HEO) is an elliptic orbit with high eccentricity with one end nearer the earth and other more distant with altitude more than 35,786 km above the Earth surface. This makes elliptical orbit useful for communications satellites. HEO is also considered for observations of Polar Regions from space (Trishchenko et al. 2019). For this purpose the Russian Arktika-M meteorological and remote sensing satellite has been launched successfully on February 28, 2021 from the Baikonur Cosmodrome in Kazakhstan.

5. *Polar orbit* (PO) passes over the Earth's Polar Regions from north to south. POs have an inclination angle between 80° and 100°. POs mainly take place at low altitudes of between 200 and 1,000 km. POs are used for reconnaissance and remote sensing. Complementing the GEO satellites are the polar-orbiting satellites known as Polar Operational Environmental Satellites (POES). The POES instruments include the AVHRR/ATOVS which provide visible, infrared, and microwave remotely sensed data. The new Capella X-SAR constellation of microsatellites also operates in PO.

6. *Sun synchronous orbit* (SSO) also called a *heliosynchronous* orbit is a nearly PO which is synchronous with the Sun. Satellites in a SSO usually operate at altitude of between 600 to 800 km with high inclination. SSO are used for remote sensing, weather forecasting, and reconnaissance. The SSOs instruments include SPOT 4-5, IKONOS, WorldView, LANDSAT 4-5, IRS, TIROS, QuickBird-1,2 (SSO/PO).

7. *Geostationary transfer orbit* (GTO), also known as a Hohmann transfer orbit, is an orbit to transfer a satellite between two circular orbits, LEO and GEO, of different radius in the same plane. GTO is a highly elliptical Earth orbit with an apogee of 42,164 km, or 35,786 km above sea level, which corresponds to the geostationary altitude. GTO is considered as a transfer trajectory for Earth-to-Moon mission (LunarSat) and also for Elon Musk's SpaceX missions.

A number of specialized orbits – *Lagrange points*, *Molniya orbits*, *Loopus orbit*, *Tundra* and *Supertundra orbits*, *Ellipso Borealis orbit*, *Galileo orbit*, and some others are used for specific satellite missions. Only a small fraction of operational satellites fall into this category.

4.7   DATA ACQUISITION

Remote sensing involves many methods and algorithms, aiming data collection, processing, assessment, analysis, interpretation, application, and management. The entire process of capturing, measuring, and exploring the collected information is known as data acquisition and analysis. An acquisition of remotely sensed data can be conducted from different platforms – ground-based, vehicles, ships, aircraft, satellites, balloons, rockets, space shuttles, etc. Therefore, there are many different data acquisition methods and analysis techniques in remote sensing dependent on sensor-platform combinations. Figure 4.7 demonstrates data acquisition chain related to remote sensing.

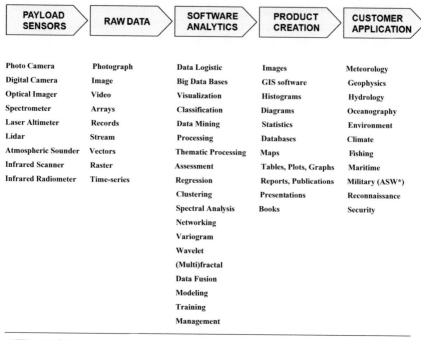

PAYLOAD SENSORS	RAW DATA	SOFTWARE ANALYTICS	PRODUCT CREATION	CUSTOMER APPLICATION
Photo Camera	Photograph	Data Logistic	Images	Meteorology
Digital Camera	Image	Big Data Bases	GIS software	Geophysics
Optical Imager	Video	Visualization	Histograms	Hydrology
Spectrometer	Arrays	Classification	Diagrams	Oceanography
Laser Altimeter	Records	Data Mining	Statistics	Environment
Lidar	Stream	Processing	Databases	Climate
Atmospheric Sounder	Vectors	Thematic Processing	Maps	Fishing
Infrared Scanner	Raster	Assessment	Tables, Plots, Graphs	Maritime
Infrared Radiometer	Time-series	Regression	Reports, Publications	Military (ASW*)
		Clustering	Presentations	Reconnaissance
		Spectral Analysis	Books	Security
		Networking		
		Variogram		
		Wavelet		
		(Multi)fractal		
		Data Fusion		
		Modeling		
		Training		
		Management		

ASW* – Anti-Submarine Warfare

FIGURE 4.7 Data acquisition chain in remote sensing.

There are basically three categories of data acquisition systems used in remote sensing: 1) image-based method, e.g., imaging spectrometer, 2) range-based method, e.g., laser altimeter, lidar, pulse radar, and 3) combined method, e.g., the use of GPS measurements and multispectral imagery for 3D digital mapping (known as *Digital Elevation Model*). The choice of an appropriate data acquisition method depends on the purposes of mission, sensor capability, observation and orbital parameters, object and/or area of the interest, the experience of the user, and the available budget.

The performance of data acquisition system has a direct impact on the retrieval of the object's properties of interest; in particular, it is of a critical importance for remote sensing of dynamical phenomena/events such as turbulent motions in the atmosphere and the ocean. Eventually, for a better performance, it will be necessary to employ *multipurpose innovative real-time acquisition* system, enabling detection and monitoring of complex dynamical phenomena/events from space.

4.8 DATA ASSESSMENT

Quantitative assessment of remotely sensed databases is an important aspect of interpretation, applications, and decision-making (Lyon and Lunetta 2005; Congalton and Green 2019). In general, the goal of quantitative accuracy assessment of data and products (e.g., GeoMaps) is the identification and measurement of errors. Figure 4.8 illustrates the possible sources of errors in remotely sensed data which are multiple and compounding. Errors can derive from the acquisition of data, their digital

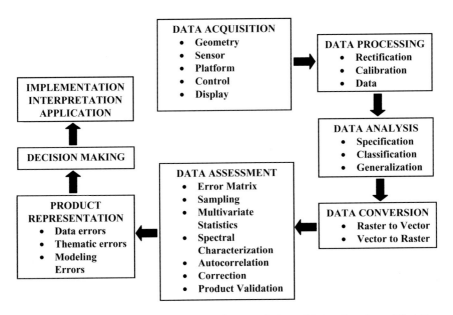

FIGURE 4.8 Sources of errors in remotely sensed data. (Updated and modified from Congalton and Green 2019).

processing, classification, and also during uncertainties of the modeling and prediction. Accuracy assessment estimates, identifies, and characterizes the impact that arises from all of the sources of error. In particular, accuracy assessment is the measurement of the rate and level to which selected image agrees with the reference image (i.e., obtained from other source).

The theory and principles of quantitative accuracy assessment have been developed in the mid 1970s with the appearance of digital technologies in remote sensing, GIS, and mapping science. At present, established accuracy assessment techniques include 1) generation of the error matrix, 2) the Kappa analysis, 3) discrete multivariate analysis, 4) sampling size and scheme, 5) spatial autocorrelation, 6) error budget analysis, 7) change detection accuracy assessment, and 8) fuzzy accuracy assessment.

Quantitative accuracy assessment methods provide a very powerful scientific basis for both descriptive and analytical study of the spatial remotely sensed data. In particular, multispectral optical imagery of oceanic and atmospheric features is definitely a subject for quantitative accuracy assessment, at the first place, due to the impact of a large number of environmental factors on an observation process and image formation. Furthermore, integration of satellite data with field observations and/or background information can help in decision-making what factor(s) could cause specific phenomenon/event.

4.9   DATA MANAGEMENT

Work with huge remotely sensed databases (called *big data*) requires a specific management in terms of data accessibility, data usability, quality of processing and data science capabilities. The term "big data" usually refers to data stores characterized by the "3 Vs": high volume, high velocity, and wide variety. Challenges include the acquisition, storage, searching, sharing, transfer, analysis, visualization, and management of big data. Typical examples of remotely sensed big data are GoogleMaps and GoogleEarth.

Big data management is a broad concept including 1) discovery and characterization of data, 2) selection of relevant information, 3) selection of the area of interest, 4) specification of geodatabases and supported information, 5) initial study of data content, and 6) data integration and distribution. Analysis of big data involves data mining, signal processing, and machine learning that can be performed automatically using mathematical modeling methods and management operations. Big data techniques and algorithms for satellite remote sensing are of a special interest (Swarnalatha and Sevugan 2018; Dey et al. 2019).

In recent years, there has been increased focus on Research Data Management (RDM). In remote sensing, effective RDM is a multiple challenge related to generating new tools, new products, and new ideas (Pryor 2012). RDM can provide research and analysis of raw data along with processed data, selection and assessment of relevant geophysical information, data sharing, data modeling expertise, and product derivation. RDM requires certain level of expertise including mathematical skills as well.

4.10 THEORETICAL MODELS

As mentioned above, the physics-based theory related to remote sensing is based on fundamental Maxwell's equations, describing propagation of electromagnetic waves in various material media including surfaces. Ishimaru (1991) considers three categories of the problem, namely 1) *"waves in random scatterers,"* 2) *"waves in random continua,"* and 3) *"rough surface scattering."* Random scatterers are random distributions of many particles; random continua are the media whose characteristics vary randomly and continuously in time and space; and rough surfaces are material surfaces with deterministic or statistical properties. There are also relevant situations when all these three categories are considered together; in this case, the radiative transfer theory is only a way to solve *remote-sensing computational* problem.

In remote sensing, major wave propagation characteristics are expressed by the reflection, transmission, absorption, scattering, and emission coefficients, which in most practical cases are setting as functions of the medium's physical parameters (at fixed observation conditions). The best option to compute these coefficients is direct numerical simulations of Maxwell's equations that provide a rigorous mathematical framework for many remote sensing applicationss. However, this may not be always optimal way for interpretation of statistical datasets, collected over a long period of time (i.e., big data) and/or large volumes of multispectral or multisensor images. Therefore, different approaches have been developed and applied to solve theoretical problems. Furthermore, inverse problems in remote sensing and GIS demand the use of whenever possible low-cost mathematical methods and algorithms, providing fast practical results. In this connection, the theory should be able to predict and explain remotely sensed data correctly, reliably, and timely.

In this section, we just outline basic theoretical concepts and selected models, underlying remote sensing of environment. References can be found in our note of the literature (Section 4.13).

4.10.1 Classification of Models

Under the term "remote sensing model," we understand any theoretical, numerical, semi-empirical, approximation model, or scaling-based mathematical relationship that provide a link between quantities measured by a remote sensor and physical quantities related to the property of the object of interest. Over the years, remote sensing models of the Earth's surfaces, oceans, and other natural media have been the subject of intensive studies and applications of many scientists and developers. Authors of books (Tsang et al. 1985; Ishimaru 1991; Fung 1994, 2015; Tsang et al. 2000; Tsang and Kong 2001; Fung and Chen 2010) have made significant contributions to electromagnetic theory of remote sensing including scattering and emission models, numerical techniques, and calculations.

The choice and the implementation of suitable models depend on the research purposes, data contents, and experiences as well. Many theoretical approaches are *congruent* in terms of physical sense and can be applied for quantitative analysis of various remote sensing problems. From this viewpoint, remote sensing models can be divided into two principle classes, referred to as

(A) *Physics-based models*, describing rigorous relationships between physical, structural, and geometrical properties of the medium and wave propagation characteristics, and

(B) *Non-physics-based models*, describing just averaged (semi)-empirical relationships between wave propagation characteristics and selected physical parameters of the interest.

Models (A) connect with rigorous or approximate solutions of electromagnetic problems, providing comprehensive computations of wave propagation characteristics (reflection, transmission, scattering, emission) and parameter dependencies. For this goal, the following methods and approximations exist:

1) Integral (Maxwell's equations),
2) Geometrical optics,
3) Kirchhoff approximation,
4) Physical optics,
5) Diffraction approximation,
6) Perturbation theory,
7) Stationary phase approximation,
8) Impedance (*Leontovich*) approximation,
9) Quasi-static approximation,

and some others more complicated (combined) methods and numerical techniques.

Models (B) operate with regressions, statistical approximations, interpolations, or correlations using bulk or spectral parameterizations. These models are usually considered in context with problems of information retrieval and deal with remotely sensed big data. Models (B) can also provide assessment and semi-empirical quantitative analysis of selected observational data in order to understand their content and practical significance better. Typical examples are empirical statistical dependencies of the backscatter coefficient and/or the brightness temperature on sea surface wind speed.

Both these classes of models, (A) and (B), are acceptable and valid anyway, e.g., in context with inverse problems of remote sensing. At the same time, remote sensing models may encounter challenges, which we refer to as the "*uncertainty of the model prediction.*" Although uncertainty-reduction methods exist and can be employed, each researcher makes own decision what kind of model is the best for the interpretation and how to handle the problem of uncertainty (if any) without big errors. In most cases, the decision making is based on his/her intuitions and/or descriptive experiences but not on the knowledge of existing *theoretical resources*. It is not a secret that even well-established models and approximations may not *always result in success* because of complexity and/or unknown factors affecting remote sensing observations.

For better understanding theoretical aspects, we suggest classification of appropriate models and techniques for remote sensing purposes (Figure 4.9). Our classification is based on *Evaluation, Validation, and Application* of remote sensing models arranged in *order of difficulty*. Some of them will be discussed in the next sections

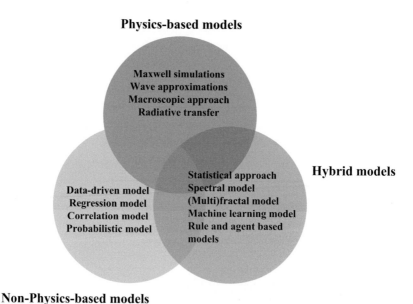

FIGURE 4.9 Classification of electromagnetic models for remote sensing.

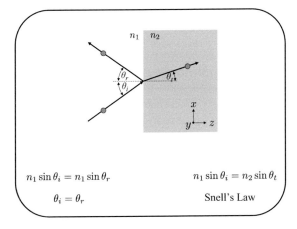

$$r_s = \frac{n_1\cos\theta_i - n_2\cos\theta_t}{n_1\cos\theta_i + n_2\cos\theta_t}$$

$$t_s = \frac{2n_1\cos\theta_i}{n_1\cos\theta_i + n_2\cos\theta_t}$$

$$r_p = \frac{n_2\cos\theta_i + n_1\cos\theta_t}{n_2\cos\theta_i + n_1\cos\theta_t}$$

$$t_p = \frac{2n_1\cos\theta_i}{n_2\cos\theta_i + n_1\cos\theta_t}$$

$$n_1 \sin\theta_i = n_1 \sin\theta_r$$

$$\theta_i = \theta_r$$

$$n_1 \sin\theta_i = n_2 \sin\theta_t$$

Snell's Law

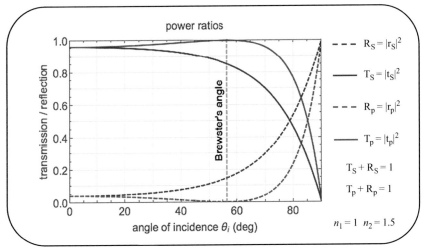

$R_S = |r_S|^2$

$T_S = |t_S|^2$

$R_p = |r_p|^2$

$T_p = |t_p|^2$

$T_S + R_S = 1$

$T_p + R_p = 1$

$n_1 = 1 \quad n_2 = 1.5$

In the case when a surface normal is pointing in the direction of gravity we have p-polarized ~ vertical polarization and s-polarized ~ horizontal polarization.

FIGURE 4.10 The Fresnel equations framework.

(*Source:* Hecht 2017).

4.10.2 Fresnel Reflection Equations

The Fresnel equations describe the reflection and transmission of electromagnetic waves at the incident onto the flat boundary between two semi-infinite media with different index of refraction (Stratton 1941; Born and Wolf 1999). The Fresnel reflection coefficients are complex quantities and function of polarization, incidence and refraction angles, and the complex index of refraction (or permittivity) of two respective media (Figure 4.10). In the case of a material half-space ($n_1 = 1$, $n_2 \neq 1$) with a flat boundary, the Fresnel equations are reduced to simple formulae

$$r_h = \frac{\cos\theta - \sqrt{\varepsilon - \sin^2\theta}}{\cos\theta + \sqrt{\varepsilon - \sin^2\theta}}, \quad r_v = \frac{\varepsilon\cos\theta - \sqrt{\varepsilon - \sin^2\theta}}{\varepsilon\cos\theta + \sqrt{\varepsilon - \sin^2\theta}}, \quad (4.1)$$

where $r_{h,v}$ are spectral complex reflection coefficients for horizontal (index "h") and vertical (index "v") polarizations, respectively, $\varepsilon = \varepsilon' - \iota\varepsilon''$ is the complex permittivity of a medium, and θ is an observation angle (incidence angle from nadir). Fresnel reflection equations in form (4.1) are involved practically in all remote sensing models of the Earth's surfaces where it is necessary to define effects of EM wave propagation in natural media.

Thermal microwave emission of a material isothermal medium is also defined through Fresnel coefficients

$$\kappa_{h,v} = 1 - |r_{h,v}|^2 = 1 - R_{h,v}, \quad T_{Bh,v} = \kappa_{h,v}T_0 = (1 - R_{h,v})T_0, \quad (4.2)$$

where $\kappa_{h,v}$ and $T_{Bh,v}$ are emission coefficient (called also "emissivity") and the brightness temperature of a medium with the surface thermodynamic temperature $T_0 = $ const in Kelvin ($R_{h,v}$ is the corresponding power reflection coefficient). Relationships (4.2) obey the Kirchhoff's law of thermal radiation that states that for thermal equilibrium for a particular body (or surface) the monochromatic emissivity equals the monochromatic absorptivity.

The fundamental application of the Fresnel equations is computing reflection $r_{h,v}$, transmission $t_{h,v}$ and emission $\kappa_{h,v}$ coefficients of a (quasi)-smooth material surface. Referring to the Rayleigh criterion of roughness, the surface is smooth if $h < \lambda/(8\cos\theta)$, a more stringer criterion is $h < \lambda/(32\cos\theta)$; otherwise, the surface is assumed to be rough (h is the height of surface elevations, λ is the electromagnetic wavelength, and θ is the incidence angle). It means that, under certain conditions, many natural surfaces may appear smooth even though the height of inherent elevations is relatively large.

There is a popular and simple formula for computing power reflection coefficient $R'_{h,v}$ of a *weakly rough dielectric surface*

$$R'_{h,v} = R_{h,v}\exp\left[-\left(k_0\sigma_h\cos\theta\right)^2\right], \quad (4.3)$$

where $R_{h,v}$ is the power Fresnel reflection coefficient, $k_0 = 2\pi/\lambda$, and σ_h is the rms surface elevation. This formula is obtained using first-order perturbation theory and valid approximately at $k_0\sigma_h \ll 1$. Equation (4.3) can be used for computing brightness temperature $T_{Bh,v} = (1 - R'_{h,v})T_0$ of a rough surface or the brightness contrast due to effects of surface roughness $\Delta T_{Bh,v} = (R_{h,v} - R'_{h,v})T_0 = R_{h,v}T_0\{1 - \exp[-(k_0\sigma_h\cos\theta)^2]\}$.

An important practical application of Equations (4.1) and (4.2) is modeling of emissivity and the brightness temperature of a smooth water surface. Complex permittivity of seawater $\varepsilon_w(\lambda;t,s)$ is a function of (t) temperature and (s) salinity (see, e.g., review Meissner and Wentz 2004; some experimental data can be found in the book Akhadov 1980). Calculations of the Fresnel coefficients $r_{h,v}(\lambda;t,s)$ with parameterization $\varepsilon_w(\lambda;t,s)$ in the ranges of $t = 0 - 40°C$ and $s = 0 - 40$psu (Practical Salinity Unit) yield strong sensitivity of $\kappa_{h,v}(\lambda;t,s)$ and $T_{Bh,v}(\lambda,\theta;t,s)$ to variations of the sea surface temperature and salinity in the wavelength range $\lambda = 0.8 - 30$ cm (see, e.g., book Ulaby and Long 2013, Chapter 18). These remarkable dependencies have been investigated and established first in the late 1960s and early 1970s. Today, it is a physical basis for the global monitoring of ocean surface salinity from space using L-band passive microwave radiometers (e.g., the SMOS mission).

The Fresnel reflection theory is successfully applied for solutions of many remote sensing problems, when it is necessary to define core values $\kappa_{Bh, v}(\lambda, \theta)$ and $T_{Bh, v}(\lambda, \theta)$ of natural media without dominated scattering effects (such as soil, sand, ice, show, water). Thus, Fresnel equation framework allows us to estimate an averaged level of the brightness temperature of the planet Earth or other planets that can be measured by microwave radio telescope.

In the Earth's science, spectral and polarization dependencies $T_{Bh, v}(\lambda, \theta)$ provide possibilities to discriminate, classify, and identify different types of natural media (e.g., Ulaby and Long 2013). An additional information can be obtained using polarization dependencies $T_{Bh}(\theta)$ and $T_{Bv}(\theta)$ and polarization ratio

$$PR(\theta) = \frac{T_{Bv}(\theta) - T_{Bh}(\theta)}{T_{Bv}(\theta) + T_{Bh}(\theta)}. \tag{4.4}$$

For example, the polarization ratio can be effectively used for the retrieval of the sea surface wind speed exponent γ from the empirical radiation dependencies $\Delta T_{Bh, v} \sim V^\gamma$ (where $\Delta T_{Bh, v}$ is the brightness temperature contrast and V is the wind speed) when polarized off-nadir radiometric measurements are available (Raizer 2017). Note that local variations of parameter γ measured by a *polarimetric* microwave radiometer can be indicators of wind turbulence over the sea surface.

The Fresnel equations in the form of Stokes vectors and Mueller matrices are used in the analysis of polarized light (Goldstein 2011). The reflection and transmission polarization matrices are also involved in the theory of polarized radiative transfer.

Ultimately, the Fresnel reflection theory, created by *Augustin-Jean Fresnel* (1788–1827) in the middle of the 19th century, has turned into favorite and vast tool in remote sensing studies of all times.

4.10.3 MACROSCOPIC APPROACH

Most natural media represent composite multiphase or multicomponent systems comprising various *microstructural inhomogeneities* (inclusions, particles, or pores). Such media often exhibit turbulent properties (e.g., bubbly flow or aeration layers at sea) which may cause measurable variations in optical and microwave signals.

In some cases, bulk electromagnetic properties and wave propagation characteristics of such media can be described using macroscopic approach known also as *effective medium approximation* or effective medium theory (Choy 2016). This approach is based on the *polarizability* and the electrostatic field solutions of macroscopic Maxwell's equations, averaged over a large volume of the medium, containing inhomogeneities (e.g., Landau, Lifshitz, and Pitaevskii 1984; Von Hippel 1995). Macroscopic approach is valid approximately at $a << \lambda$, where a is a characteristic size of inhomogeneities and λ is the electromagnetic wavelength.

In common case, electromagnetic properties of inhomogeneous and anisotropic macroscopic random medium are described by the vector electromagnetic wave equation

$$\Delta \mathbf{E} + k^2 \mathbf{E} + \mathrm{grad}\left(\mathbf{E} \frac{1}{\varepsilon} \mathrm{grad}\,\varepsilon \right) = 0, \quad k^2(\mathbf{r}) = \frac{\omega^2}{c^2} \varepsilon(\mathbf{r}) \tag{4.5}$$

with complex permittivity $\varepsilon(\mathbf{r}) = <\varepsilon> + \Delta\,\varepsilon(\mathbf{r})$, where $<\varepsilon>$ is the mean, ensemble averaged part of the permittivity, and $\Delta\varepsilon(\mathbf{r})$ is the fluctuation parts which is a function of spatial coordinates $\mathbf{r} = \{x, y, z\}$, $k(\mathbf{r})$ is the propagation constant, and ω is the frequency of the EM wave. Equation (4.5) is valid as for continuous as well as discrete macroscopic medium depending on the distribution of permittivity $\varepsilon(\mathbf{r})$. Mean and fluctuation parts, $<\varepsilon>$ and $\Delta\varepsilon(\mathbf{r})$, can be associated with spatial variations of medium's macroscopic properties, motivated, e.g., by changes in structure, phase content, density, or scattering properties of the particles. Methods of the solution of stochastic wave Equation (4.5) have been considered in Chapter 3 (see also book Tsang et al. 2000).

In our interpretation, Equation (4.5) describes wave propagation effects associated with so-called *macroscopic turbulence* which occurs due to fluctuations of microstructural properties of the medium. The classical case study is turbulence in porous media. Environmental examples are a multiphase flow of complex geometry such as dust jets or ash plumes or simply densely packed systems of particles in the atmosphere or oceans. Electromagnetic properties of such media can be specified through parameterization of permittivity $\varepsilon(\mathbf{r})$.

In remote sensing, the macroscopic or quasi-electrostatic models of composite inhomogeneous media usually are connected with the theory of heterogeneous dielectric mixtures, operated with effective dielectric constant. This theory includes ~10+ different mixing rule formulae to compute effective complex permittivity $\varepsilon_{\mathrm{eff}} = f(\varepsilon_i, \upsilon_i)$ for so-called "matrix" and "statistical" dielectric systems (here ε_i and υ_i are the permittivity and the volume fraction of the i-th phase component). The most known are Bruggeman, de Loor, Clausius–Mossotti, Lorentz–Lorenz, Maxwell-Garnet, Polder-van Sanden, Wiener, Odelevskii, and Landau-Lifshitz mixing formulas (see also book Sihvola 1999).

Well-known example is two-phase air-water mixture like sea foam, dense spray, or aeration layer (bubbly flow). In this case, effective complex permittivity $\varepsilon_{\mathrm{eff}}(\lambda) = \varepsilon'_{\mathrm{eff}}(\lambda) - \iota \varepsilon''_{\mathrm{eff}}(\lambda)$ at microwave frequencies can be represented by the Cole–Cole diagram dependent on the void fraction (Cherny and Raizer 1998). Graph, $\varepsilon''_{\mathrm{eff}}(\lambda)$ versus $\varepsilon'_{\mathrm{eff}}(\lambda)$, computed in a wide range of $\lambda = 0.3 - 30$ cm is useful for analysis of microwave radiometric data and the retrieval of air-water mixture parameters (Raizer 2017).

The simplest way to compute reflection and emission coefficients $r_{h, v}$, $\kappa_{h,v}$, and brightness temperature $T_{Bh, v}$ of a composite multiphase medium is to replace complex permittivity $\varepsilon \rightarrow \varepsilon_{\mathrm{eff}}$ in Fresnel Equations (4.1) and then to use formulae (4.2). The choice of appropriate mixing formula for computing $\varepsilon_{\mathrm{eff}}(\lambda)$ is a critical issue. Note that Fresnel-based reflection and emission models of the effective medium may not always explain experimental data, obtained at grazing observation angles. Dielectric mixture formulae are very popular (because of simplicity) and widely used in many remote sensing applications related to microwave radiometry, polarimetry, and spectroscopy of the Earth's surfaces.

Other relevant case study is propagation (reflection and transmission) of EM waves through a *multilayer dielectric medium* with flat (or quasi-flat) interfaces. It can be as random as well deterministic time-variable multilayer structure with fluctuations of input parameters (i.e., kind of macroscopic "layer" turbulence). Here we have a lot of options for electromagnetic modeling of radiation characteristics, involving different (discrete or continuous) profiles of the complex permittivity $\varepsilon_m(z)$ as a function of the media depth z. Some analytical solutions for specified types of profiles $\varepsilon_m(z)$ and a number of numerical algorithms are available in the literature (for microwaves see, e.g., book Ulaby et al. 1981, Vol. 1).

More complicated variant is a multiphase stratified dielectric system with *vertical profile* of effective complex permittivity $\varepsilon_m(z; \varepsilon_i, \upsilon_i)$ dependent on the fractions of the phase components and their dielectric properties. In this case, the reflection coefficient of the system $R_{h,v}(\lambda, \theta; \varepsilon_m)$ is computed numerically using a method of layer recursions, operated with discrete complex profile $\varepsilon_m(z)$ of any configuration. The brightness temperature of the system is defined by standard formula $T_{Bh,v}(\lambda, \theta; \varepsilon_m) = [1 - R_{h,v}(\lambda, \theta; \varepsilon_m)]T_0$. This method is very flexible and can be applied for analysis of microwave emission of various natural stratified multiphase media (e.g., thick sea foam and whitecap, Raizer 2007, 2017; Anguelova 2008).

Finally, several approaches exist to define the effective complex propagation constant $k_{eff}(\lambda)$ of a nonuniform dielectric medium, containing densely packed particles or other inclusions. In this case, cooperative (or coherent, or multiple scattering) effects due to interactions between particles (mostly spherical form) can be taken into account and calculated numerically using some approximations. The initial theory of multiple scattering is based on the Foldy–Twersky's integral equation; the corresponding formulations can be found in books (Ishimaru 1991; Tsang and Kong 2001). However, remote sensing applications of this particular theory are limited.

4.10.4 WAVE APPROACH: SCATTERING AND EMISSION

Wave approach is a rigorous theoretical framework, operated with the vector wave equation for *inhomogeneous* medium, Equations (3.10) and (3.11), or with the familiar Helmholtz equations for homogeneous medium, Equations (3.20) and (3.21). The corresponding applied electromagnetic theory, remote sensing models, and numerical data are discussed in many publications; we refer the reader to books (e.g., Fung 1994; Tsang et al. 2000, 2001; Tsang and Kong 2001; Fung and Chen 2010).

The most common theories of remote sensing consider two major topics: 1) scattering and emission from random rough surfaces and 2) wave propagation through scattering media, containing distributed particles. In both cases, interactions of electromagnetic fields with medium's nonuniformities (which we specify as "surface" or "volume") can lead to various scattering and emission phenomena.

Scattering on a random rough surface is a wide area of research, where wave propagation characteristics are analyzed for different statistical parameters such as surface height distribution, correlation function, spatial spectrum, or fractal dimension. Several excellent books (Bass and Fuks 1979; Beckmann and Spizzichino

1987; Voronovich 1999; Franceschetti and Riccio 2007; Fung 2015) provide rigorous mathematical treatment of the scattering on random rough surfaces.

For example, small perturbation theory yields simple formulae for computing radar backscatter coefficient (e.g., Valenzuela 1978)

$$\sigma_{0pq}(\lambda;\theta) = 16\pi k_0^4 \cos^4\theta \left|g_{pq}(\theta)\right|^2 W(K,0), \qquad (4.6)$$

where $k_0 = 2\pi/\lambda$ is radar wavenumber, K is the surface wavenumber, $W(K,0)$ is the wavenumber spectrum of the surface with integration over all wavenumbers where $\iint W(K_x, K_y)dK_x dK_y = \langle\xi^2\rangle$ is the surface displacement variance; subscript p and q demote polarization (p = h, v) or (q = v, h). In Equation (4.6), factor g_{pq} depends on the polarization of the transmitted and received signals and the backscattering coefficient $\sigma_{0pq}(\lambda;\theta)$ is defined by only Bragg resonance component $K = 2k_0 \sin\theta$ in the spectrum, i.e., $W(K = 2k_0 \sin\theta, 0)$. To the first order, for cross-polarization factors $g_{h,v} = g_{v,h} = 0$ (that is $\sigma_{0h,v} = \sigma_{0v,h} = 0$); for co-polarization they are

$$g_{hh}(\lambda;\theta) = \frac{\varepsilon - 1}{\left[\cos\theta + \left(\varepsilon - \sin^2\theta\right)^{1/2}\right]^2}, \qquad (4.7)$$

$$g_{vv}(\lambda;\theta) = \frac{(\varepsilon-1)\left[\varepsilon\left(1+\sin^2\theta\right) - \sin^2\theta\right]}{\left[\varepsilon\cos\theta + \left(\varepsilon - \sin^2\theta\right)^{1/2}\right]^2}, \qquad (4.8)$$

where $\varepsilon(\lambda)$ is the complex permittivity of the medium. Equations (4.6)–(4.8) represent well-known Bragg theory of radar backscatter from the ocean surface with permittivity $\varepsilon = \varepsilon_w(\lambda;t,s)$. In particular, these formulae have been used extensively in computations of the dependency of the backscattering coefficient on ocean surface wind speed in context with interpretation of recent satellite SAR observations (e.g., Hwang et al. 2010).

Spectral emissivity of a rough surface can be defined as

$$\kappa_p(\lambda;\theta_0,\varphi_0) = 1 - \frac{1}{4\pi}\iint[\sigma_{pp}(\lambda;\theta_0,\phi_0;\theta_s,\varphi_s) + \sigma_{pq}(\lambda;\theta_0,\phi_0;\theta_s,\phi_s)]d\Omega_s, \qquad (4.9)$$

where σ_{pp} and σ_{pq} are the bistatic scattering coefficients at co- and cross-polarizations (p = h, v) or (q = v, h); θ_0, ϕ_0; θ_s, φ_s are the angular coordinates for incident (emitted) and scattered radiation, and $d\Omega_s = \sin\theta_s d\theta_s d\varphi_s$ is the elementary solid angle. Note that computed by integral Equation (4.9) emissivity is much less sensitive to surface roughness variations than the input scattering coefficient σ_{pp}.

Since the 1990s, direct numerical simulations of Maxwell's equations have been proposed and developed for the problems of scattering and emission from a random rough surface. These works included a number of integral methods, providing computations of electromagnetic fields, interacted with the surface. The discussion is beyond the scope of this chapter and we encourage the interested reader to find out the corresponding literature data. Our understanding is that numerical Maxwell's simulations enable to provide more accurate analysis of scattering and emission fields especially related to time-varying complex systems. As a relevant application, such computations would make considerable contributions to remote sensing of turbulent media.

Wave propagation through a scattering medium is often associated with the Mie (1908) theory of scattering and absorption of electromagnetic waves for a single sphere of different size. The classical Mie theory is discussed in monographs (van de Hulst 1957; Kerker 1969; Deirmendjian 1969; Bohren and Huffman 1983; Shifrin 1988; Hergert and Wriedt 2012). The theory has also been developed for scattering by inhomogeneous and anisotropic particles (e.g., Babenko et al. 2003) as well as non-spherical particles (Schuerman 1980; Mishchenko et al. 2000; Rother and Kahnert 2014). The subject has also wide engineering applications (Dombrovsky and Baillis 2010). Theoretically, scattering depends upon the wavelength of radiation being scattered and the size of the particles causing the scattering. Spectral characterization of scattering for atmospheric particles is shown in Figure 4.11.

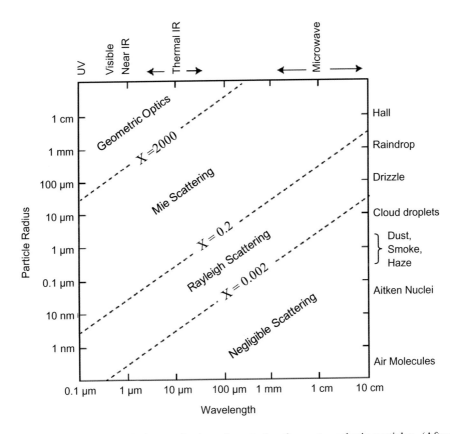

FIGURE 4.11 Spectral characterization of scattering from atmospheric particles. (After Petty 2006).

According to the Mie theory, scattering efficiency factor Q_s, extinction efficiency factor Q_e, and absorption efficiency factor Q_a of a spherical particle are defined as

$$Q_s(a) = \frac{2}{x^2} \sum_{n=1}^{\infty} (2n+1)\left(\left|a_n\right|^2 + \left|b_n\right|^2\right), \tag{4.10}$$

$$Q_e(a) = \frac{2}{x^2} \sum_{n=1}^{\infty} (2n+1)\,\mathrm{Re}\left(a_n + b_n\right), \tag{4.11}$$

$$Q_a(a) = Q_e - Q_s, \tag{4.12}$$

where $x = ka = 2\pi a/\lambda$ is the diffraction (or size) parameter ($0.1 < x < 100$), a is the radius of spherical particle, and λ the electromagnetic wavelength. The Mie scattering complex coefficients are given by (van de Hulst 1981)

$$a_n = \frac{\psi_n(x)\,\psi_n'(mx) - m\psi_n(mx)\,\psi_n'(x)}{\zeta_n(x)\,\psi_n'(mx) - m\psi_n(mx)\,\zeta_n'(x)}, \tag{4.13}$$

$$b_n = \frac{m\psi_n(x)\,\psi_n'(mx) - \psi_n(mx)\,\psi_n'(x)}{m\zeta_n(x)\,\psi_n'(mx) - \psi_n(mx)\,\zeta_n'(x)}, \tag{4.14}$$

where $\psi_n(x) = \sqrt{\pi x/2} \cdot J_{n+1/2}(x)$ and $\zeta_n(x) = \sqrt{\pi x/2} \cdot H_{n+1/2}^{(1)}(x)$ expressed in terms of the Bessel function and the Hankel function of the first kind; $m = \sqrt{\varepsilon}$ is the complex refractive index of a particle material. An example of calculations $Q_{s,e,a}(a)$ for single water droplets at optical wavelengths is shown in Figure 4.12. Resonance effects are revealed very well. The corresponding spectral cross sections of a spherical particle are given by $\sigma_{s,e,a} = \pi a^2 Q_{s,e,a}$.

Numerous calculations using Mie theory have shown that there are several regions of the scattering, defined by the value of diffraction parameter x and complex refractive index m of the spherical particle. Correspondingly, the following approximations are known:

- Rayleigh approximation: $x \ll 1$ and complex refractive index m is arbitrary (applies to scattering by small particles).
- Rayleigh–Gans approximation: $2x|m-1| \ll 1$ and $|m-1| \ll 1$ (applies to scattering by non-spherical particles).
- Perelman approximation (also called S-approximation): x arbitrary but $m \to 1$ (x region is wider than by Rayleigh and Rayleigh–Gans approximations).
- Resonant Mie scattering: $1 \le x \le 20$, ($a \sim \lambda$), both parameters x and m are arbitrary.
- van de Hulst approximation (also called the anomalous diffraction approximation): $x \to \infty$ and $|m-1| \ll 1$ (applied to scattering by large soft particles).
- Geometrical optics approximation: $x \ge 20$ and refractive index m is real (applies to scattering by large particles).

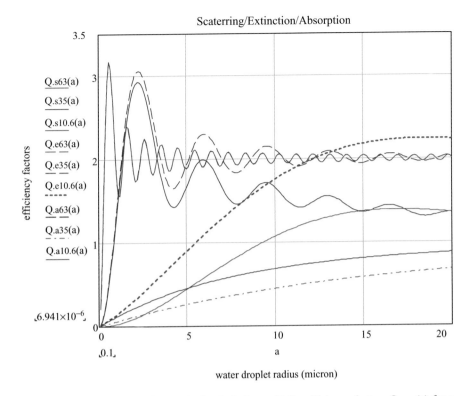

Scaterring/Extinction/Absorption

efficiency factors

Q.s63(a)
Q.s35(a)
Q.s10.6(a)
Q.e63(a)
Q.e35(a)
Q.e10.6(a)
Q.a63(a)
Q.a35(a)
Q.a10.6(a)

$6.941×10^{-6}$

0.1

a

water droplet radius (micron)

FIGURE 4.12 Example of numerical calculations of Mie efficiency factors $Q_{s,\,e,\,a}(a)$ for a single spherical particle of radius a. Data correspond to scattering, extinction, and absorption of laser light by water droplets (fine fog) at wavelengths 630 nm, 3500 nm, and 10.6 μm (10600 nm). Note: $Q_{s63}(a) \approx Q_{e63}(a)$.

The spectral volume coefficients for a system of particles of different size (known as *polydispersed* system of particles) are given by

$$\langle \sigma_{s,e,a} \rangle = \pi \int_0^\infty Q_{s,e,a}(a) a^2 f(a) da, \qquad (4.15)$$

where f(a) is the size distribution probability density function ($\int_0^\infty f(a) da = 1$). The size distribution can be described by the two-parameter gamma function

$$f(a) = \frac{A^{B+1}}{\Gamma(B+1)} a^B \exp(-Aa), \qquad (4.16)$$

where A and B are empirical coefficients. Electromagnetic scattering on polydispersed systems of spherical particles is considered in great detail in the book (Deirmendjian 1969).

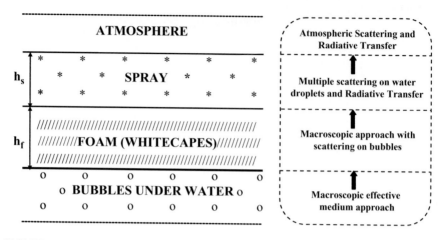

FIGURE 4.13 Combined electromagnetic microwave model "*Sandwich*" of the ocean surface at very high winds and hurricanes. Thickness of the layers: $h_s \approx 0.1 - 10m$, $h_f \approx 10^{-3} - 10^{-1}$ m. (Based on Raizer 2017.)

An interesting case is light scattering from a bubble in water (Jonasz and Fournier 2007).

The scattering by particles is central problem in atmospheric optics, aerosol optics, and radiative transfer (e.g., Kokhanovsky 2008). Numerous theoretical studies and computations have been made in this regard over the years. Model data vary largely depending on geometry and structure of scattering media. In particular, atmospheric turbulence due to scattering reduces contrast of optical images and changes spectral signatures of ground objects.

Real-world media often exhibit turbulent properties associated with the *fluctuations of scatterer density* resulting in the changes of volume scattering and emissivity. In the case of dispersed systems with small particles, these effects can be taken into account using the effective medium approximation, e.g., the Lorentz–Lorenz equation (Kerker 1969).

4.10.5 Radiative Transfer Theory

The radiative transfer theory (RTT), also called radiation transport, is a phenomenological and well-established mathematical theory, describing the interaction of radiation with matter. RTT is considered in classical books (Chandrasekhar 1960; Sobolev 1963; Preisendorfer 1965; Ishimaru 1991; Apresyan and Kravtsov 1996). The physical basis of RTT is the law of conservation of energy. Applicability of this law to propagation of radiance through scattering material media is formulated as the following: the difference between the intensity that leaves the unit volume and the intensity that enters it must be equal to the algebraic sum of the power loss due to absorption and scattering (extinction) and the total multiple scattering contribution within the unit volume.

The RTT is governed by the integro-differential equation known as radiative transfer equation (RTE). In vector notation, the radiative transfer equation for a plane-parallel scattering medium has the following form:

$$\mu \frac{d\mathbf{I}(\tau,\mu,\phi)}{d\tau} = \mathbf{I}(\tau,\mu,\phi) - \frac{\omega_0(\tau)}{4\pi} \int_0^{2\pi} \int_{-1}^{1} \mathbf{P}(\mu,\phi,\mu',\phi') \mathbf{I}(\tau,\mu',\phi') d\mu' d\phi' + \mathbf{F}(\tau,\mu,\phi), \quad (4.17)$$

where $\mathbf{I} = (I,Q,U,V)^T$ is the specific four-component intensity defined by the Stokes parameters (I,Q,U,V), $\mathbf{P}$ is the phase matrix, $\mathbf{F}$ is the source vector, τ is the optical depth, $\mu = \cos\theta$, θ is the zenith angle ($\theta < 90°$ for upward and $\theta > 90°$ for downward), ϕ is the azimuthal angle, and ω_0 is the single scattering albedo.

Commonly, the RTE is solved using numerical methods, e.g., with the finite volume method or with Monte Carlo simulations. Chandrasekhar (1960) in his book provides an extensive treatment of the RTE for a finite plane-parallel layer under a variety of conditions involving scattering and absorption. Analytic solutions are considered for special cases of diffuse and isotropic reflection and scattering as well as for Rayleigh's phase function. The most common methods of the RTE solution are based on the two-stream and diffusion approximations (Ishimaru 1991).

The RTE constitutes the physical basis of optical and microwave remote sensing techniques due to its relative simplicity and capability to deal with multiple-scattering effects. In the case of geophysical source, RTE can be represented in terms of emissivity and the brightness temperature. For example, the brightness temperature of the Earth's surface, observed by satellite-borne microwave radiometer $T_{Bh,v}(\lambda,\theta)$ is given by

$$\begin{aligned} T_{Bh,v}(\lambda,\theta) &= \kappa_{h,v}(\lambda,\theta) T_0 \exp(-\tau_0/\mu) + T_{atm}^{\uparrow}(\lambda,\theta) + T_{atm}^{\downarrow}(\lambda,\theta)\left[1 - \kappa_{h,v}(\lambda,\theta)\right] \\ &\quad \exp(-\tau_0/\mu) + T_c\left[1 - \kappa_{h,v}(\lambda,\theta)\right]\exp(-2\tau_0/\mu), \end{aligned} \quad (4.18)$$

where $\kappa_{h,v}(\lambda,\theta)$ is spectral emissivity of the surface, T_0 is thermodynamic temperature in Kelvin, $T_{atm}^{\uparrow}(\lambda,\theta)$ and $T_{atm}^{\downarrow}(\lambda,\theta)$ are the upwelling and downwelling brightness temperatures of atmosphere, respectively, $T_c = 2.7K$ is the brightness temperature of cosmic background radiation, and τ_0 is the opacity of the atmosphere (total absorption coefficient). Equation (4.18) is widely used in global modeling of microwave emission from the ocean-atmosphere system. Simplified retrieval algorithms based on Equation (4.18) have been developed as well (see, e.g., papers Guissard 1998; Mitnik and Mitnik 2003).

As a whole, the RTT/RTE is a complex mathematical framework having various options and modifications for analyses, solutions, and applications in atmospheric physics, astrophysics, applied physics, optics, planetary sciences, atmospheric sciences, meteorology, climatology, and remote sensing (Goody and Yung 1995; Liou 2002; Petty 2006; Kuznetsov et al. 2012; Stamnes et al. 2017).

Wave propagation through turbulent continuous/discrete medium with scattering and absorption is a common problem in many remote sensing applications. In this case, combination of phenomenological RTT and macroscopic/wave electromagnetic theory is an appropriate option to provide relevant practical results. Various model configurations with particle and medium properties can be considered in this regard. One realistic example related to the ocean, is bubble-foam-spray "sandwich"

(Figure 4.13), representing complex nonstationary *turbulent fluid flow of different particles and water jets* – dispersed two-phase system, formed on the sea surface at very high winds and/or hurricanes. Complicated interplay between multiple scattering, radiative transfer, and macroscopic (dielectric) effects take place that yields specific and sometimes unexpected spectral signatures in thermal microwave emission (Cherny and Raizer 1998, Raizer 2007) and in radar backscatter (Raizer 2012, 2013) as well.

4.10.6 A Simple Image Formation Model

Formation of images is one of the most important aspects in remote sensing. Historically, this subject is overlapped from physical optics and digital signal processing. Today we can distinguish several types of remotely sensed images: optical/IR (multispectral, hyperspectral, lidar), radar (SAR, Doppler, polarimetric), microwave radiometer (multiband, polarimetric), and some others. Although physical principles of remote sensing imaging are different for various instruments; however, a simple image formation model can be generally formulated in terms of 2D spatial convolution

$$g(x,y) = \iint f(x',y')h(x'-x,y'-y,)dx'dy' + \eta(x,y),$$ (4.19)

where $g(x,y)$ is the observed image or signal, $f(x,y)$ is the actual image, $h(x,y)$ is the impulse response which is often called the point spread function (PSF, see also Section 3.9), and $\eta(x,y)$ is additive noise. In digital domain, the 2D convolution is performed by multiplying and accumulating the instantaneous values of the overlapping samples corresponding to two input signals, one of which is flipped twice

$$g(i,j) = \sum_{m=-\infty}^{\infty}\sum_{n=-\infty}^{\infty} f[m,n] \cdot h[i-m,j-n],$$ (4.20)

where matrix $g(i,j)$ is output or filtered image, matrix $f[m,n]$ is actual input image, and matrix $h(i,j)$ is the filter kernel. The range of the convolution is determined by both data array and the kernel size. Here elements m and n are concerned with the image matrixes while those of elements i and j are related to the kernel (noise is omitted). Discrete convolution (4.20) is a basic concept in digital image processing (e.g., Pratt 2007).

The image formation process usually results in a blurring of the actual image; fine details maybe lost. In Fourier domain, this leads to degradation of high frequency spatial components. For example, the resulting grayscale image becomes *bandlimited*. More information about remote sensing image formation can be found in book (Blahut 2004).

As mentioned in Section 3.9, a convolution model (4.19) describes an *image restoration* process mostly used in optics. Restoration methods require complete knowledge of the spatiotemporal PSF which involves both instrument and medium parameters (called also statistical or turbulent PSF). The task then is to estimate the signal f(x,y), and perhaps h(x,y) from the observation g(x,y).

The noise $\eta(x,y)$ is unknown, but its mean value and 2D correlation function are usually known.

The convolution model (4.19) can also be applied for spaceborne passive microwave imagery; in this case, the resulting microwave image can be represented in terms of the brightness temperature field

$$T_B(x,y) = \iint T_{B0}(x',y')G(x'-x,y'-y,)dx'dy' + noise, \quad (4.21)$$

where $T_B(x,y)$ is the measured radio brightness image, $T_{B0}(x,y)$ is actual brightness temperature of the scene, and $G(x,y)$ is the spatial response function dependent on both radiometer antenna gain pattern and observation geometry. This relation is well known from radio astronomy; however, in the case of multiband scanning microwave radiometer of high resolution, accurate reconstruction of radio brightness image $T_{B0}(x,y)$ for nonstationary scene is not a trivial problem (e.g., Long and Brodzik 2016). For stationary scenes, image reconstruction algorithm is based on a Fourier synthesis.

4.11 CONCLUSIONS

This chapter is a brief overview of remote sensing. The author discussed only those topics which he believed can deserve more attention in context with the title of this book and related problems. Here we focused on satellite technology because it is the most promising option for exploring our dynamical turbulent environment that can be applied aslocally as well globally. Moreover, satellite optical-microwave observations provide great visualization of geophysical flows of various geometries and scales and this is very important circumstance for detection of 2D/3D turbulence in view of topological changes and time *reversal symmetry*.

Remote sensing is a vast area of scientific and engineering activity. Today comprehensive remote sensing is considered as the ultimate tool for exploration of our environments from space. It is quickly becoming the new frontier in education, economics, and politics.

Remote sensing is qualified as multistage *cumulative IT process*, involving data acquisition, analysis, utilization, management, and application. Miscellaneous aspects are data assessment, theoretical modeling, statistical treatment, interpretation, and retrieval and mapping. In this context, satellite remote sensing is the state-of-the art technology, where applied science and innovative developments play a key role in obtaining new data and knowledge. Remote sensing thereby enables future exploration success.

Satellite remote sensing has advantages and certain disadvantages associated with *overall difficulty* in providing relevant data utilization solutions and *quantitative interpretation* (QI) of phenomena/events of the interest. There is also a tension between the observational aspects and the environmental benefits. Advantages can be formulated by three sentences: global observing, great capabilities, and multipurpose applicability. Typical technical disadvantages include payload limitations, low operational flexibility, high cost and complexity of the implementation, and possible measurement biases as well.

> *Perhaps the greatest limitation is that it is often oversold. Remote sensing is not a panacea that will provide all the information needed to conduct physical, biological, or social science research. It simply provides some spatial, spectral, and temporal information of value in a manner that we hope is efficient and economical.*

(Jensen 2014, page 8)

This is not exactly true because remote sensing inspires a real *breakthrough in turbulence science*.

4.12   NOTES ON THE LITERATURE

Books (Slater 1980; Joseph 2005; Elachi and van Zyl 2006; Jin 2006; Sabins 2007; Olsen 2007; Campbell and Wynne 2011; Rees 2013; Jensen 2014; Njoku 2014; Lavender and Lavender 2015; Khorram et al. 2016) provide basic information on various remote sensing methods, instruments, theory, models, and applications.

Microwave remote sensing is considered in fundamental 3-volume book set (Ulaby et al. 1981, 1982, 1986) and also in comprehensive books (Shutko 1986; Janssen 1993; Sharkov 2003; Woodhouse 2005; Matzler 2006; Ulaby and Long 2013; Karmakar 2014; Weng 2017).

Remote sensing studies of the oceans are reviewed in books (Cherny and Raizer 1998; Robinson 2010; Hou 2013; Martin 2014; Grankov and Milshin 2015; Raizer 2017, 2019).

Book (Skou and Le Vine 2006) addresses the design and construction of microwave radiometer systems for remote sensing.

Recent books (Emery and Camps 2017; Chuvieco 2020) are perfect introduction to satellite remote sensing.

Satellite payloads, missions, and global observations are reviewed in books (Kramer 2002; Tan 2014; Qian 2016; Kelkar 2017; Ilčev 2019). Satellite orbits and parameters are described in books (Montenbruck and Gill 2000; Capderou 2005).

REFERENCES

Akhadov, Y. Y. 1980. *Dielectric Properties of Binary Solutions: A Data Handbook.* Pergamon, Oxford, England.

Anguelova, M. D. 2008. Complex dielectric constant of sea foam at microwave frequencies. *Journal of Geophyical Research* 113, C08001. doi: 10.1029/2007JC004212.

Apresyan, L. A. and Kravtsov, Y. A. 1996. *Radiation Transfer: Statistical and Wave Aspects (translated from Russian by M. G. Edelev).* Gordon and Breach Publishers, Amsterdam, The Netherlands (reprinted by CRC Press, Boca Raton, FL, 2019).

Babenko, V. A., Astafyeva, L. G. and Kuzmin, V. N. 2003. *Electromagnetic Scattering in Disperse Media: Inhomogeneous and Anisotropic Particles.* Springer – Praxis Publishing, Chichester, UK.

Bass, F. G. and Fuks, I. M. 1979. *Wave Scattering from Statistically Rough Surfaces (translated from Russian and edited by C. B. Vesecky and J. F. Vesecky).* Pergamon, Oxford, New York.

Beckmann, P. and Spizzichino, A. 1987. *The Scattering of Electromagnetic Waves from Rough Surfaces*. Artech House, Norwood, MA.

Blahut, R. E. 2004. *Theory of Remote Image Formation*. Cambridge University Press, Cambridge, UK.

Bohren, C. F., and Huffman, D. R. 1983. *Absorption and Scattering of Light by Small Particles*. John Wiley & Sons, New York.

Born, M. and Wolf, E. 1999. *Principles of Optics*, 7th (expanded) edition. Cambridge University Press, Cambridge, UK.

Campbell, J. B. and Wynne, R. H. 2011. *Introduction to Remote Sensing*, 5th edition. The Guilford Press, New York.

Capderou, M. 2005. *Satellites: Orbits and Missions (translated by S. Lyle)*. Springer, Paris, France.

Chandrasekhar, S. 1960. *Radiative Transfer*. Dover Publications, New York.

Cherny, I. V. and Raizer, V. Yu. 1998. *Passive Microwave Remote Sensing of Oceans*. Wiley, Chichester, England.

Choy, T. C. 2016. *Effective Medium Theory: Principles and Applications (International Series of Monographs on Physics 165)*, 2nd edition. Oxford University Press, Oxford, UK.

Chuvieco, E. 2020. *Fundamentals of Satellite Remote Sensing: An Environmental Approach*, 3rd edition. CRC Press, Boca Raton, FL.

Congalton, R. G. and Green, K. 2019. *Assessing the Accuracy of Remotely Sensed Data: Principles and Practices*, 3rd edition. CRC Press, Boca Raton, FL.

Deirmendjian, D. 1969. *Electromagnetic Scattering on Spherical Polydispersions*. American Elsevier Publishing Company, New York.

Dey, N., Bhatt, C., and Ashour, A. S. (Eds.). 2019. *Big Data for Remote Sensing: Visualization, Analysis and Interpretation: Digital Earth and Smart Earth*. Springer International Publishing, Switzerland.

Dombrovsky, L. A. and Baillis, D. 2010. *Thermal Radiation in Disperse Systems: An Engineering Approach*. Begell House Publishers Inc., Redding, CT.

Elachi, C. and van Zyl, J. 2006. *Introduction to the Physics and Techniques of Remote Sensing*, 2nd edition. John Wiley & Sons, Hoboken, NJ.

Emery, W. and Camps, A. 2017. *Introduction to Satellite Remote Sensing: Atmosphere, Ocean, Land and Cryosphere Applications*. Elsevier, Amsterdam, The Netherlands.

Franceschetti, G. and Riccio, D. 2007. *Scattering, Natural Surfaces, and Fractals*. Elsevier Academic Press, San Diego, CA.

Fung, A. K. 1994. *Microwave Scattering and Emission Models and Their Applications*. Artech House, Norwood, MA.

Fung, A. K. 2015. *Backscattering from Multiscale Rough Surfaces with Application to Scatterometry*. Artech House, Norwood, MA.

Fung, A. K. and Chen, K.-S. 2010. *Microwave Scattering and Emission Models for Users*. Artech House, Norwood, MA.

Goldstein, D. H. 2011. *Polarized Light*, 3rd edition. CRC Press, Boca Raton, FL.

Goody, R. M. and Yung, Y. L. 1995. *Atmospheric Radiation: Theoretical Basis*, 2nd edition. Oxford University Press, New York.

Grankov, A. G. and Milshin, A. A. 2015. *Microwave Radiation of the Ocean-Atmosphere: Boundary Heat and Dynamic Interaction*, 2nd edition. Springer International Publishing, Cham, Switzerland.

Guissard, A. 1998. The retrieval of atmospheric water vapor and cloud liquid water over the oceans from a simple radiative transfer model: Application to SSM/I data. *IEEE Transactions on Geoscience and Remote Sensing*, 36(1):328–332. doi: 10.1109/36.655346.

Hecht, E. 2017. *Optics*, 5th edition. Pearson Education Limited, Harlow, UK.

Hergert, W., and Wriedt, T. (Eds.). 2012. *The Mie Theory: Basics and Applications*. Springer-Verlag, Berlin, Heidelberg.

Hou, W. W. 2013. *Ocean Sensing and Monitoring: Optics and Other Methods.* SPIE Press, Bellingham, WA.

Hwang, P. A., Zhang, B., Toporkov, J. V., and Perrie, W. 2010. Comparison of composite Bragg theory and quad-polarization radar backscatter from RADARSAT-2: With applications to wave breaking and high wind retrieval. *Journal of Geophysical Research,* 11(C8) C08019. doi:10.1029/2009jc005995.

Ilčev, S. D. 2019. *Global Satellite Meteorological Observation (GSMO) Applications: Volume 2.* Springer Nature, Switzerland.

Ishimaru, A. 1991. *Electromagnetic Wave Propagation, Radiation, and Scattering.* Englewood Cliffs, Prentice Hall, NJ.

Janssen, M. A. 1993. *Atmospheric Remote Sensing by Microwave Radiometry.* John Wiley & Sons, New York.

Jensen, J. R. 2014. *Remote Sensing of the Environment: An Earth Resource Perspective,* 2nd edition. Pearson Education Limited, England.

Jin, Ya-Q. 2006. *Theory and Approach of Information Retrievals from Electromagnetic Scattering and Remote Sensing.* Springer, Dordrecht, The Netherlands.

Jonasz, M. and Fournier, G. 2007. *Light Scattering by Particles in Water: Theoretical and Experimental Foundations.* Elsevier – Academic Press, London, UK.

Joseph, G. 2005. *Fundamentals of Remote Sensing,* 2nd edition. University Press, (India) Private Limited, Hyderguda, Hyderabad, India.

Karmakar, P. K. 2014. *Ground-Based Microwave Radiometry and Remote Sensing: Methods and Applications.* CRC Press, Boca Raton, FL.

Kelkar, R. R. 2017. *Satellite Meteorology,* 2nd edition. PS Publication, Hyderabad, India – CRC Press, Boca Raton, FL.

Kerker, M. 1969. *The Scattering of Light and Other Electromagnetic Radiation.* Academic Press, New York.

Khorram, S., van der Wiele, C. F., Koch, F. H., Nelson, S. A. C., and Potts, M. D. 2016. *Principles of Applied Remote Sensing.* Springer Science, New York, Dordrecht.

Kokhanovsky, A. A. 2008. *Aerosol Optics: Light Absorption and Scattering by Particles in the Atmosphere.* Springer, Praxis Publishing, Chichester, UK.

Kramer, H. J. 2002. *Observation of the Earth and Its Environment: Survey of Missions and Sensors,* 4th edition. Springer, Berlin, Germany.

Kupperberg, P. 2003. *Spy Satellites.* The Rosen Publishing Group, New York.

Kuznetsov, A., Melnikova, I., Pozdnyakov, D., Seroukhova, O., and Vasilyev, A. 2012. *Remote Sensing of the Environment and Radiation Transfer: An Introductory Survey.* Springer, Berlin, Germany.

Landau, L. D., Lifshitz, E. M., and Pitaevskii, L.P. 1984. *Electrodynamics of Continuous Media. Course of Theoretical Physics, Vol. 8 (translated from Russian by J. B. Sykes, J. S. Bell, and M. J. Kearsley),* 2nd edition. Elsevier – Butterworth-Heinemann, Burlington, MA.

Lavender, S. and Lavender, A. 2015. *Practical Handbook of Remote Sensing.* CRC Press, Boca Raton, FL.

Lillesand, T., Kiefer, R. W., and Chipman, J. 2015. *Remote Sensing and Image Interpretation,* 7th edition. John Wiley & Sons, Hoboken, NJ.

Liou, K.-N. 2002. *An Introduction to Atmospheric Radiation,* 2nd edition. Academic Press, San Diego, CA.

Long, D. G. and Brodzik, M. J. 2016. Optimum image formation for spaceborne microwave radiometer products. *IEEE Transactions on Geoscience and Remote Sensing,* 54(5):2763–2779. doi: 10.1109/TGRS.2015.2505677.

Lyon, J. G. and Lunetta, R. S. (Eds.). 2005. *Remote Sensing and GIS Accuracy Assessment(Mapping Science).* CRC Press, Boca Raton, FL.

Maini, A. K. and Agrawal, V. 2014. *Satellite Technology: Principles and Applications,* 3rd edition. John Wiley & Sons, Chichester, UK.

Martin, S. 2014. *An Introduction in Ocean Remote Sensing*, 2nd edition. Cambridge University Press, Cambridge, UK.

Matzler, C. (Ed.). 2006. *Thermal Microwave Radiation: Applications for Remote Sensing*. The Institution of Engineering and Technology, London, UK.

Meissner, T. and Wentz, F. J. 2004. The complex dielectric constant of pure and sea water from microwave satellite observations. *IEEE Transactions on Geoscience and Remote Sensing*, 42(9):1836–1849. doi:10.1109/TGRS.2004.831888.

Mie, G. 1908. Beitrage zur Optik truber Medien, speziell kolloidaler Metallosungen. *Annalen Der Physik*, 330(3):377–445. doi:10.1002/andp.19083300302.

Mitnik, L. M., Kuleshov, V. P., Mitnik, M. L., Streltsov, A. M., Chernyavsky, G. M., and Cherny, I. V. 2017. Microwave scanner-sounder MTVZA-GY on new Russian meteorological satellite Meteor-M No. 2: Modeling, calibration, and measurements. *IEEE Journal of Selected Topics in Applied Earth Observations and Remote Sensing*, 10(7):3036–3045. doi:10.1109/JSTARS.2017.2695224.

Mitnik, L. M., and Mitnik, M. L. 2003. Retrieval of atmospheric and ocean surface parameters from ADEOS-II Advanced Microwave Scanning Radiometer (AMSR) data: Comparison of errors of global and regional algorithms. *Radio Science*, 38(4) 8065. doi:10.1029/2002rs002659.

Mishchenko, M. I., Hovenier, J. W., and Travis, L. D. (Eds.). 2000. *Light Scattering by Nonspherical Particles: Theory, Measurements, and Applications*. Academic Press, San Diego, CA.

Montenbruck, O. and Gill, E. 2000. *Satellite Orbits: Models, Methods and Applications*. Springer-Verlag, Berlin.

Njoku, E. G. (Ed.). 2014. *Encyclopedia of Remote Sensing (Encyclopedia of Earth Sciences Series)*. Springer, New York.

Qian, S.-E. (Ed.). 2016. *Optical Payloads for Space Missions*. John Wiley & Sons, Chichester, UK.

Olsen, R. C. 2007. *Remote Sensing from Air and Space*. SPIE Press, Bellingham, WA.

Petty, G. W. 2006. *A First Course in Atmospheric Radiation*, 2nd edition. Sundog Publishing, Madison, WI.

Pratt, W. K. 2007. *Digital Image Processing*, 4th edition. John Wiley & Sons, Hoboken, NJ.

Preisendorfer, R. W. 1965. *Radiative Transfer on Discrete Spaces*. Pergamon Press, New York.

Pryor, G. (Ed.). 2012. *Managing Research Data*. Facet Publishing, London, UK.

Raizer, V. 2007. Macroscopic foam-spray models for ocean microwave radiometry. *IEEE Transactions on Geoscience and Remote Sensing*, 45(10):3138–3144. doi:10.1109/TGRS.2007.895981.

Raizer, V. 2012. Microwave scattering model of sea foam. In *IEEE International Geoscience and Remote Sensing Symposium*, 22–27 July 2012, Munich, Germany, pp. 5836–5839. doi:10.1109/IGARSS.2012.6352282.

Raizer, V. 2013. Radar backscattering from sea foam and spray. In *IEEE International Geoscience and Remote Sensing Symposium*, 21–26 July 2013, Melbourne, VIC, Australia, pp. 4054–4057. doi:10.1109/IGARSS.2013.6723723.

Raizer, V. 2017. *Advances in Passive Microwave Remote Sensing of Oceans*. CRC Press, Boca Raton, FL.

Raizer, V. 2019. *Optical Remote Sensing of Ocean Hydrodynamics*. CRC Press, Boca Raton, FL.

Rees, W. G. 2013. *Physical Principles of Remote Sensing*, 3rd edition. Cambridge University Press, Cambridge, UK.

Robinson, I. S. 2010. *Discovering the Ocean from Space: The Unique Applications of Satellite Oceanography*. Springer, Berlin, Germany.

Rother, T. and Kahnert, M. 2014. *Electromagnetic Wave Scattering on Nonspherical Particles: Basic Methodology and Simulations*, 2nd edition. Springer-Verlag, Berlin, Heidelberg.

Sabins, F. F. 2007. *Remote Sensing: Principles and Interpretations*, 3rd edition. Waveland Press, Long Grove, IL.

Schuerman, D. W. (Eds.). 1980. *Light Scattering by Irregularly Shaped Particles*. Plenum Press, New York.

Sharkov, E. A. 2003. *Passive Microwave Remote Sensing of the Earth: Physical Foundations*. Springer Praxis Books, Chichester, UK.

Shifrin, K. S. 1988. *Physical Optics of Ocean Water (translated from Russian by D. Oliver)*. AIP Translation Series. American Institute of Physics, Woodbury, NY.

Sihvola, A. 1999. *Electromagnetic Mixing Formulas and Applications (IEE Electromagnetic Waves Series 47)*. The Institution of Electrical Engineers, London, UK.

Shutko, A. M. 1986. *Microwave Radiometry of a Water Surface and The Ground*. Nauka, Moscow (in Russian: Шутко А. М. СВЧ-радиометрия водной поверхности и почвогрунтов. Москва, Наука, 1986).

Skou, N. and Le Vine, D. M. 2006. *Microwave Radiometer Systems: Design and Analysis*, 2nd edition. Artech House, Norwood, MA.

Slater, P. N. 1980. *Remote Sensing: Optics and Optical Systems*. Addison-Wesley Publishing Company, Reading, MA.

Sobolev, V. V. 1963. *A Treatise on Radiative Transfer (translated from Russian by S. I. Gaposchkin)*. Van Nostrand, Princeton, NJ.

Stamnes, K., Thomas, G. E., and Stamnes, J. J. 2017. *Radiative Transfer in the Atmosphere and Ocean*, 2nd edition. Cambridge University Press, Cambridge, UK.

Stratton, J. A. 1941. *Electromagnetic Theory*. McGraw-Hill Book Company (reprinted by John Wiley & Sons– IEEE Press, 2007).

Swarnalatha, P. and Sevugan, P. (Eds.). 2018. *Big Data Analytics for Satellite Image Processing and Remote Sensing*. IGI Global Engineering Science Reference, Hershey, PA.

Tan, S.-Y. 2014. *Meteorological Satellite Systems*. International Space University – Springer, New York.

Trishchenko, A. P., Garand, L., and Trichtchenko, L. D. 2019. Observing Polar Regions from Space: Comparison between Highly Elliptical Orbit and Medium Earth Orbit Constellations. *Journal of Atmospheric and Oceanic Technology*, 36(8):1605–1621. doi:10.1175/jtech-d-19-0030.1.

Tsang, L. and Kong, J. A. 2001. *Scattering of Electromagnetic Waves: Advanced Topics*. John Wiley & Sons, New York.

Tsang, L., Kong, J. A., and Ding, K.-H. 2000. *Electromagnetic Waves: Theories and Applications*. John Wiley & Sons, New York.

Tsang, L., Kong, J. A., and Shin, R. T. 1985. *Theory of Microwave Remote Sensing*. Wiley-Interscience, New York.

Tsang, L., Kong, J. A., Ding, K.-H. and Ao, C. O. 2001. *Scattering of Electromagnetic Waves: Numerical Simulations*. John Wiley & Sons, New York.

Ulaby, F. T. and Long, D. G. 2013. *Microwave Radar and Radiometric Remote Sensing*. University of Michigan Press, Ann Arbor, Michigan.

Ulaby, F. T., Moore, R. K., and Fung, A. K. 1981, 1982, 1986. *Microwave Remote Sensing. Active and Passive* (vol. I, II, III). Artech House, Norwood, MA.

Valenzuela, G. R. 1978. Theories for the interaction of electromagnetic and oceanic waves – A review. *Boundary-Layer Meteorology*, 13(1–4):61–85. doi:10.1007/bf00913863.

van de Hulst, H. 1957. *Light-Scattering by Small Particles*. Wiley, New York (reprinted by Dover, New York, 1981).

Von Hippel, A. R. 1995. *Dielectrics and Waves*, 2nd edition. Artech House, Boston.

Voronovich, A. 1999. *Wave Scattering from Rough Surfaces (Springer Series on Wave Phenomena)*, 2nd updated edition. Springer-Verlag, Berlin, Heidelberg.

Weng, F. 2017. *Passive Microwave Remote Sensing of the Earth: For Meteorological Applications*. Wiley-VCH, Weinheim, Germany.

Woodhouse, I. H. 2005. *Introduction to Microwave Remote Sensing*. CRC Press, Boca Raton, FL.

5 Turbulence Observations

One never notices what has been done; one can only see what remains to be done.

Marie Curie

In this chapter, we will discuss satellite remote sensing capabilities for turbulence detection. Airborne and ground-based atmospheric measurement techniques, operated with scanning or non-scanning (weather) radars and lidars, are well known and provide, as a rule, local measurements with relatively low level of predictability or forecasting. We refer the interested reader to the corresponding literature (e.g., Weitkamp 2005; Doviak and Zrnić 2006; Banakh and Smalikho 2013; Sharman and Lane 2016). Unlike atmospheric turbulence, methods of non-acoustic detection of oceanic turbulence are still under development, and available today remotely sensed data may not fully reflect underlying problems.

There are a number of reasons to study Earth's turbulence from space.

First, this is the only option to observe global and local turbulent motions worldwide.

Second, by studying *Dynamic Earth Environments*, we can highlight the role of turbulence in weather forecast and prediction of hazard events.

Third, turbulence affects the propagation of EM waves in the atmosphere that influences significantly on optical and microwave communications especially related to Naval Network and Space Operations.

Fourth, extreme turbulence (e.g., CAT) and wind shear are two of the most important hazards for aviation. Satellite data can help diagnose and predict these meteorological hazards.

Fifth, measuring properties of turbulence at the ocean-atmosphere interface, we can understand better heat and mass transfer processes.

Finally, remote sensing of natural turbulence allows us to improve geophysical fluid dynamics (GFD) models and create *predictive* analytics techniques.

Observationally, it is possible to identify three main classes of Earth's turbulence: 1) the surface turbulence, 2) the ABL turbulence, and 3) the upper atmospheric or tropospheric turbulence. All these stochastic motions are characterized by fluctuations of physical parameters (velocity, density, temperature, etc.), exhibiting deferent spatiotemporal scales and distributions. It is well known that even for fair weather conditions, *in situ* measurements are insufficient to examine the spatial and temporal evolution of turbulent flows in the ocean and the atmosphere. Alternatively, satellite remote sensing is capable of capturing spatial structures of various scales; however, the study requires the development of certain observation methodology and optimization of remote sensing techniques as well.

Since the late 1990s, satellite microwave instruments such as radars, altimeters, radiometers, scatterometers, and recently GPS have been designed and employed to

DOI: 10.1201/9781003217565-5

collect previously inaccessible data concerning global variability of the ocean and the atmosphere for ecological, meteorological, and climatological studies (Chuvieco 2020). Commercial space technologies are also applied widely for monitoring and conducting assessments of the environment and natural resources (Razani 2019). Today, several new NASA and NOAA geophysical programs (e.g., GOES-R series or JPSS series) continue this effort and may include remote sensing of turbulence although this particular task is beyond the scope of these missions. The possible reason is the overall difficulty to resolve natural turbulent flows due to their complexity, chaotic behavior, and nonlinearity. Furthermore, apparent uncertainty in the information content as the amount and variety of satellite data increases can lead to a lack of confidence in the resulting analytics process and decisions made thereof.

For example, detailed thematic interpretation of satellite optical images often requires expertise and routine work (microwave data are more complicated for analysis) and/or invoking additional geophysical information which may not be available at time. Comprehensive model studies are necessary as well in order to predict parameter dependencies and/or determine possible inherent biases and errors. In this regard, the choice of adequate and objective criteria for detection and recognition of relevant turbulent signatures becomes a key question in the satellite remote sensing observations.

5.1   HOW AND WHAT TO OBSERVE?

Most often, Earth's environments are found in a turbulent state like, for instance, rough sea surface or circulation of wind in the atmosphere. Models of isotropic incompressible turbulence partly can describe properties of turbulence although turbulent flows in the ocean and the atmosphere are not ideals nor incompressible. Intermittency is a fundamental characteristic of variability associated with emergence of disturbances, stochastic motions, and appearances of coherent structures that lead to dissipation, heating, transport, and/or inverse cascade of energy. Most conducted *in situ* investigations of turbulence and intermittency with a large portfolio of data are based on the statistical approach or higher-order analyses, such as probability distribution functions, structure functions, spectral power laws, and multifractals.

As mentioned earlier, traditionally, the presence of turbulence is indicated by a characteristic power-law behavior of the energy spectra which could obey the Kolmogorov "5/3" law (that is a commonly used criterion and measure of turbulence for many); however, the turbulence spectra describe only energy cascades across the inertial interval of scales but not an entire turbulence statistics. Usually, the problem of turbulence is considered on a case-by-case basis with most cases still unsolved.

To narrow the gap between observations and analysis, a concept of turbulence detection from space is suggested. The concept is based on the following remote sensing information principles and objectives:

1. Provide microwave/optical sounding – the measurement of vertical profiles of atmospheric physical parameters. Detect anomalous features in the fields of atmospheric pressure, temperature, water vapor, and various trace

gases. Turbulence can be associated with localized fluctuations of data profiles.

2. Provide measurements of sea surface parameters (the near surface wind, surface displacements, temperature, salinity, color, bioproductivity) using high-resolution radar, altimeter, optical sensors, and/or multiband microwave radiometer-imager. Detect any stochastic changes or fluctuations in physical parameters. Provide spectral, correlation, and scaling analysis of remotely sensed data/images. Turbulence can be associated with localized stochastic variations of physical parameters (and/or their correlations) being measured.

3. Provide global aerosol mapping in the atmospheric boundary layer (ABL) using optical/lidar techniques. Define sources of aerosol production. Detect regions with anomalous concentrations of aerosol particles and map their spatial distributions. Define origin and type of aerosol: sea-salt, gaseous-formed, dust, biological, pollution, volcanic, or anthropogenic. Determine anomalous fluctuations of aerosol concentrations in the presence of stochastic (turbulent) airflows. Turbulence can be induced by interactions of aerosol and airflows even at clear atmosphere.

4. Detect localized features at the clear air: jets, wakes, or plumes. For this goal, use remote sensors of any type, e.g., sounder, Doppler or laser profilograph, high-resolution optical/IR imager or radar. Provide structural and spectral analysis of the collected data/images. Define geometrical properties of possible turbulence signatures. The Doppler shift of the radio signals can also be used to characterize the winds and turbulence levels in the atmosphere.

5. Detection of ocean subsurface turbulent events – submarine wakes, jets, wave-induced turbulence, bubbly flows, and/or cavitational effects –requires the use of sophisticated remote sensing techniques and enhanced data processing. High-resolution multisensor active/passive optical/microwave imagery (which perhaps currently unavailable for space missions) is preferable instrumentation. Some possibilities, however, can be considered today, e.g., high-resolution optical multispectral (with spatial resolution ~ 1 m) imagery and/or lidar imagery. Briefly speaking, specific variations of *spatial statistical characteristics* of localized 2D image patterns can be perceived as potential turbulent signatures.

6. Space-based radars enable observing surface ship wakes, fronts, jets, vortexes, currents, internal gravity waves (or solitons), and oil spills; all of them are potential sources of turbulence at the air-sea interface. However, it does not mean that space-based radars are capable for detecting chaotic fluid motions, localized instabilities, and/or induced turbulent flows at the *ocean interior*. The solving problem still requires serious experimental evidence.

Convection, orography, and tropospheric baroclinic instability can also lead to significant turbulence in both cloudy and clear air atmospheric conditions.

The observed fluid motions and/or perturbations may exist in several scales: 1) mesoscale, with length scales of hundreds of kilometers and lifetimes of the order of 10 hours that is related to local climate conditions, 2) large scale, with length scales of several kilometers caused, for instance, by atmospheric gravity waves, and 3) small scale and/or microscale with length scales of tens of meters and less, caused by atmospheric turbulence including CAT. Indeed, typical atmospheric observations and weather models enable prediction randomized perturbations at large-area domains, whereas the measurement and modeling of local turbulence is still a big concern for many civil and military industries.

Tropospheric turbulence presents a significant aviation hazard, associated with CAT and jet streams. This type of turbulence is difficult to measure and predict; however, satellite observations could identify favorable for turbulence production areas. For example, meteorological satellite instruments such as GOES, MODIS, AVHRR, METEOSAT, NPOESS, and GMS enable providing observations of large Earth's areas that potentially may contain turbulence, e.g., over mountains, or ocean regions (Kelkar 2017). Hyperspectral sounding satellite systems of very high resolution are also capable of resolving the onset and development of turbulence of different scales (Qian 2020). Satellite altimetry may also help to select regions with enhanced turbulence over the oceans and land (Stammer and Cazenave 2018). Thus, current satellite missions potentially ready for detecting turbulence signatures of the interest at various conditions. Prospective ideas today are integrating satellite data with other observational and model insights and creating an operational analytic predictive product, allowing automated identification of potentially hazardous turbulence in the Earth's environments (Mecikalski et al. 2007; Sharman and Lane 2016).

5.2 STATE-OF-THE-ART TECHNOLOGY

Space-based sensors can be an attractive way to undertake Earth science particularly associated with natural hazards and turbulent features. The state-of-the-art program, aiming this study, may have the following three primary objectives:

1. Develop new space-based technology to enhance or expand the capabilities of satellite observations of turbulence. It could be either new technology that is revolutionary efficient and innovative or incremental improvements of existing satellite remote sensing systems. High-resolution multiband (multisensor) imagery is preferable.
2. Demonstrate new remote sensing technology and applications with near real-time capability for turbulence detection from airspace. It could be technology that has not previously been used for turbulence studies. The concept of small automated satellites as a paradigm shift for NASA and NOAA and the larger space community can be promoted as well.
3. Use advanced payload platforms and technology solutions enabling a scientific prediction and analysis of turbulent phenomena/events of the interest. The implementation challenges may include reduction of the cost, risk, complexity, and time required to turbulence detection. Apply an improved methodology of data acquisition, utilization, and interpretation.

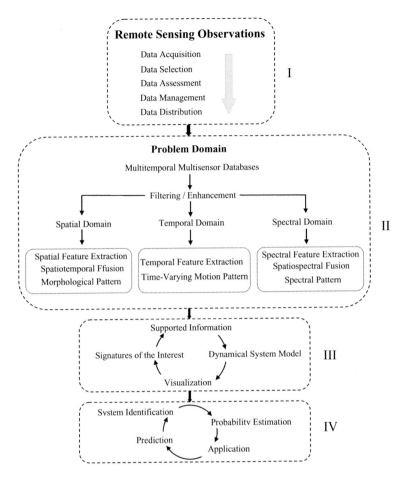

FIGURE 5.1 Conceptual framework for remote sensing of dynamical (turbulent) signatures (RSTF). I – Basic part, II – Implementation, III – Decision-making part, IV – Thematic part.

The purpose of this program is to develop the state-of-the-art remote sensing technology and to perform detailed airspace observations of turbulence. The program suggested may require more attention to scientific research, filed experiments, and numerical modeling in order to provide an appropriate remote sensing product and increase observational capabilities in a pooled and highly efficient way. Moreover, planning for infusion of breakthrough technological solutions and capabilities into satellite missions will be an important factor in program implementation.

Specifically, a conceptual framework for remote sensing of turbulent flows or other dynamical features in our environment (RSTF) can be developed and implemented on the basis of the critical factors and potential measures of the physical parameters. A possible version of the RSTF is shown in Figure 5.1; it consists of several parts with a number of operations: basic part, implementation part, decision-making part, and thematic part. Basic part consists of acquisition, assessment, analysis, interpretation, management, and distribution of data – all standard operations of any satellite observational technology

(Macdonald and Badescu 2014). Thematic part consists of joint analysis and fusion of multitemporal and/or multisensor data, enhanced data/image processing, selection of relevant signatures, and supported numerical modeling. Two last parts are decision-making and thematic analysis, involving modeling, interpretation, parameter estimation, and application. They can be delivered in terms of probability distribution functions of the main physical parameters (velocity, temperature, pressure, etc.), spectral and correlation characteristics, geomaps, or GIS products. The proposed solution to addressing turbulence detection problem is a participatory observation and implementation process that brings various satellite, ground-based, and model data together. Their integrated utilization creates databases allowing decision-making with certain but highest probability. Note that the RSTF should be closely connected with *satellite network architecture* which provides communication, application, and distribution services (Elbert 2004).

The challenges of the suggested framework are reducing false alarm, increasing information capacity, and improving detection performance at overall measurements. For example, some of the principal components of RSTF may include:

- Managing satellite data at the region scale.
- Increasing efficiency of distributed database system.
- Optimizing thematic processing by conducting enhanced assessments of satellite data.
- Providing real-time access to collected data resources through their visualization.
- Operational data exchange capabilities (specifically related to aviation meteorology).
- Establishing integrated automated detection tools, based on artificial neural network (ANN), support vector machine (SVM), or other state-of-the-art algorithms.

The selection and use of the state-of-the-art RSTF in some cases may require supporting information or specific knowledge from other available frameworks. In this respect, we have to cautiously apply RSTF and be aware of its ability to create particular results that will not disadvantage some other structures (e.g., integrated space-terrestrial network). Undoubtedly, the space technologies will continue to influence on remote sensing community and resources management.

The increased availability of RSTF connected with regional measurements is essential and may have a huge potential impact on turbulence prediction at locations where environmental changes are most favorable for formation of weather hazards. The state-of-the-art technology for turbulence detection would require some form of adaptation to the multi-disciplinarily level of expertise to make it available for users of different background. In this context, we believe that the RSTF capabilities could play a more central role in the observation process and decision-making, representing an art tool for comprehensive research and robust forecasting. In recent years, the development and availability of (very) high-resolution imagery from low-orbit satellites have offered another opportunity for advanced studies and applications. In next sections, we will explore the most relevant and meaningful information concerning observations of geophysical turbulence from space.

5.3   GEOPHYSICAL ASSESSMENT

Although Earth observation techniques were first used for military and security applications, in the Cold War, nowadays they are available for a wide range of multipurposeful applications in the fields of human activities, science, society, and politics (Brünner et al. 2018). Since the 1970s, satellite imagery has become a powerful tool for discovery of the Earth's environments, but it had typically been of relatively poor spatial resolution. According to the U.S. Geological Survey (USGS) publication *"Aerial Photographs and Satellite Images"* (1997), in the earliest days, satellite photographs were taken only over limited areas of the Earth on NASA's Gemini (1965–66) and Apollo (1968–69) missions. Beginning in 1972, Landsat data kick-started the big-data epoch by capturing imagery of the whole Earth's surface every two weeks. Then three Skylab missions in 1973 and 1974 have resulted in more than 35,000 photographs. Astronauts aboard the Space Shuttle, which began flying in 1981, have taken many photographs of the Earth with hand-held cameras. These photographs documented sites of scientific interest around the world and depict geophysical hazard phenomena such as hurricanes, tropical cyclones, and erupting volcanoes (ash plumes). Several color pictures of Earth from space can be found in books (Grant 2016; Nataraj 2017).

In the late 20th century, modern high-quality digital information technology (DIT) has brought natural science into a new challenging era. Digital revolution has provided 3D imagery, reconstruction, retrieval, restoration, and projection of environmental scenes and complex objects.

Digital refers to more than the electronic format of the data in bits and bytes or the automated workflow used to manage the data. The Digital Era encompasses the much wider and greater societal and technological transformations facing humans.

(Guo et al. 2020)

Nowadays, electro-optical and IR remote sensing systems of high resolution provide space-based imagery suitable for geographical intelligence and control of natural and/or man-made events. In particular, multispectral (hyperspectral) observations allow users to analyze intelligence imagery using *Big Data Analytics* technologies and computer tools based on digital data transformation (Swarnalatha and Prabu 2018). Big data analytics examines large amounts of data to uncover hidden patterns, correlations, and other insights.

Satellite earth-observing and GIS technologies emphasize on a broad *quantitative geophysical assessment* for many scientific and civil applications. Remotely sensed data are integrated into the assessment process at various levels depending on the scale or desired resolution. The most interesting and potentially far-reaching trends in space-based missions are identification and assessment of disasters (earthquake, flood, avalanche, forest fire, and pollution damage) and hydro-meteorological hazards (hurricane, cyclones, tsunami, and storm surges impact). All these events in one way or another exhibit turbulent properties. The ability to identify these natural hazards and/or detect turbulence depends not only on the satellite acquisition system (e.g., quality and resolution of remotely sensed images) but also on digital signal processing capability and robust *digital analytics* strategy. Moreover, the availability of geophysical models of the scene being studied can greatly enhance analysis and interpretation of satellite data.

A critical application of (very) high-resolution satellite imagery is the identification of natural hazards, vulnerability, disasters, management, and risk assessment (Dalezios 2017). Many regions in the world are exposed to several types of natural calamities or *geohazards*, each with their own spatiotemporal characteristics and huge environmental impacts (Smith 2013). The most prominent examples of natural disasters are the Indian Ocean tsunami (2004) and recent hurricanes of category 5 in the Caribbean and the USA ("Irma" and "Maria" in 2017 and "Dorian" in 2019). Such catastrophic events are always followed by *severe or extreme turbulence* induced by giant atmospheric vortices, extreme wind, heavy rain or squalls, even hailstorm, and high gale forces.

Potentially, *hazardous turbulence* situations occur at all thunderstorms and hurricanes and their avoiding is the best policy in commercial aviation. Turbulence occurs in complex wind fields where numerous vortices are generated due to significant differences of adjacent air streams velocities. An example of satellite image of hazardous turbulence is shown in Figure 5.2, which is GOES-16 merged visible and IR animation of *aboveanvil cirrus plume* (AACP).The presence of an AACP is a strong indicator that a storm may produce a tornado, large hail, or powerful wind. This hazardous turbulence is typically generated in environments with strong storm-relative wind shear. Researchers have identified hundreds of plumed storms over the U.S. using high-resolution imagery from GOES-16.

The application of satellite images in environmental hazard identification and risk assessment now seems inevitable if consider possibilities of data acquisitions with a

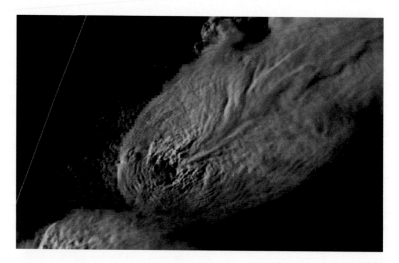

FIGURE 5.2 Satellite images of hazardous turbulence. This GOES-16 "sandwich" imagery shows a supercell storm over Texas on May 5, 2019. The green, yellow, and red areas show the temperatures of cloud tops within the storm. Red represents colder cloud tops, which indicate areas of greater storm updraft intensity. Green areas downwind of the updraft indicate anomalously warm temperatures. (Image from NOAA).

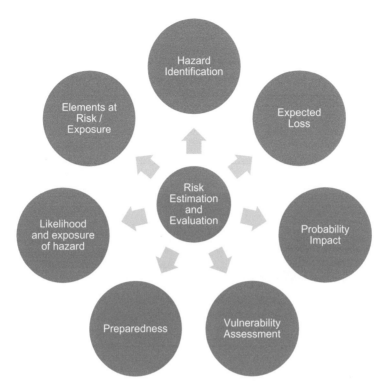

FIGURE 5.3   Hazard risk assessment and management framework. (Based on Bobrowsky 2013; Dalezios 2017).

wide swath and rapid revisit of remote sensing satellites. The advantages of high-resolution satellite imagery in hazard risk assessment are obvious: 1) low-cost and high accuracy measure, 2) global view and real-time mapping, and 3) inspiring more insights than the traditional methods for hazard risk assessment. As illustrated in Figure 5.3, risk assessment and management is a function of many factors (Bobrowsky 2013). Disaster risk cannot be eliminated completely, but it can be assessed and managed in order to reduce the impact of disasters.

Risk assessment is a term used to describe the overall process of hazard identification, risk analysis, and risk evaluation. Risk assessment is effective but typically tough statistical technique in which many aspects are not fully quantifiable or have a very large degree of uncertainty. This is because of the difficulty to define or predict hazard scenarios, characterize the elements-at-risk, or define the vulnerability. In order to overcome these problems, the risk is often assessed using so-called *risk matrices* which permit to classify risks based on expert knowledge with limited quantitative data (Haimes 2016). In this connection, greater access to insight would greatly facilitate detection, forecasting, and risk assessment capabilities.

5.4 SATELLITE IMAGE GALLERY

Earth is the greatest artwork as is evident from numerous satellite images. In this gallery, we are looking especially for distinct turbulence phenomena/events evolved on the Earth's environment by various geophysical factors and flow effects. These images are chosen because they improve our understanding of the Nature of Turbulence and offer us a new vision that is constantly changing and endlessly fascinating. In fact, satellite images provide contributions to scientific discoveries. Brief description of each image gives the interested reader the chance to explore selected features in more detail. Here is an annotated list of the case studies we have presented and discussed.

5.4.1 HURRICANE IRMA

Hurricane Irma (August 30, 2017–September 12, 2017) intensified into a strong and "potentially catastrophic" category 5 storm (Figure 5.4). By definition, category 5 storms deliver maximum sustained winds of at least 157 miles (252 km) per hour. Irma's winds, September 5, approached 180 miles (290 km) per hour – the strongest ever measured for an Atlantic hurricane outside of the Gulf of Mexico or north of the Caribbean. Huge vortex around the eye of the hurricane was captured in a short video taken by the NOAA GOES-13 and GOES-16 Earth-observing satellites on September, 2–5 (GOES stands for "Geostationary Operational Environmental Satellite"). The NASA-NOAA Suomi NPP (National Polar-orbiting Partnership) satellite captured a high-resolution, visible-light image of Hurricane Irma on September 4, before it was upgraded to a category 5 storm. The Visible Infrared Imaging Radiometer Suite (VIIRS) instrument on the satellite also captured images showing cloud-top temperatures above the storm. These images reveal the rising cloud temperatures near the eye of the hurricane (Figure 5.4a). Measuring the size of Irma, based on the densest cloud bands, it shows ~ 645 km (400 miles) in diameter from outer dense cloud bands through the eye.

Hurricane Irma is the perfect example of large-scale *helical* turbulent motions (or simply *helical turbulence*) in the atmosphere when air spiraling into the center, and then rising up and out the middle. Supposedly, genesis and further vortex intensification of the strongest hurricane like Irma represent a process of self-organization, in which several rotating mesovortices merge and form specific coherent structures known as *vortical hot tower* (VHT) There is some evidence for the presence of VHTs and convective burst vortices within the Irma eye wall (Figure 5.4b).

In the case study of Irma, helical scenario is more preferable for the explanation of the large-scale long-lived vortex formation and evolution, accompanied by self-organization of helical convective turbulence – mechanism, known today as *turbulent vortex dynamo in tropical cyclogenesis* (Moiseev et al. 1988; Levina 2018; see also book Sharkov 2012). The theory, introduced some 30 years ago, emphasizes the role of integrated helicity $H = \int \mathbf{u} \cdot \text{curl} (\mathbf{u}) d\mathbf{r}$ of the velocity vector field $\mathbf{u}$ in the evolution of large-scale vortex instability resulting to the formation of intense spiral vortices in tropical cyclones. Preliminary, Irma's helical structure can be investigated using an image processing (Figure 5.4c), which reveals a set of internal spirals. In this case study, turbulent momentum flux, turbulent kinetic energy (TKE), and vertical eddy diffusivity can be estimated for the hurricane eyewall. (The author pays

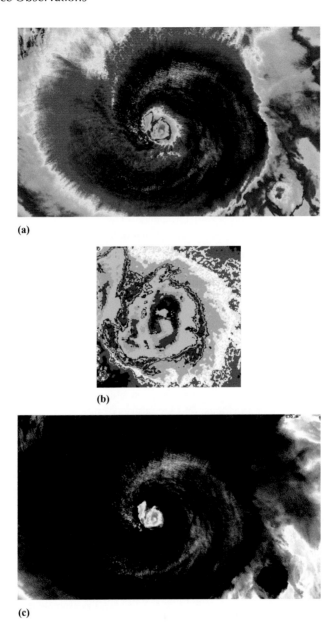

FIGURE 5.4 Satellite image of hurricane Irma. (a) Original image, September, 5 2017, VIIRS-1 – band 5 – 11 μm (Image from NASA/NOAA/ Suomi NPP). Processing: (b) vortical hot tower VHT (filtered), and (c) helical structure.

special attention to this extremely powerful event because he has experienced Irma category 5 at Caribbean island, Saint Martin, September 5–6, 2017). Satellite images of hurricane Irma available on the Internet websites: https://www.space.com/38043-hurricane-irma-satellite-images-approaching-caribbean.html and https://www.nasa.gov/feature/goddard/2017/irma-atlantic-ocean.

5.4.2 JET STREAMS

Jet streams are fast-moving air currents found high in the atmosphere, near the tropopause, which is a boundary between the troposphere and the warmer layer of the
stratosphere. The width of jet streams is typically a few hundred kilometers and each
one is only a few kilometers deep. They extend for thousands of kilometers around
the planet and are usually continuous, but meandering, over these long distances. On
average, jet streams move at about 110 miles (177 km) per hour. Dramatic temperature differences between the warm and cool air masses can cause jet streams to move
at much higher speeds, about 250 miles (~ 400 km) per hour or faster. Such high
speeds usually happen in polar jet streams in the winter time.

Strong vertical wind shear, which is prevalent within the jet stream, is known as one
of the main causes of sudden severe turbulence or aviation turbulence (e.g, CAT).
According to NOAA information (https://www.nesdis.noaa.gov/GOES-R-Series-
Satellites), monitoring and mapping of jet streams cover of more than half the globe –
from the west coast of Africa to New Zealand and from near the Arctic Circle to the
Antarctic Circle.

Large-scale atmospheric jet streams can be detected by various satellite instruments. Those include all three types of imagery: 1) visible, 2) infrared, and 3)
water vapor. Weather satellites, such as the GOES-R (Geostationary Operational
Environmental Satellites-R Series) use infrared radiation to detect water vapor in
the atmosphere. With this technology, meteorologists can define the location of
the jet stream. Recently designated NOAA's GOES-17 provides advanced high-
resolution visible and infrared imagery of atmospheric jets and so-called "atmospheric rivers," troughs and ridges, and signatures of potential turbulence among
other weather phenomena (hazards and lightning). An atmospheric river is a long,
narrow, and transient corridor of strong horizontal water vapor transport that is
typically associated with a low-level jet stream ahead of the cold front of an
extratropical cyclone. A well-known example of a strong atmospheric river is
called the "Pineapple Express" because moisture builds up in the tropical Pacific
around Hawaii and can wallop the U.S. and Canada's West Coasts with heavy
rainfall and snow.

Water vapor images, however, depict atmospheric flows much better than visible
or infrared images. In Figure 5.5 the dark, dry zones located near jet streaks usually
correspond to the cold side of the jet axis. In the case of a classic cold front, the jet
axis is oriented along the back of the frontal zone, showing as a moisture boundary
within the cold, sinking air mass. The highest wind speeds in the upper troposphere
are located at these moisture boundaries. The reasons behind an atmospheric stream
becoming turbulent are manifold, reaching from thermal instability and orography to
strong wind shear.

5.4.3 MOUNTAIN TURBULENCE AND LEE WAVES

Mountain turbulence originates when air flows over mountain ridges within a stably
stratified atmosphere and directly connected to orography and wind field. MODIS
image of mountain waves is shown in Figure 5.6 (https://cimss.ssec.wisc.edu/). On

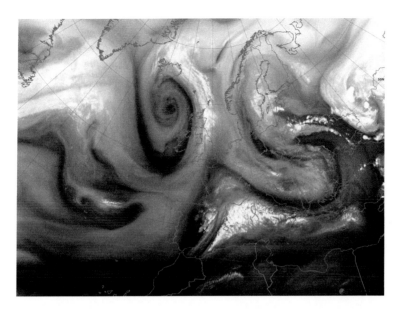

FIGURE 5.5 Satellite image of atmospheric jet stream. Water vapor satellite image also shows the "atmospheric river."

(*Source:* Ralph et al. 2020).

this image, lee waves represent quasi-periodic streamline wave patterns, extending downwind from the mountains for some distance. Sometimes, these waves propagate over long distances from mountain ridge and form "lee wave trains" at regular intervals of several kilometers.

Breaking of mountain waves can cause hazard weather events such as strong to extreme turbulence, "rotor or roll clouds," and damaging downslope windstorms on the leeward side of a mountain barrier. The roll or rotor cloud looks like a line of cumulus clouds parallel to the ridge line (Figure 5.6). They are produced by the rotor circulation of air on the leeward side of the mountain below the wave crests. This is the area of the greatest danger due to the extreme turbulence in and below rotor clouds. GOES-16/17, Himawari-8/9, and VIIRS can resolve mountain turbulence signatures very well (Figure 5.7). Thus, satellite observations provide unique possibilities for detection of turbulence in local mountain regions.

5.4.4 Atmospheric Gravity Waves

Atmospheric gravity waves (AGW), also called atmospheric internal waves, are generated by air flow over orography depending on the atmospheric Froude number, stratification, and topographic forcing. The downstream waves are often standing waves and are known as lee waves. They commonly occur in the leeside of terrain barriers and are supported either by stably stratified lower troposphere or by vertical wind shears upstream the barrier.

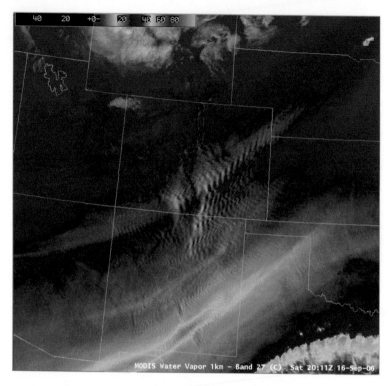

FIGURE 5.6   Satellite image of mountain waves. The AWIPS MODIS 6.7 µm "water vapor channel" image depicts mountain wave signatures over parts of the southern and central Rocky Mountains region, USA. Strong winds associated with a jet stream axis interact with the rugged terrain. (Image from NASA).

According to existing SAR observations, the horizontal length scales (wavelengths) of AGW range from a few to several tens of kilometers. In the satellite images, AGW usually represent quasi-periodic band pattern in form of either transverse waves or divergent waves over the terrain obstacles. Other abundant types – concentric AGWs, are often generated in offshore zones and forced by thunderstorms. Note that AGWs should not be mistaken with oceanic IWs which are detected in the SAR images as *"a series of alternating light/darklinear or curvilinear bands that represent the crests and troughs of the waves. The light/dark signatures are the result of variations in sea surface roughness"* (as mentioned John Apel in his book 1985). AGW form when buoyancy pushes air up and gravity pulls it back down.

AGW are perfectly observed by various satellite sensors – SAR, optical/IR imager, or even high-resolution microwave radiometer (from aircraft). Periodic cloud band patterns in the images frequently occur as a result of the updraughts imposed by AGW (Figure 5.8). Large-scale AGW appear in the form of a wave packet containing several waves located between 50 and 200 km offshore. The NASA MODIS captured this image of a large-scale, overlapping AGW pattern in the sunglint region. The long, dark bands show where the troughs of AGW have roughened the surface. The brighter bands show the crests of the AGW. In addition

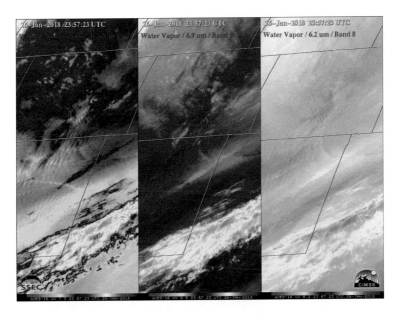

FIGURE 5.7   Satellite image of mountain wave turbulence. GOES-16 7.3 µm, 6.9 µm and 6.2 µm. Water Vapor images revealed widespread mountain wave turbulence in Colorado and New Mexico. (Image from NASA).

to the AGW patterns, several much smaller waves appear on the water surface as well. However, sudden severe turbulence (such as CAT) induced by breaking AGW is difficult to detect from space.

AGW are fundamental in Earth's atmospheric dynamics and other planetary atmospheres because they link vertical layers and have no preferred horizontal direction. AGW can propagate vertically from low ABL to stratosphere and mesosphere up to several hundred kilometers where space weather processes primarily influence the upper atmosphere (Yiğit and Medvedev 2019).

5.4.5 Atmospheric von Kármán Vortex Streets

Atmospheric von Kármán vortex street (AVS) is a classical example of large-scale turbulent vortex motions in the atmosphere or even in oceans. These vortexes bear a striking resemblance to those seen in high Reynolds number airflow around a bluff body. In nature, AVS appear in any region where fluid flow is disturbed by topography – mountains or rising islands. For example, when wind-driven clouds meet an island they flow around it clockwise and anticlockwise and form spinning eddies. That oscillation creates alternating vortices around the obstacle.

Various satellite imagers regularly capture AVS around the globe since the late 1960s (e.g., Tsuchiya and Tokuno 1989; Young and Zawislak 2006; Horváth et al. 2020). Meteorological NASA and NOAA observations show that the most common places in the world where AVS occur are where the Trade Winds are prevalent. These include the Canary Islands, Madeira Island, Cape Verde Islands, as well as Guadalupe and Socorro

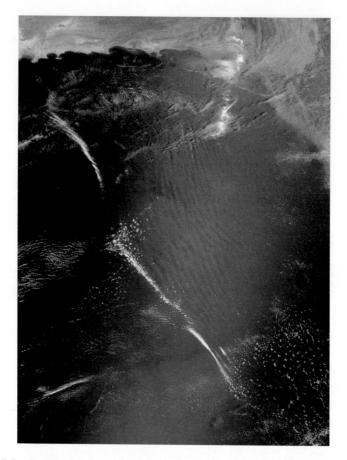

FIGURE 5.8   Satellite image of atmospheric gravity waves (AGW). On May 23, 2005, the MODIS on NASA's Terra satellite captured this image of a large-scale, overlapping wave pattern in the sunglint region of an image of the Arabian Sea. (Image from NASA).

Islands west of Baja California and the Juan Fernandez Islands off the Chilean coast. AVS cloud patterns have also been spotted in the Greenland Sea, in the Arctic, and even next to a tropical storm. The world's largest von Kármán atmospheric vortices tend to form near the Hallasan Volcano on Jeju Island, off the coast of South Korea.

The spatial coverage and resolution of satellite (e.g., SAR) imagery provides observations of AVSs whose spatial scale ranges from 100 km to 400 km. There are also a number of color composite images of AVS obtained from MODIS and GOES-16 satellites. Spectacular AVS patterns have been captured by the VIIRS on the Suomi NPP satellite and MODIS-Aqua (Figure 5.9) (https://visibleearth.nasa.gov). In these pictures, we see a vortex wake behind an asymmetric island. The wake consists of a number of cyclonic eddies having larger peak vorticities. Vorticity generally has decreased with downstream distance, perhaps, due to viscous diffusion. According to the common theory, evolution of von Kármán vortex streets and such an inherent vortex shedding can occur only under the influence of moderate background rotation when the Rossby number is equal to or larger than unity.

Geometrical properties of AVS in the satellite images can be characterized by the following metrics:

- The aspect ratio, h/a,
- The dimensionless width ratio, h/a,
- The vortex shedding frequency, $f_{vs} = U_{down}/a$ (or period $T_{vs} = 1/f_{vs}$),
- The Strouhal number for normalized shedding frequency, $St = D_{inv}/T_{vs}U_{up}$,

where h is the distance between the two counter-rotating vortex trains, a is the distance between two vortices in the same vortex train, d is the diameter of the obstacle, U_{down} is defined as the downstream eddy propagation speed, U_{up} is the upstream velocity, D_{inv} is the crosswind island diameter at inversion base, and T_{vs} is the shedding period between two consecutive like-rotating vortices. Supposedly, these metrics may characterize AVS at various environmental conditions; however, this doesn't necessarily imply that these metrics will be valid in overall cases of atmospheric turbulent vortex shredding. Moreover, the accuracy of the computations depends on spatial resolution of satellite data. Most known estimates (Young and Zawislak 2006) give values of the aspect ratio in the range 0.37–0.42 and dimensionless width ratio in the range 1.61–1.83. These data are based on a 30-case sample of MODIS satellite images of well-developed AVS.

As seen from space, atmospheric vortex streets can differ from their laboratory analogs (see Figure 1.5 in Section 1.2.2). This is because AVS occur in a stratified rotating fluid in the presence of "non-ideal" topographic obstacles such as mountains or islands. Furthermore, atmospheric vortexes are usually controlled by several parameters including the Froude number, the Rossby number, and the Strouhal number besides the Reynolds number, which is only criterion for generation of turbulent flows in laboratory experiments with axisymmetric bluff bodies.

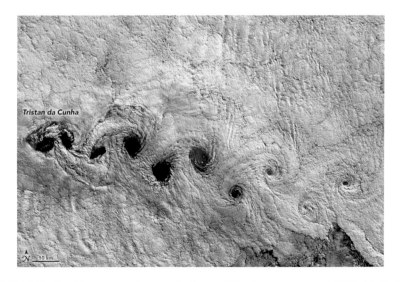

FIGURE 5.9 Satellite image of atmospheric von Kármán vortex street (AVS). (a) On May 24, 2017, the (VIIRS) on the Suomi NPP satellite captured a natural-color image of AVS on the lee side of Guadalupe Island.

(Continued)

FIGURE 5.9 **(Continued)** (b) On 5 July, 2002, MODIS-Aqua satellite image (visible) capturing an intense AVS forming leeward of Madeira Island, and extending its influence south of Canaries Archipelago. (Images from NASA).

Nevertheless, there is some superficial morphological (structural or geometrical) similarity or even statistical similarity between AVS and idealized laboratory vortex streets although the size of the observed atmospheric eddies is several orders of magnitude larger than that in the laboratory. The apparent similarity allows the researchers to model AVS and turbulent phenomena in the same way using both CFD methods and laboratory tests.

5.4.6 CLEAR AIR TURBULENCE (CAT)

CAT usually occurs in the lower stratosphere and the upper troposphere. CAT or out-of-cloud *convectively induced turbulence* (CIT) is impossible to observe either with the naked eye or on-board aviation radar. Traditionally, optical ground-based techniques such as scintillometers or Doppler lidar are used locally for turbulence detection and forecasting. Possibilities to evolve extremely sensitive microwave systems (e.g., high-resolution radar or even radiometer) for CAT detection are considered as well.

Satellite-based methods look very attractive because they provide a global coverage of the phenomenon. However, the spatial and temporal resolutions of available satellite instruments are not sufficient to detect CAT and/or CIT directly. Nowadays, some new, indirect methods are developed to approach the problem.

An idea is based on the measurements of water vapor in the upper atmosphere. According to recent studies (Wimmers and Moody 2004a, 2004b), the upper-tropospheric boundary between air masses – region, called *tropopause fold* (TPF) – can be detected from satellites. It is assumed that TPF is one of the possible mechanisms that can initiate dynamical turbulence including CAT. The project was named as the "tropopause folding turbulence product" (Wimmers and Feltz 2010).

The TPFs (described first in the early 1930s) led to the identification of multiple stable layers in the troposphere and stratosphere. These stable layers were thought to represent the same folded discontinuity surface at different heights (i.e., TPF). TPF is a mesoscale feature, which extends from this boundary to a limited distance into and underneath the weather air mass. TPFs are located by their association with gradients in moisture, which are evident in satellite images sensitive to upper tropospheric water vapor.

Satellites GOES series, MODIS, and CIMSS (NOAA's Cooperative Institute for Meteorological Satellite Studies) with 6.5 μm, 7.0 μm, and 7.4 μm wavelength channels provide water vapor imagery, resolving specific humidity gradients in the TPF. In satellite images (Figure 5.10), these features are revealed as the large-scale variations of brightness, which define boundaries between the air masses. Analysis of GOES/MODIS satellite data in spectral, temporal, and spatial domains provides insight to the identification of potential regions of interest (or predecessor) associated with CAT and CIT phenomena.

In any case, precise satellite-based measurements of main atmospheric parameters (air pressure, temperature, humidity, velocity, etc.) or their profiles would be necessary in order to determine microstructure and velocity fluctuations that are a preliminary indicator of the presence of turbulence including CAT and CIT. We believe that multisensor visible/IR and microwave satellite imaging capabilities will bring a new perspective to the solution of this challenging problem.

5.4.7 AIRCRAFT CONTRAILS

Aircraft contrail (*condensation trail*) forms in the atmosphere when the mixture of water vapor in the aircraft exhaust and the air condenses and freezes. The temperature and humidity of the air affects life time of contrails. At dry air, contrails last just seconds or minutes. At humid air, contrails can be long-lived and spread outward until they become difficult to distinguish from naturally occurring cirrus clouds. Persisting contrails can spread into extensive cirrus clouds and become distorted by wind shear. The spreading occurs due to the air mass being unstable or turbulent. This turbulence dissipates the dense contrail lines and spreads them over a wide area, giving them a more cloud-like appearance. Contrail formation and evolution are clearly visible on the satellite images (Figure 5.11). NASA's Terra satellite captured

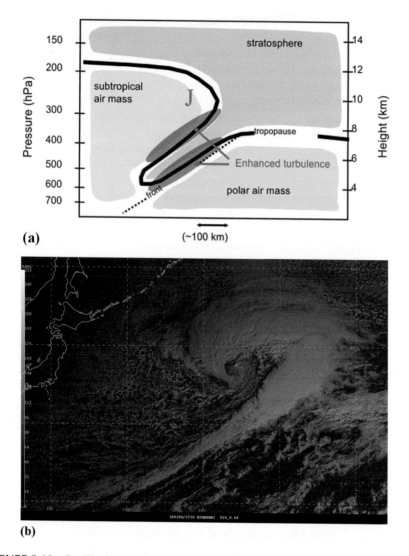

(a)

(b)

FIGURE 5.10 Satellite image of tropopause fold (TPF) region as a potential predecessor of CAT. (a) Schematic illustration of TPF. (Based on Wimmers and Feltz 2010) (b) The Himawari-8 visible and RGB air mass satellite image of TPF area. A vertical cross-section over the low center indicates a stratospheric TPF to the south of the low center. (Image from NOAA).

these two amazing images of aircraft contrails, spreading across the sky off the coast of Newfoundland, Canada. Satellites have observed clusters of contrails lasting as long as 14 hours, though most remain visible for four to six hours (https://earthobser-vatory.nasa.gov/images/78154/the-evolution-of-a-contrail).

5.4.8 Ocean Mesoscale Eddies, Swilling Flow, and Spirals

Ocean mesoscale eddies (OME) are considered as the "weather of the ocean" with cyclonic and anticyclonic eddies, having significant impact on the large-scale ocean circulation, the air-sea interactions, transport processes, and thereby the Earth climate as a whole. OME are formed under the influence of interactions of currents and winds. Typical horizontal scales of OME are less than 100 km and timescales on the order of a month. In general, the eddy field is a typical turbulent dynamical system, characterized by both a high-dimensional phase space and a large number of instabilities. OME comprise rich cascades of coherent structures such as vortices, rings, filaments, squirts, and spirals; some of them are closely related to Lagrangian coherent structures (LCS).

Satellite observations of the last several decades have demonstrated the existence of a strongly energetic mesoscale turbulent eddy field in all the oceans. The most active regions of oceanic eddies are the Kuroshio Extension region, the Subtropical Countercurrent zone, the Northeastern Tropical Pacific, and the Southeast Pacific (Talley et al. 2011). Large-scale spiral eddies have been detected also in the Black, Caspian, and Baltic seas (Karimova 2012).

Recent studies have revealed new properties of OME associated with sea surface temperature (SST) and salinity (SSS) anomalies and ocean color as well. Geostrophic

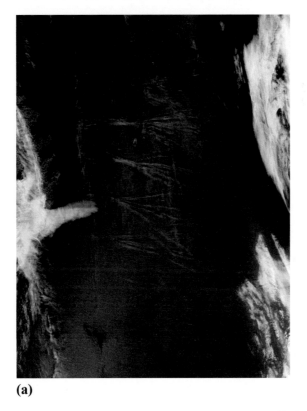

(a)

FIGURE 5.11   Satellite image of aircraft contrails. The condensation trails that form behind high-altitude aircraft, or contrails, are one of the most visible signs of the human impact on the atmosphere. (a) On May 26, 2012, the MODIS on NASA's Terra satellite captured multiple contrails, criss-crossing the Atlantic.

(Continued)

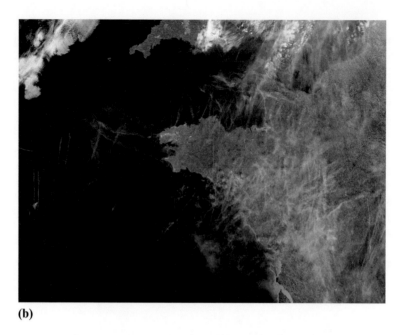

(b)

FIGURE 5.11   (Continued) (b) Contrails above Western France (Aqua 2006.11.04. 12:42). (Images from NASA).

turbulence in the ocean is now known to involve a much broader range of scales from 1 to 5,000 km including fine-scale and submesoscale currents and eddies, baroclinic and barotropic instabilities, and so-called *centrifugal instabilities*. Computer scientific visualization of the "Perpetual Ocean" created by NASA, Figure 5.12 (https://svs.gsfc.nasa.gov/3827), shows the swirling turbulent flows of tens of thousands eddies and currents of various geometry. NASA believes that this visualization offers a realistic study in both the order and the chaos of the circulating waters that populate Earth's ocean. However, high-resolution space observations are needed for detecting multiscale eddy flows and validate these model data.

OMEs have been observed by satellite SAR since the 1980s (e.g., Ivanov and Ginzburg 2002). Series of photographs of oceanic eddies were taken by Commander Richard H. Truly during NASA Space Shuttle mission STS-8 (August, 30–September, 5 1983). These images gave the first information about geometry and size of OME. Some data are available on the Internet https://www.lpi.usra.edu/publications/slidesets/oceans/oceanviews/oceanviews_index.shtml.

Later, the existence of sea spiral eddies was confirmed and investigated by Australian-American astronaut and oceanographer Paul Scully-Power during NASA Space Shuttle mission STS-41-G (October 5–13, 1984). Large volume of color photographs of sea eddies at clear air was taken with handheld Hasselblad 6×6 cm cameras using 100 mm and 250 mm focal length lenses. The Shuttle flew at altitudes ranging from 200 km to 400 km. At a typical altitude of 300 km, the resolution of these photographs is 50 m or 20 m for the two respective focal lengths. These materials have aroused great research interest and discussions (Munk et al. 2000; McWilliams 2016).

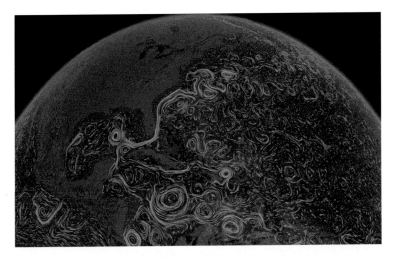

FIGURE 5.12 Computer scientific visualization of the "Perpetual Ocean."
(*Source:* NASA).

Nowadays, ocean eddies can be observed by various active and passive satellite remote sensors. SAR data become available for automatic detection of OME using computer vision algorithms (e.g., machine learning and neural networks). Eddy-induced sea surface height anomalies can be detected by altimeter, whereas eddy-related patterns of SST and SSS can be mapped by a passive microwave radiometer-imager. Ocean eddies are also tracked using multispectral optical ocean color sensors.

Here we present two examples: satellite images of OME, captured by the NASA ERS-1 SAR (Figure 5.13) and completely cloud-free photography from Space Shuttle (Figure 5.14). Last one covers an area of approximately 50 kilometers on a side. These images illustrate complex structure of the swirling flow, involving individual and connected eddies of various size and shape. For example, enhancement and segmentation of the selected SAR image fragment reveal fine structure of the eddy pair in the form of concentric spirals similar to the classical Lorenz attractor. Created for weather prediction, Lorenz attractor (Equation 1.5) describes convective turbulence in a simple mathematical fashion, and therefore, this model seems to be an appropriate for evaluation of chaotic eddies observed in the swirling flow. Furthermore, we may consider short-lived fine-scale or submesoscale eddies as the response of local instabilities, and mesoscale long-lived eddies as a response of geostrophic circulation. The Lorenz model may be feasible for simulation and prediction of individual OME and/or spiral eddies as well as overall swirling turbulent flow in the first place.

On the other hand, chaotic structure and behavior of most natural turbulent flows have tendency spontaneously form interconnected multiple features and/or complex patterns that may not be predicted or explained not by attractors (because of high sensitivity to initial conditions), nor by classical fluid mechanics (because of deterministic description). Affordable quantitative interpretation can be based on (multi) fractal analysis of the visualization data. This approach allows us to explore scaling and self-similarity of the OME flow, at least in statistical sense. In this particular case

FIGURE 5.13   Satellite ERS-2 SAR image (September 19th, 1993) of ocean mesoscale eddies (OME) in the Mediterranean Sea (image size 15 km ×13 km). (Image from ESA).

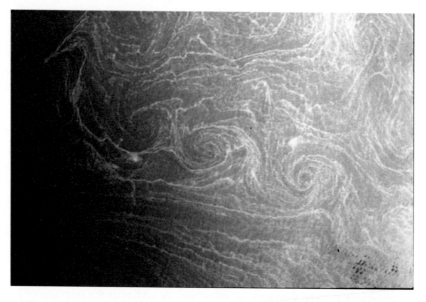

FIGURE 5.14   Photography of ocean mesoscale eddies (OME) from Space Shuttle Challenger (mission STS-41G, October 5–13, 1984).

(*Source:* NASA).

study, (multi)fractal is a measure of the change in topological properties of overall turbulent swirling flow which can be detected from space in one way or the other.

Supposedly, both presented images (Figures 5.13 and 5.14) illustrate topological chaos and coherent structures, which highly likely dominated by rotating flow that can be associated with a number of geophysical factors: 1) strongly horizontally sheared motions, 2) boundary currents, 3) complex bottom topography, 4) baroclinic instability, and 5) combination of Coriolis forces and vorticity produced by wind shear. In many cases, the ambient stratification is significant and its effects may change flow characteristics. So far, a more detailed look at the problem is needed.

5.4.9 OCEAN WHIRLPOOL

Turbulent environment poses many violent vortexes; one of them is known as *whirlpool*.

According to the Encyclopaedia Britannica,

> *Whirlpool, rotary oceanic current, a large-scale eddy that is produced by the interaction of rising and falling tides. Similar currents that exhibit a central downdraft are termed vortexes and occur where coastal and bottom configurations provide narrow passages of considerable depth. Slightly different is vortex motion in streams; at certain stages of turbulent flow, rotating currents with central updrafts are formed. These are called kolks, or boils, and are readily visible on the surface.*

A large, powerful, or violent water whirlpool is also defined as maelstrom. Maelstroms typically form in the ocean near narrow straights as a result of interacting tides. Notable oceanic whirlpools include Moskstraumen (Norway), Corryvreckan (Scotland), Naruto Straits (Japan), Old Sow (Canada). These whirlpools are long-lasting, some having been on maps for centuries.

The NASA satellite – Terra (EOS AM-1) at December 26, 2011 recorded a giant whirlpool near the coast of South Africa (Figure 5.15). This whirlpool, known as *The*

FIGURE 5.15 Satellite image of oceanic whirlpool near the coast of South Africa. The MODIS on NASA's Terra satellite captured this natural-color image of a deep-ocean whirlpool on December 26, 2011. (Image from NASA).

FIGURE 5.16   Fragment of satellite image of the Great Whirlpool (GW) and plankton-fueled Agulhas Current. (Image from NASA).

Great Whirl (GW), has been observed *in situ* by oceanographers for years (e.g., Wirth et al. 2002). It was found that the clockwise-spinning GW forms every spring off the east coast of Africa when winds, blowing across the Indian Ocean change direction from west to east. The close-up image (Figure 5.16) shows the GW's structure, traced in light blue by the plankton blooming (also called *plankton-fueled Agulhas Current*) in the 150-kilometer-wide swirl. This Agulhas whirlpool reaches as broad as 500 km and its circular currents extend hundreds of meters downward and can go farther than 1 km (0.6 miles) deep in some areas.

More detailed investigations have been conducted recently by (Melzer et al. 2019) using multi-year remotely sensed data. They examined 22 years of general circulation model reanalysis data and 23 years of satellite altimetry and found interannual variability of the GW life cycle in both timing and duration. In particular, the average area of the GW over 23 years was 275,000 square kilometers (106,000 square miles) and persists for about 200 days out of the year.

In view of fluid dynamics, as a first approach, ocean whirlpool is considered as a material vortex in 2D turbulent flow closely related to the rotating Lagrangian coherent structures (LCS). Coherence occurs in the invisible barrier that surrounds a vortex – the "coherent boundary," in which water circles, but does not fall in or break away boundary region, creating so-called *coherent material belts* (Haller and Beron-Vera 2013).

As seen in the picture (Figure 5.16), a fast-rotating spiraling whirlpool closely resembles "black hole" which is well known in astrophysics. (Haller and Beron-Vera 2013) put this similarity on a formal footing by describing the behavior of vortices in turbulent fluids using the same mathematical tool that describe black holes.

More realistic scenario, however, can be associated with the development of *hydrodynamic helical turbulence* under the influence of nonlinear quasi-resonant tide interactions (triads) in rotating and stratified flow. Inverse cascades play a central role in this process providing a stable mechanism for the formation of coherent structures and for the self-organization of disorganized fluid flows. This idea (which actually is not novel) can be adapted for a number of oceanographic problems and practical solutions including phenomena such as whirlpools, rotary currents, internal waves, and even turbulent wakes.

5.4.10 INTERNAL WAVE-INDUCED TURBULENCE

Oceanic internal waves (IWs), their generation, dynamics, and parameters have been in the focus of satellite oceanography for quite a long time. Pioneering remote sensing studies of oceanic IWs and solitons were conducted by J. Apel and W. Aplers in the mid 1970s and 1980s. Today, we know that surface manifestations of IWs are well detected by satellite radar (SAR), by optical imager, and even by sensitive microwave radiometer (Raizer 2017). However, turbulence itself is difficult to identify among complex IW field that is usually observed by satellite SAR (Figure 5.17). In most stratified ocean interior, turbulence is driven by IW breaking, resulting to intensive mixing of water masses and upwelling that can create localized areas of enhanced turbulence at the subsurface layer. Therefore, it is clearly desirable to be able to distinguish between the effects of IW turbulence, ambient (or wave) turbulence, and their interactions as well.

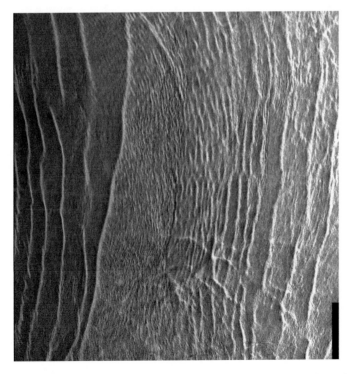

FIGURE 5.17   Satellite ERS-2 SAR image of oceanic internal waves (IWs) in Andaman Sea (February 11, 1997). (Image from ESA).

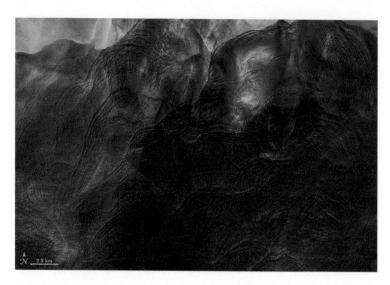

FIGURE 5.18 Satellite optical image of oceanic internal waves (IWs). The reflection of the Sun on the ocean (sunglint) makes the IWs visible. (Image from NASA).

In the optical image (Figure 5.18), a number of hydrodynamic effects can be revealed. The Operational Land Imager (OLI) on Landsat 8 satellite captured this image of the Andaman Sea on November 29, 2019. The reflection of the Sun on the ocean surface makes visible IWs and some turbulent features – swirls, small-scale eddies, fronts, and stochastic wave patterns. The lighter and darker tones reflect the depth of the water (darker is deeper) and turbulent features fill out areas of variable brightness. This 30-m resolution true color image demonstrates great sensitivity of optical radiance to ocean hydrodynamics, particularly in the case of large-scale intensive nonlinear IWs, generated by tide-topography on the continental shelf.

Other satellite image (Figure 5.19) presents more evidence that detection of IW turbulence is quite realistic (https://visibleearth.nasa.gov/images/3586/internal-waves-sulu-sea). This true-color image was acquired by NASA Aqua MODIS on April 8, 2003 in the Sulu Sea between the Philippines and Malaysia. According to the description

> *internal waves alter surface currents, changing the overall sea surface roughness. Where these currents converge, the sea surface is more turbulent, and therefore brighter. Where the currents diverge, the surface is smoother and darker creating zones called "slicks".*

These slicks appear as dark bands in the center of the image.

The characterization of IW turbulence as an oscillation process in the field of surface is more recent. Viewed from this perspective, turbulence can be regarded as a response of a fluid to an internal shock front moving across the ocean. Front transfers mass and momentum in both its mean and its fluctuations to the surface that may

FIGURE 5.19 Satellite optical image of turbulence, induced by oceanic internal waves (IWs) in Sulu Sea between the Philippines and Malaysia. In this true-color Aqua MODIS image acquired on April 8, 2003, the slicks created by IWs appear as dark bands in the center of the image. (Image from NASA).

trigger turbulent transport and mixing. Moreover, the tidal cycle and Coriolis effects may have significant influence on the onset of IW turbulence. The exact mechanisms for generation of IW turbulence are not yet established, but the two main hypotheses, known as leewave formation and barotropic–baroclinic scattering are the most important for remote sensing.

Ultimately, the problem of ocean turbulence detection (induced or natural) requires the application of sophisticated observational technology and enhanced analysis. For example, the detection of the hydrodynamic effects associated with surface turbulent-laminar patterns (e.g., ocean wakes) can be made through the concept of "spectral portrait," delivered from high-resolution optical imagery (Raizer 2019). The idea of ocean turbulence-related spectral Fourier signatures has also been considered and tested during field remote sensing experiments, specially organized in Hawaiian Ridge (Gibson et al. 2008).

5.4.11 SUBMARINE VOLCANO ERUPTION

Submarine volcanoes are underwater vents or fissures on certain zones of the ocean floor. The erupted magma and lava create the edges of new oceanic plates and supply heat and chemicals to some of the Earth's most unusual and rare ecosystems. The most productive submarine volcanoes are hidden under the surface at an average of 8,500 feet (2,600 m) deep. However, even very large, deep-water eruptions may not appear in the surface. This makes *in situ* observations of submarine volcano eruptions very difficult.

Nowadays, satellite data provide unlimited possibilities to identify the location of volcanic source and monitor undersea eruptions. Advanced Spaceborne Thermal Emission and Reflection Radiometer, ASTER VNIR satellite captured eruption of volcano Kavachi in 24 March 2006 (Figure 5.20). Kavachi is an undersea volcano on the southern edge of the Solomon Islands in the western Pacific Ocean. It erupted dozens of times in the 20th century, often breaking the water surface, only to be

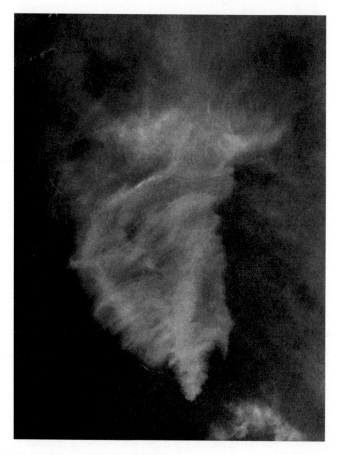

FIGURE 5.20   Satellite image of submarine volcano eruption. There appears to be turbulence at the ocean surface and a possible line of pumice along the lower left edge of the discolored area. (Image from NASA).

eroded back below the water line within a few months. This ASTER image shows turbulent ash-and-steam eruption plume and pulses of discolored water originating from the vent, confirming ongoing volcano activity. Bright aquamarine water indicates high concentrations of volcanic material. Extended area of brown water corresponds to thermo-hydrodynamic surface disturbances, associated with *turbulent thermal mixing* in the water column over the vent due to powerful volcano eruption.

The surface turbulent plume detected in the image (Figure 5.20) resembles (multi) fractal object very much. We can assume that morphology and structure of the plume depend largely on buoyancy force and density gradients that define wind-driven dynamics and variability of plume's position, shape, and surrounding area. More detailed interpretation can be made using fractal analysis of the satellite data and Lagrangian dynamic model of the plume as well.

As follows from observations, submarine volcano eruption results in significant changes in subsurface turbulent layer and generation of additional instabilities (i.e., stochastic signatures), while ambient sea has more stable and homogenous structure. Thus, submarine volcanic processes, in which there are localized, intense releases of energy, can lead to the occurrence of turbulent buoyant plumes in the surface that is perfectly detected by a satellite sensor. The same possibility exists also in the case of underwater (nuclear) explosion and/or seismic activity.

5.4.12 Plumes at Sea

Plumes refer to common features in environmental fluids in which a body of one fluid moving through another. More specifically (in view of remote sensing), a plume is defined as a fluid enriched in sediment, ash, biological or chemical matter that enters another fluid (Shanmugam 2018). Plumes occur whenever a persistent source of buoyancy creates motions of the fluid away from the source. In aquatic environments, in particular oceans, estuaries, and rivers, the combination of density stratification and velocity shear results in active mixing of different fluids and creation of the *turbulent buoyant plume* which changes the near-surface properties and hydrodynamics. Usually turbulent plumes grow vertically, spreading radially as a result of entraining ambient fluid into the plume at the subsurface boundary layer (or air-water interface).

Satellite remote sensing has been successfully used for mapping plumes at sea since the 1970s. Recent studies have demonstrated possibilities for detection and recognition of the following types of plumes: 1) density plume, 2) sediment plume, 3) dredge plume, 4) hydrothermal plume, 5) hypopycnal plume, and 6) oil plume.

Most remotely sensed data refer to natural and anthropogenic plumes of category 1)–3), which are determined by the hydrodynamic, thermodynamic, and morphological features inherent to coastal zones and river estuaries (Klemas 2011). Coastal plumes (also known as *turbidity plumes*) are characterized by their higher concentration of suspended matter giving them a distinct visibility contrast against clear waters. Sediments are unconsolidated organic and inorganic particles that accumulate on the seafloor and rise to the surface. A typical sediment plume at the sea surface represents a nonuniform thick layer consisting of large amounts of sediments, suspended in water. Such plume covers large spaces along the shore and exhibits distinct from pure water optical (and sometimes microwave) properties. This provides opportunities to define physical and structural characteristics of complex plumes using remote sensing measurements (by analogy with detection of oil spills).

Plumes at sea, their evolution, structure, and impact on marine ecology are of considerable interest in geosciences and remote sensing. The problem connects with remote sensing of coastal environments where plumes occur due to direct impacts of natural habitats and human-related activity. The coastal zone is the most dynamic interface between land and sea, and therefore, detection and identification of plumes at coastal aquatic environments requires remote sensing data acquisition with highest resolution (temporal, spatial, spectral, and radiometric).

At present, the MODIS on TERRA and AQUA optical sensors as well as Sentinel-2 with the MultiSpectral Imager (MSI) provide medium-resolution maps of coastal zones worldwide. These data are suitable for identification and morphological analysis of plumes in coastal waters (Figure 5.21). In some cases, satellite SAR imagery can be used for automatic detection of the plume's distinct boundaries such of those in coastal oil plume (Figure 5.22). As a whole, satellite-based operational control and assessment of coastal waters on a regular basis is still a challenge. More information about coastal remote sensing can be found in the book (Wang 2010).

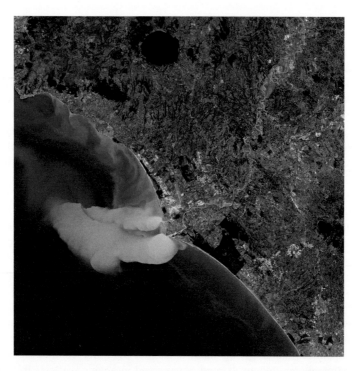

FIGURE 5.21 Satellite multispectral image of sediment plume at sea in coastal waters. On February 5, 2019 the Copernicus Sentinel-2B satellite captured this true-color image of sediment gushing into the Tyrrhenian Sea, part of the Mediterranean Sea. (Image from ESA).

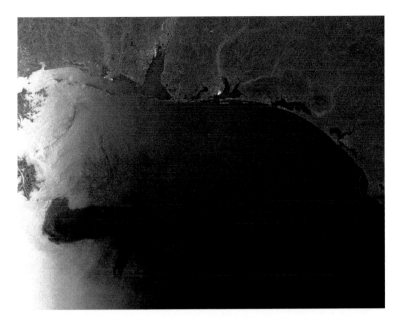

FIGURE 5.22 ENVISAT Advanced Synthetic Aperture Radar (ASAR) image of oil plume in Gulf of Mexico captured on April 28, 2010. The oil spill is visible as a lighter grey whirl on the left side of the large black pattern stretching across the Gulf. (Image from ESA).

5.5 CONCLUSIONS

The primary goal of this chapter is to highlight satellite capabilities in turbulence detection that may provide cumulative conception of scientific progress in this field. This chapter also aims to identify the current challenges and opportunities in satellite imagery and interpretation efforts concerning turbulence. Several important points can be noted.

First, presented in this chapter data and materials establish, in our opinion, measurement strategies and specify priorities and accessibility to the research community, interested to the subject matter. A secondary, but nonetheless important, outcome of this chapter is to reaffirm the utility of current remote sensing observations for addressing key science goals and tasks in turbulence detection from space. Third, we believe that advances in ability to observe turbulent environment from space and to assimilate remote sensing data into scientific product are already achieved today.

One question we have not considered in this regard is an *instrument optimization* that is absolutely necessary to provide a high-quality assessment of turbulence (in a broad sense) from space-based observations. In general, instrumentation needs to be in a sufficient technological innovative level, which could demonstrate reliability and relevance of the sophisticated remotely sensed information for current and future applications.

Much progress has been made in satellite capabilities in recent decades, especially in remote sensing of the atmosphere and oceans. Yet the ability to obtain the necessary data for understanding turbulent and convective environments has stagnated, because capacity of satellite platforms and sensors have not grown in proportion to technological advances and scientific efforts.

We believe that more highly resolved remote sensing observations are needed for scientific advancement in understanding and prediction of turbulent and convective processes and their impacts. Detailed measurements are required to provide comprehensive study of dynamical characteristics in the Earth's environments ranging from micro- to organized mesoscale convective systems, and supercell storms to tropical cyclones. There are many constraints to observing these processes from satellite-based platforms. These include, first of all, instrument performance limitations and difficulty in obtaining sufficient temporal and spatial resolution, and, ultimately, a lack of resources to maintain observing capabilities and the state-of-the art scientific research. Remotely sensed data, obtained occasionally from various satellite platforms may not (and do not so far) reflect spatiotemporal variability and stochastic nature of turbulent environment as a whole dynamical system.

For example, techniques for automatic identification and tracking of ocean eddies are largely based on image processing algorithms. Such approaches rely primarily on proxy variables such as ocean color, SST and SSS, or the near-surface wind vector field. These physical fields and/or variables should be measured with the highest accuracy in order to distinguish relevant turbulent signatures from dynamical background which also exhibits stochastic motions. On the other hand, a gap between physics-based models and observations can be reduced using the state-of-the-art CFD methods and scientific visualization techniques. Elaborating computer science capabilities and scientific expertise provide successful data analytics.

The most valuable information concerning geophysical turbulence can be gained from multisensor observations. Multisensor concept is based on the integration or *synergy* of different types of sensors (active, passive, optical, and microwave) into a unit observational payload platform which should be able to provide synchronized data acquisition and combined data analysis. Multisensor remote sensing is preferable option in enhanced observations of rare events which we call *weakly emerged turbulence or mild turbulence*, natural or human-made, such as induced turbulent wakes in the ocean or contrails and jets in the atmosphere. Multisensor imagery is also an optimal technique for investigating spatially statistical characteristics of turbulent phenomena, their origin and evolution. Finally, multisensor observations increase the detection capability especially in the case of *stochastic targets*.

Specialized methods of turbulence detection can be developed on the basis of *laser topographic mapping* – technique, which is widely used in remote sensing of Earth's surfaces (Dong and Chen 2018; Shan and Toth 2018). More advanced technique known as *laser reflection tomography* is a long-range active super-resolution imaging technology allowing accurate targeting and reconstruction of 3D objects. (Tomography refers to imaging by sections, or sectioning, and is usually the representation of a 3D object by means of its 2D cross sections, see, e.g., book Herman 2009). Tomography is well-established technology, commonly used in geophysical and medical imaging, although the application to remote sensing of environment is still in its infancy. In particular, laser tomography of the ocean-atmosphere interface from space will lead to a complete breakthrough of major applications. If this ever happens, nothing much more should be desired, nor expected.

5.6   NOTES ON THE LITERATURE

Several perfect books (Dowman et al. 2012; Tupin et al. 2014; Saha 2015; Zhao et al. 2016; Eugenio and Marcello 2019) provide comprehensive treatment of principles, methods, instruments, and applications of high-resolution remote sensing imagery.

Hyperspectral satellite imagery is discussed in recent books (Manolakis et al. 2016; Pu 2017).

Recent advances in radar (SAR) imagery and polarimetry are reviewed in books (Lee and Pottier 2009; Richards 2009; Berizzi et al. 2016; Zhao et al. 2016; Yamaguchi 2021).

Books (Nayak and Zlatanova 2008; Li 2017; Petropoulos and Islam 2018; Singh and Bartlett 2018; Maggioni and Massari 2019) describe various methods of observing natural hazards, disasters, and hurricanes.

REFERENCES

Banakh, V. and Smalikho, I. 2013. *Coherent Doppler Wind Lidars in a Turbulent Atmosphere.* Artech House Publishers, Boston, MA.

Berizzi, F., Martorella, M., and Giusti, E. (Eds.) 2016. *Radar Imaging for Maritime Observation (Signal and Image Processing of Earth Observations).* CRC Press, Boca Raton, FL.

Bobrowsky, P.T. (Ed.). 2013. *Encyclopedia of Natural Hazards.* Springer Science, Dordrecht, The Netherlands.

Brünner, C., Königsberger, G., Mayer, H., and Rinner, A. (Eds.). 2018. *Satellite-Based Earth Observation: Trends and Challenges for Economy and Society.* Springer Nature, Switzerland.

Chuvieco, E. 2020. *Fundamentals of Satellite Remote Sensing: An Environmental Approach,* 3rd edition, CRC Press, Boca Raton, FL.

Dalezios, N. R. 2017. *Environmental Hazards Methodologies for Risk Assessment and Management.* IWA Publishing, London, UK.

Dong, P. and Chen, Q. 2018. *LiDAR Remote Sensing and Applications.* CRC Press, Boca Raton, FL.

Doviak, R. J. and Zrnić, D. S. 2006. *Doppler Radar and Weather Observations,* 2nd edition. Dover Publications, Mineola, New York.

Dowman, I., Jacobsen, K., Konecny, G., and Sandau, R. 2012. *High Resolution Optical Satellite Imagery.* Whittles Publishing, Dunbeath, UK.

Elbert, B. R. 2004. *The Satellite Communication Applications Handbook,* 2nd edition. Artech House, Norwood, MA.

Eugenio, F. and Marcello, J. (Eds.). 2019. *Very High Resolution (VHR) Satellite Imagery: Processing and Application.* MDPI, Basel, Switzerland.

Gibson, C. H., Bondur, V. G., Keeler, R. N., and Leung, P. T. 2008. Energetics of the beamed Zombie turbulence maser action mechanism for remote detection of submerged oceanic turbulence. *Journal of Applied Fluid Mechanics,* 1(1):11–42.

Grant, B. 2016. *Overview: A New Perspective of Earth Hardcover.* Amphoto Books, Berkeley, CA.

Guo, H., Goodchild, M. F., and Annoni, A. (Eds.). 2020. *Manual of Digital Earth.* Springer Nature, Singapore.

Haimes, Y. Y. 2016. *Risk Modeling, Assessment, and Management,* 4th edition. John Wiley & Sons, Hoboken, NJ.

Haller, G., and Beron-Vera, F. J. 2013. Coherent Lagrangian vortices: The black holes of turbulence. *Journal of Fluid Mechanics,* 731 R4. doi:10.1017/jfm.2013.391.

Herman, G. T. 2009. *Fundamentals of Computerized Tomography: Image Reconstruction from Projections (Advances in Computer Vision and Pattern Recognition)*, 2nd edition. Springer, London, UK.

Horváth, Á., Bresky, W., Daniels, J., Vogelzang, J., Stoffelen, A., Carr, J. L., Wu, D. L., Seethala, C., Günther, T., and Buehler, S. A. 2020. Evolution of an atmospheric Kármán vortex street from high-resolution satellite winds: Guadalupe Island case study. *Journal of Geophysical Research: Atmosphere*, 125(4) e2019JD032121. doi:10.1029/2019JD032121.

Ivanov, A. Y. and Ginzburg, A. I. 2002. Oceanic eddies in synthetic aperture radar images. *Journal of Earth System Science*, 111(3):281–295. doi:10.1007/bf02701974.

Karimova, S. 2012. Spiral eddies in the Baltic, Black and Caspian seas as seen by satellite radar data. *Advances in Space Research*, 50(8):1107–1124. doi:10.1016/j.asr.2011.10.027.

Kelkar, R. R. 2017. *Satellite Meteorology*, 2nd edition. CRC Press, Boca Raton, FL.

Klemas, V. 2011. Remote sensing techniques for studying coastal ecosystems: An overview. *Journal of Coastal Research*, 27(1):2–17. doi:10.2112/jcoastres-d-10-00103.1.

Levina, G. 2018. On the path from the turbulent vortex dynamo theory to diagnosis of tropical cyclogenesis. *Open Journal of Fluid Dynamics*, 8:86–114. Available on the Internet https://www.scirp.org/pdf/OJFD_2018032815215345.pdf.

Lee, J.-S. and Pottier, E. 2009. *Polarimetric Radar Imaging: From Basics to Applications*. CRC Press, Boca Raton, FL.

Li, H. (Ed.). 2017. *Hurricane Monitoring with Spaceborne Synthetic Aperture Radar*. Springer Nature, Singapore.

Macdonald, M. and Badescu, V. (Eds.). 2014. *The International Handbook of Space Technology*. Springer-Verlag, Berlin, Heodelberg.

Maggioni, V. and Massari, C. (Eds.). 2019. *Extreme Hydroclimatic Events and Multivariate Hazards in a Changing Environment: A Remote Sensing Approach*. Elsevier, Amsterdam, Netherlands.

Manolakis, D., Lockwood, R., and Cooley, T. 2016. *Hyperspectral Imaging Remote Sensing: Physics, Sensors, and Algorithms*. Cambridge University Press, Cambridge, UK.

McWilliams, J. C. 2016. Submesoscale currents in the ocean. *Proceedings of the Royal Society A: Mathematical, Physical and Engineering Science*, 472(2189) 20160117. doi:10.1098/rspa.2016.0117.

Mecikalski, J. R., Feltz, W. F., Murray, J. J., Johnson, D. B., Bedka, K. M., Bedka, S. T., Wimmers, A. J., Pavolonis, M., Berendes, T. A., Haggerty, J., Minnis, P., Bernstein, B., and Williams, E. 2007. Aviation applications for satellite-based observations of cloud properties, convection initiation, in-flight icing, turbulence, and volcanic ash. *Bulletin of the American Meteorological Society*, 88(10):1589–1607. doi:10.1175/bams-88-10-1589.

Melzer, B. A., Jensen, T. G., and Rydbeck, A. V. 2019. Evolution of the Great Whirl using an altimetry-based eddy tracking algorithm. *Geophysical Research Letters*, 46:4378–4385. doi:10.1029/2018gl081781.

Moiseev, S. S., Rutkevich, P. B., Tur, A. V., and Yanovskii, V. V. 1988. Vortex dynamo in a convective medium with helical turbulence. *Soviet Physics – JETP*, 67(2):294–299. Available on the Internet http://www.jetp.ac.ru/cgi-bin/dn/e_067_02_0294.pdf.

Munk, W., Armi, L., Fischer, K., and Zachariasen, F. 2000. Spirals on the sea. *Proceedings of the Royal Society A: Mathematical, Physical and Engineering Sciences*, 456(1997):1217–1280. doi:10.1098/rspa.2000.0560.

Nataraj, N. 2017. *The Planets: Photographs from the Archives of NASA*. Chronicle Books, San-Francisco, CA.

Nayak, S. and Zlatanova, S. (Eds.). 2008. *Remote Sensing and GIS Technologies for Monitoring and Prediction of Disasters*. Springer-Verlag, Berlin, Heidelberg.

Qian, S.-E. 2020. *Hyperspectral Satellites and System Design*. CRC Press, Boca Raton, FL.

Petropoulos, G. P. and Islam, T. 2018. *Remote Sensing of Hydrometeorological Hazards*. CRC Press, Boca Raton, FL.

Pu, R. 2017. *Hyperspectral Remote Sensing: Fundamentals and Practices (Remote Sensing Applications Series)*. CRC Press, Boca Raton, FL.

Raizer, V. 2017. *Advances in Passive Microwave Remote Sensing of Oceans*. CRC Press, Boca Raton, FL.

Raizer, V. 2019. *Optical Remote Sensing of Ocean Hydrodynamics*. CRC Press, Boca Raton, FL.

Ralph, F.M., Dettinger, M. D., Rutz, J. J., and Waliser, D. E. (Eds.). 2020. *Atmospheric Rivers*. Springer Nature, Switzerland.

Razani, M. 2019. *Commercial Space Technologies and Applications: Communication, Remote Sensing, GPS, and Meteorological Satellites*, 2nd edition. CRC Press, Boca Raton, FL.

Richards, J. A. 2009. *Remote Sensing with Imaging Radar*. Springer-Verlag, Berlin, Heidelberg.

Saha, S. K. (Ed.). 2015. *High Resolution Imaging: Detectors and Applications*. CRC Press, Boca Raton, FL.

Shan, J. and Toth, C. K. (Eds.) 2018. *Topographic Laser Ranging and Scanning:Principles and Processing*, 2nd edition. CRC Press, Boca Raton, FL.

Shanmugam, G. 2018. A global satellite survey of density plumes at river mouths and at other environments: Plume configurations, external controls, and implications for deep-water sedimentation. *Petroleum Exploration and Development*, 45(4): 640–661. doi:10.1016/s1876-3804(18)30069-7.

Sharkov, E. A. 2012. *Global Tropical Cyclogenesis*, 2nd edition. Springer-Verlag – Praxis, Berlin, Heidelberg.

Sharman, R. and Lane, T. (Eds.). 2016. *Aviation Turbulence: Processes, Detection, Prediction*. Springer International Publishing, Switzerland.

Singh, R. and Bartlett, D. (Eds.). 2018. *Natural Hazards: Earthquakes, Volcanoes, and Landslides*. CRC Press, Boca Raton, FL.

Smith, K. 2013. *Environmental Hazards: Assessing Risk and Reducing Disaster*, 6th edition. Routledge Taylor & Francis Group, New York.

Stammer, D. and Cazenave, A. (Eds.). 2018. *Satellite Altimetry Over Oceans and Land Surfaces*. CRC Press, Boca Raton, FL.

Swarnalatha, P. and Prabu, S. 2018. *Big Data Analytics for Satellite Image Processing and Remote Sensing*. IGI Global, Hershey, PA.

Talley, L. D., Pickard, G. L., Emery, W. J., and Swift, J. H. 2011. *Descriptive Physical Oceanography: An Introduction*, 6th edition. Elsevier – AcademicPress, London, UK.

Tsuchiya, K. and Tokuno, M. 1989. Analysis of Karman vortex clouds revealed by satellite images. *Advances in Space Research*, 9(7):425–431. doi:10.1016/0273-1177(89)90195-6.

Tupin, F., Inglada, J., and Nicolas, J.-M. (Eds.). 2014. *Remote Sensing Imagery*. ISTE – John Wiley & Sons, Hoboken, NJ.

Wang, Y. (Ed.). 2010. *Remote Sensing of Coastal Environments*. CRC Press, Boca Raton, FL.

Weitkamp, C. (Ed.). 2005. *Lidar: Range-Resolved Optical Remote Sensing of the Atmosphere*. Springer Science, New York.

Wimmers, A. and Feltz, W. 2010. Tropopause folding turbulence product. Algorithm Theoretical Basis Document (ATBD). *NOAA NESDIS Center for Satellite Applications and Research*. Available on the Internet https://www.goes-r.gov/products/ATBDs/option2/Aviation_Turbulence_v1.0_no_color.pdf.

Wimmers, A. J. and Moody, J. L. 2004a. Tropopause folding at satellite-observed spatial gradients: 1. Verification of an empirical relationship. *Journal of Geophysical Research, Atmospheres*, 109(D19) D19306. doi:10.1029/2003jd004145.

Wimmers, A. J. and Moody, J. L. 2004b. Tropopause folding at satellite-observed spatial gradients: 2. Development of an empirical model. *Journal of Geophysical Research, Atmospheres*, 109 (D19) D19307. doi:10.1029/2003jd004146.

Wirth, A., Willebrand, J., and Schott, F. 2002. Variability of the Great Whirl from observations and models. *Deep Sea Research Part II: Topical Studies in Oceanography*, 49(7–8):1279–1295. doi:10.1016/s0967-0645(01)00165-5.

Yamaguchi, Y. 2021. *Polarimetric SAR Imaging: Theory and Applications (SAR Remote Sensing)*. CRC Press, Boca Raton FL.

Yiğit, E. and Medvedev, A. S. 2019. Obscure waves in planetary atmospheres. *Physics Today*, 72(6):40–46. doi:10.1063/pt.3.4226.

Young, G. S. and Zawislak, J. 2006. An observational study of vortex spacing in island wake vortex streets. *Monthly Weather Review*, 134(8):2285–2294. doi:10.1175/mwr3186.1.

Zhao, Y., Yi, C., Kong, S. G., Pan, Q., and Cheng, Y. 2016. *Multi-band Polarization Imaging and Applications*. National Defense Industry Press Beijing – Springer, Berlin, Germany.

Appendix

LYAPUNOV EXPONENT

Lyapunov exponents measure exponential rates of separation of nearby trajectories in the flow of a dynamical system. In mathematics, the Lyapunov exponent or Lyapunov characteristic exponent of a dynamical system is a quantity that characterizes the rate of separation of infinitesimally close trajectories (Figure A1). Quantitatively, two trajectories in phase space with initial separation vector $\delta\mathbf{Z_0}$ diverge (provided that the divergence can be treated within the linearized approximation) at a rate given by

$$\left|\delta\mathbf{Z}(t)\right| \approx e^{\lambda t}\left|\delta\mathbf{Z_0}\right|, \tag{A1}$$

where λ is the Lyapunov exponent.

The rate of separation can be different for different orientations of initial separation vector. Thus, there is a spectrum of Lyapunov exponents – equal in number to the dimensionality of the phase space. It is common to refer to the largest one as the Maximal Lyapunov exponent (MLE), because it determines a notion of predictability for a dynamical system. A positive MLE is usually taken as an indication that the system is chaotic (provided some other conditions are met, e.g., phase space compactness). The MLE is defined as follows:

$$\lambda = \lim_{t \to \infty} \lim_{|\delta\mathbf{Z_0}| \to 0} \frac{1}{t} \ln \frac{\left|\delta\mathbf{Z}(t)\right|}{\left|\delta\mathbf{Z_0}\right|} \tag{A2}$$

The MLE defines stability and type of dynamics:

$\lambda > 0$ – chaos, unstable dynamics. The trajectory is unstable and chaotic. Nearby points, no matter how close, will diverge to any arbitrary separation and unstable

$\lambda = 0$ – regular dynamics. The trajectory is a neutral fixed point. A physical system with this exponent exhibits Lyapunov stability. The trajectories in this situation usually maintain a constant separation.

$\lambda < 0$ – fixed-point dynamics. The trajectory attracts to a stable fixed point or stable periodic orbit. Negative Lyapunov exponents are characteristic of dissipative or non-conservative systems. A system exhibits asymptotic stability; the more negative the exponent, the greater the stability.

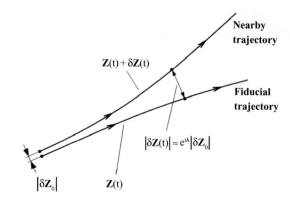

FIGURE A1 The sketch demonstrated definition of the Lyapunov exponent.

Index

Page numbers in **bold** indicate tables, page numbers in *italic* indicate figures.